The Social Life of Water

THE SOCIAL LIFE OF WATER

∞

Edited by

John Richard Wagner

berghahn

NEW YORK · OXFORD

www.berghahnbooks.com

Published in 2013 by

Berghahn Books

www.berghahnbooks.com

Library of Congress Cataloging-in-Publication Data

The social life of water / edited by John Richard Wagner.
    pages cm
   Includes bibliographical references and index.
   ISBN 978-0-85745-966-4 (hardback : alk. paper) — ISBN 978-1-78238-910-1 (paperback :
alk. paper) — ISBN 978-0-85745-967-1 (ebook)
   1. Water--Social aspects. 2. Water—Symbolic aspects. 3. Water—Economic aspects. I. Wagner,
John R. (John Richard)
   GB665.S59 2013
   304.2—dc23

                                                                                      2013005538

British Library Cataloguing in Publication Data

A catalogue record for this book is available from the British Library

Printed on acid-free paper

ISBN: 978-0-85745-966-4 hardback
ISBN: 978-1-78238-910-1 paperback
ISBN: 978-0-85745-967-1 ebook

To water

CONTENTS

ILLUSTRATIONS

Figures

Tables

ACKNOWLEDGMENTS

My interest in water issues began the year I returned to the Okanagan Valley of British Columbia to take up a position at Okanagan University College, an institution that has since become a second campus of the University of British Columbia. A series of forest fires devastated the valley just after my arrival in Kelowna in July 2003, forcing the evacuation of thousands of residents from their homes. Many homes were lost to the fires but just as startling for me was a controversy that broke out between the Town of Summerland and the federal Department of Fisheries. Fisheries staff accused the Town Council of contravening fisheries regulations by withdrawing too much water from Trout Creek, the town's main supply of water. In a very public and dramatic fashion, the valley thus came face to face with something most wished to ignore—water scarcity had become a fact of life. The Okanagan watershed has the lowest precipitation per capita in all of Canada but is also blessed with several large valley-bottom lakes and most residents had not previously perceived themselves as living in a water-scarce region. But by 2003 the valley had experienced several consecutive years of below-average precipitation, which led to the fires and the unusually low volumes of water in local creeks. Having lived in the Okanagan for several years during the 1970s and 1980s, I was as startled as most long-term residents by the events of 2003, and like many others was mobilized to take action.

I recognized from the outset that Okanagan water issues, though severe by Canadian standards, are far less severe than those that occur in many other regions of the world. I suspected, however, that what I learned about water governance here would apply to many other settings; seven years of research have confirmed that opinion. I have not included a detailed description of my Okanagan research in this volume, but the theoretical approach I describe in the introduction arises directly out of that research. I therefore wish to acknowledge the contribution of all those who have participated over the past few years in my Okanagan Valley research. They include the many orchardists, grape-growers, ranchers, and water managers—too many to mention by name—who I have interviewed and whose meetings I have attended as observer, participant, and occasionally as guest speaker.

I also wish to acknowledge the contributions of Jeannette Armstrong, executive director of the En'owkin Centre, with whom I collaborated on my first water research project in 2004; and Marlowe Sam and Rose Caldwell, who worked with us as undergraduate research assistants. Nelson Jatel and Anna Warwick Sears of the Okanagan Basin Water Board have also become valued colleagues in water research over the past few years, as have many members—once again too many to mention—of the Okanagan Water Stewardship Council. I would also like to thank several other undergraduate and graduate students who have worked with me on Okanagan research projects that have informed this volume: Angele Clarke, Amanda Daignault, Lindsay Eason, Mohammad Anwar Hossen, Moira McNairnay, Jennifer Mooney, and Kasondra White. These projects were funded by the Social Sciences and Humanities Research Council of Canada and by internal grants provided by Okanagan University College and the University of British Columbia Okanagan.

This volume owes particular thanks to all of those who participated in the dedicated meeting on the "Anthropology of Water" in Philadelphia, especially to Ben Orlove for coorganizing the session with me, and to Lyla Mehta, who traveled all the way from England to join us. Lyla also introduced me to the work of exceptional scholars from India, including Amita Baviskar, Rita Brara, and Rohan D'Souza, all of whom I was able to meet and learn much from during a subsequent trip to India. John Donahue and Bryan Bruns, both scholars with many more years of experience on water issues than me, have provided invaluable help through their contributions to this volume, their previous publications, and their advice as editors of previous volumes on water issues. I also owe a huge debt of gratitude to all the other contributors to this volume, and to the publisher, Marion Berghahn, for sticking with this project as I struggled to bring it to a successful completion.

Finally, I would like to thank family, friends and especially Jennifer Gustar, my wife, for tolerating, usually with good humor, my growing obsession with water over the past several years, an obsession that seems unlikely to diminish any time soon. She has provided me with sage counsel on many occasions and when I succumb to the unfortunate tendency to think that all my opinions on water are "expert" opinions, she finds ways to remind me, as scholar and life partner, that expertise requires constant effort and accommodation of other points of view.

INTRODUCTION

John Richard Wagner

∞

> In the beginning, there is nothing but water and darkness. God changes all this by this thoughts, creating the sun, causing a bush to grow up from a vast expanse of water, its blossoms transforming into the first people.
> —Syilx oral narrative, as told by Harry Robinson,
> Robinson and Wickwire, *Write It on Your Heart:*
> *The Epic World of an Okanagan Storyteller*

> We made from water every living thing.
> —Qur'an Surah 21: 30

> In the beginning, God created the heavens and the earth. The earth was without form and void, and darkness was over the face of the deep. And the Spirit of God was hovering over the face of the waters.
> —Bible Genesis I: 1–2

It is hard to conceive of any substance more fundamental to life than water. Scientists tell us that the cells that constitute all life forms are, on average, 70 percent water by weight. Even more impressive is the fact that 99 percent of all human molecules are water molecules (Freitas 1999). Three-quarters of the earth's surface is covered by either saltwater or freshwater. Human beings can survive for several weeks without food, but only a few days without water. We should not be surprised, then, to discover that water also occupies a place of privilege in the human imagination. According to the Qur'an, every living thing is made from water. According to the Bible, water existed before God created the heavens and the earth. The oral narratives of many indigenous people, such as the Syilx of North

America, also stress the primacy of water at the beginning of time (Robinson and Wickwire 1989: 31). The very name of the Hindu religion derives from the Sanskrit name for the Indus River (Sindu). It is also worth noting that religious accounts of the emergence of life from water converge, metaphorically at least, with scientific theories about the emergence of early life forms from a prebiotic ocean.

The primordial and spiritual qualities of water are not the main topic of this introduction nor of the chapters that follow, but it seems appropriate to acknowledge them at the outset of a book focused on the social life of water—a term I use here, in its broadest sense, to encompass all domains of human relationships including the political, economic, technological, and spiritual. Conflict over water tends to evoke primordial responses in affected populations, even when it arises as the cumulative result of mundane political and economic activities. As Orlove and Caton (2010: 402) argue, employing Mauss' concept of the "total social fact," the human relation to water finds expression within institutions and sets of behavior that operate in all social domains, including the moral, juridicial, aesthetic, political, and economic (Mauss 1990: 3). While Orlove and Caton (2010: 402) emphasize the ways in which "water connects different domains of social life to each other," I also wish to emphasize that, as water circulates continuously through complex hydrological cycles, it establishes connections among all life forms and between animate and inanimate worlds. The human social connections brought into play through water thus constitute one subset of a broader set of ecological relations, and from that perspective we can understand water not only as a "total social fact," but also as a "total ecological fact," in which the social is separable from the ecological only by semantic convention.

It is now impossible to think globally about the human relation to water without recognizing the extent to which that relation is conditioned by a sense of crisis. Millions of people face water crises on a daily basis as they struggle to make do in urban shantytowns that have no piped water or sanitation services and no nearby source of clean water for drinking, cooking, or bathing. Others experience periodic droughts and floods and become water refugees. An increasing world population means we need to increase agricultural production, which requires more use of water for irrigation in a world where agriculture consumes more than 70 percent of all the freshwater used by humans (UNESCO 2006: 247). The World Health Organization reports that insufficient and unclean water supplies are the direct or indirect cause of 20 percent of all deaths of children under fourteen years of age, and that about one-tenth of the global disease burden could be prevented through better management of water supplies and sanitation (Prüss-Üstün et al. 2008: 10). The list of crises is long and growing, and so are the many popular books and documentary films that now chronicle them.[1] These works describe the depletion of aquifers around the world, contamination of water supplies by sewage and industrial and agricultural pollutants, battles over water privatization and commodification, environmental degradation due to damage to riparian habitat and wetlands by dam construction, river channelization, urban development, and the impacts of climate change that include the melting of glaciers and reduced flows of water in glacier-fed rivers, as well as the more frequent occurrence of extreme weather events.

Contributors to this volume were asked to provide case studies that shed light on how a sense of crisis informs and shapes the human relation to water in the twenty-first century. They also were asked to contextualize their studies in terms of the most important issues occurring in their country or region; a special effort was made to ensure that all continents, and a diverse range of geographical regions, were included. The polar regions are not represented, unfortunately, and no true desert region is included, though several arid regions are (in British Columbia, Canada; Kenya; Palestine; and the Southwest United States). A diverse range of elevations and biomes are included and a balance of sites in tropical (Bolivia, Cambodia, El Salvador, Fiji, Ghana, India, Indonesia, Kenya, and Peru) and temperate zones (Argentina, Australia, Canada, Chile, Ireland, New Zealand, Palestine, and the United States). A deliberate effort was made as well to include authors from every continent, especially those who live and work presently in countries outside Europe and North America. This attempt to achieve a truly global account of human–water relations, though limited by time constraints and my own professional research networks, will hopefully prove informative for a broad audience and assist the process of developing improved water policies in diverse settings. The book has been written for an audience that includes water practitioners and the public, as well as undergraduate and graduate students, and academic researchers.

Although the idea of crisis provides essential context for the chapters that follow, contributors were not asked to evaluate the legitimacy or severity of the various claims of water crises we hear about on a daily basis, nor were they asked to theorize about crisis, disaster, and risk as an emergent research paradigm in the social sciences. They were invited to theorize about the social life of water in settings where they have conducted ethnographic research, with the goal of demonstrating the immense variety of ways in which social practices shape and are shaped by water. Before discussing contributors' diverse approaches to the social life of water in more detail, however, I would like to describe some important contributions to theory about crisis, disaster, and risk, in order to provide additional context for the chapters that follow.

Water Crises

Kirsten Hastrup, editor of a recent collection of papers on climate change, notes in her introduction that the volume is organized on the basis of three "regions of disaster": the melting of glaciers and the arctic ice cap, the rise in sea levels that is occurring in response to melting ice, and the desertification that is occurring in many regions, most notably in Africa, as a result of shifts in precipitation patterns (Hastrup 2009: 16). Following the arguments of Wisner and colleagues (2004), she notes that these types of events are often classified as natural hazards, but that human actions are a major contributor to global climate change and thereby to most of the specific, localized events attributed to climate change. It is also the case that disasters like desertification, in Africa and elsewhere, are not caused by climate change alone, but by climate change together with forms of agricultural intensification that have reduced the resilience of local ecosystems (Blaikie and Brookfield 1987; Franke and Chaisin 1980; Wisner et al. 2004: 135). As Wisner and colleagues (2004: 6) emphasize, however, media representations tend to focus on the natural causes of these disasters, and the public is left with the impression that there are no solutions other than technical, one-off responses to isolated events. This perception works against our ability to recognize the root causes of disasters and modify the human behavior that creates or worsens them.

Although the contributors to the Hastrup volume recognize climate change is a socionatural hazard, they do not focus on how socionatural hazards are produced. They concentrate instead on understanding how communities adapt to the many uncertainties generated by climate change in varied settings. Gregory Button, by contrast, in studies that range from Love Canal, to the Exxon–Valdez oil spill, to the impact of Hurricane Katrina on New Orleans, emphasizes the "production of uncertainty as an ideological tactic" (2010: 16) and the role of "informational vacuums" in generating uncertainty (2010: 17). His goal, Button writes, "is to examine how the conscious tendency to manufacture, revise, or withhold knowledge politicizes the discourse in the wake of disasters" (Button 2010: 16). His approach converges with that of Hastrup, however, when he argues that a disaster should not be analyzed as an "isolated event" but rather as one in a series of events that began many years or decades previously and that will generate "cascading events" for years to come (2010: 17). Wisner and colleagues (2004: 20) similarly emphasize the importance of understanding disasters in relation to the normal, day-to-day social and economic behavior that gives rise to them.

If we classify approaches to crisis and disaster in two general schools, those that emphasize the socionatural production of disaster and those that document the responses and adaptations of specific communities, cities, and regions, then the influential work of Ulrich Beck (1992, 1995, 1999) falls clearly in the former category. Beck argues that processes of modernization have created new forms of risk on a global scale and that the institutions on which we depend to control or limit these risks are ineffective. Government agencies are poorly equipped, for instance, to deal with nuclear disasters on the scale of what occurred in 1986 at Chernobyl, in the Ukraine, or more recently, in 2011, at the Fukushima power station in Japan. Bioengineering technologies in medical and agricultural industries also introduce new, globally distributed, and poorly understood levels of risk. Global climate change also generates hazards and risks for which no single institution or set of institutions is responsible. According to Beck, the level and scale of risk in the world today is unprecedented, thus justifying his claim that we now live in a world risk society and that much of modern life is shaped by our efforts to cope with this new reality (Beck 1999). While Beck has been criticized for overstating the negative aspects of risk (c.f. Giddens 2003), and for not providing sufficient empirical evidence to back up his claims (see Mythen 2004), his theory has been extremely influential.

Beck's theory is driven largely by his analysis of environmental risk (Beck 1995); from that perspective his work is informative for this study. Water crises, as a global phenomenon, fit neatly within the typology of risk he identifies. Problematic, however, from the perspective of this volume, is the overly sharp distinction he draws between the natural hazards he claims dominated the world until very recently, and the new forms of manufactured risk that dominate our lives today. While the extent of the human impact on natural systems has clearly expanded during the past half century, the boundary between natural hazard and manufactured risk is very difficult to establish in the case of contemporary water crises. As a result of global warming, for instance, we know that severe weather events are becoming increasingly common, and it is no longer possible to classify flooding and drought simply as natural hazards. We also know that human actions have been generating or worsening natural water hazards like flooding and drought for thousands of years, and not just for the past half century. As I will argue in more detail in the next section of this introduction, solutions for the water crises we face today require that we theorize water and humans as simultaneously social and ecological, not just as one or the other.

Contributors to this volume focus mainly on people's day-to-day lives, noting the ways in which the routine decisions of governmental and nongovernmental agencies, private individuals, and corporations generate hazard and risk that may erupt in disaster or, just as often, generate less-extreme forms of hardship that come to be viewed as part of normal life. Many of the crises described here do not attract headlines; it is only through their intractability and wide distribution, especially among the poor and marginalized, that they come to be recognized as crises at all. We focus on crisis as context, then, in recognition of the fact that crisis, both chronic and exceptional, is often a catalyst for social transformation, and that "the arrangements of a society become most visible when challenged by crisis" (Wolf 1990, as cited by Button 2010: 11).

Actors, Agency, and Things

We do not have to look far to find examples of economic activities, like agriculture and fisheries, that are entirely water-dependent. But water is also essential to most manufacturing processes, to hydropower generation, and to the technologies we use to carry sewage and industrial waste away from human settlements. The location of all human communities, until fairly recently, could be predicted by the location of water, whether in the form of rivers and lakes, or underground springs and aquifers; historically, the places where people go to get their water are also places where they go to socialize. The influence of water on our social lives reaches well beyond these obvious examples, however, to our political institutions, class relations, and landscape aesthetics (Wagner 2008). Irrigation technology was an essential prerequisite for the development of early state-level societies around the world since it enabled food surpluses, labor specialization, and the growth of urban centers. And as this volume will demonstrate, the hierarchical political and economic structures that dominate the modern state and today's global economy also owe a great deal to the human–water relationship. For many, perhaps most people in the world, access to water can be predicted on the basis of class distinctions.

The claim that human actions in relation to water have important social consequences is not the same thing, however, as the claim that water itself has a social life. *The Social Life of Things*, edited by Arjun Appadurai and published in 1986, provides a valuable starting point for thinking about water from this perspective. Though now somewhat dated within the world of commodity and exchange theory, the authors' assertion that things, and not just people, have social lives, lays some of the groundwork we need to build a unified approach to studying socioecological systems. It thus anticipates some of the more recent criticisms leveled against conventional approaches to studying water systems (see especially Swyngedouw 1999, 2004), and converges with theory developed by Bruno Latour and others concerning the agency of objects and quasi-objects within social systems (Bennett 2010; Latour 2005). It is partly in appreciation of this particular contribution of Appadurai and his coauthors that I have titled this volume *The Social Life of Water.*

In his introduction to *The Social Life of Things*, Appadurai (1986) outlines a methodology for studying the ways in which things acquire value as objects of exchange. He begins by acknowledging that the exchange value of objects is constructed by humans, and is not inherent in the objects themselves, but proposes that we should "follow the things themselves," in order to "illuminate" their "human and social context" (Appadurai 1986: 5). The conceit that things have social lives, he argues, is justified as a methodological strategy even though it must be rejected as a theoretical argument (1986: 3). In the same volume, Kopytoff (1986) similarly proposes that social scientists can study cultural behavior by constructing biographies of things rather than, or in addition to, biographies of people, but he also stops short of assigning agency to the things themselves. One of the contributors to this present volume, Swathi Veeravalli, addresses this issue directly, asserting in her chapter on water and identity in Kenya that water has agency within social systems just as it has agency within ecological systems. While other contributors to this volume do not directly address this issue, most include detailed descriptions of the objects that populate the water worlds they describe, objects such as water pumps and pipes, dams, storage tanks, irrigation equipment, and sanitation facilities. A few specifically emphasize the importance of these objects in shaping social systems, most notably Hugo De Burgos in his historical analysis of public water storage tanks in El Salvador, Rita Brara in her study of irrigation practices in India, and Nefissa Naguib in her study of piped water in Palestine. While I cannot speak for all contributors on this issue, I believe their case study materials provide readers with enough data to decide for themselves whether the agency of things is a valid theoretical concept or merely a useful methodology.

The works of Bruno Latour and other actor network theorists provide, I believe, an opportunity for some further development and refinement of our approaches to studying human–water relations (Wagner 2012). If our goal is to live sustainably within the watersheds we inhabit, then we need to understand watersheds as socioecological systems, as whole systems, not systems that are sometimes social and sometimes ecological, and not always both at the same time. Socioecological resilience theorists have attempted to move us in this direction, and the emphasis by many such theorists on social learning and adaptation seem useful and promising (see, e.g., Anderies and Walker 2006; Folke et al. 2005; Holling 1973; Pahl-Wostl et al. 2007; Schlüter and Pahl-Wostl 2007). I see little evidence, however, that they have made progress toward breaking down the mental divide between the social and the ecological, between nature and culture, a habit of mind that, arguably, lies at the very root of contemporary, unsustainable industrial economies and lifestyles (Latour 2004; Swyngedouw 1999, 2004).

The particular contribution of actor network theory that I wish to emphasize here insists that the important actors within complex actor networks are nonhuman as well as human, and that we should undertake our research projects with the goal of determining empirically which are the most important actors in a given system and which are less important. The durability of social systems depends as much, perhaps more, Latour tells us, on the objects included within them than on the social ties among the human actors within those systems. Many types of objects thus populate and shape our social systems. They include inanimate objects such as the oil, coal, gold, and uranium that we extract from nature in order to construct other objects such as knives, bowls, houses, cars, schools, cell phones, and hospitals. They also include animate objects such food plants, trees, and animals. And they include water, a clearly inanimate substance according to normal definitions, but such a vital, moving, and ubiquitous part of all biological and ecological life—both the mother of life and its inmost cellular constituent—that it *seems* animate and is often venerated as such. And if we agree with Latour that the social is not something given and preordained but rather something we assemble on a daily basis, then it makes sense to treat water as an actor within this process of assembly and to assign to it the same capacity for agency that we assign to ourselves as intentional human beings.

If the actors within a given social system are human, nonhuman, animate, and inanimate, then agency does not have to be defined exclusively in terms of intentionality but can be defined in terms of the impact of a given actor within the network. While most people would quite rightly reject the idea that a water pipe exercises agency when it is an entirely passive object within an irrigation system, it is much easier to accept the notion that water itself exercises agency. Veeravalli, for instance, describes how fluctuations in rainfall in East Africa affect the degree of inequality of access to water. Since only the most prosperous households own water storage tanks, only they have ready access to water for domestic and agricultural needs during times of ample rainfall. During prolonged periods of little or no rainfall, however, the source of inequality disappears as both rich and poor families obtain water in the same way from the same alternative sources. Periods of little or no rainfall thus flatten the social landscape by reducing inequalities of access to water. We cannot assign intentionality to water as we conventionally define that term, but we can assign agency to it in ecological terms and therefore in socioecological terms. Just as the social behavior of other species constitutes a domain of ecological interaction, so also does human behavior.

Organization of the Book

Contributors to the volume were not asked to adopt or respond directly to the work of Latour or Appadurai, or to the theoretical approach I outline above. They were simply invited to describe the social meaning of water and the relation of water to ongoing social, political, economic, and ecological transformations in their fieldwork setting. They were also told that the ultimate goal of the publication is to contribute to the development of more-effective governance policy for the challenges we are now facing on a worldwide basis. I communicated to all contributors about my theoretical orientation during the early stages of the project but they were left entirely free to organize their materials along whatever lines seemed most appropriate to them, and the result is a diverse and creative set of readings that sometimes converge in productive and fascinating ways but other times diverge in equally productive ways. Some authors share my interest in the work of Latour (see especially Brara) and all are aware of the relevance of *The Social Life of Things* to this volume, but few have referenced the latter publication and the perspectives of those who did were quite different from my own (see the chapter by Veronica Strang).

Contributors were also made aware of the fact that political ecology approaches based on first-hand ethnographic research would dominate the volume, but that other approaches were also welcome. In the end, only about three-quarters of the contributors employed an ethnographic approach and only half can unequivocally be classified as political ecology. Despite the diversity of approaches, however, all chapters include some political analysis and most include some ethnographic content. Sarah Smith, for instance, follows what she refers to as a "critical-interpretive medical anthropology perspective" (this volume, p. 180), with much of her chapter devoted to a technical description of the conditions under which the dengue vector thrives and to a critique of the biomedical approach taken by local health authorities to control the vector in urban Cambodia. But her critique is grounded in ethnographic description of local perceptions of water quality—perceptions that bring the local population into conflict with the biomedical approach being applied. Although Liam Leonard's account of a water crisis in Ireland could be classified as political ecology, he relies entirely on media and government reports and his own insider status as a permanent resident in the study area rather than on ethnographic data. As indigenous scholars and activists, Marlowe Sam and Jeannette Armstrong write from a Syilx perspective that does not fit within any Western intellectual tradition, though it does inform and overlap with several including political ecology. Hugo De Burgos' approach is informed by his perspective as a medical anthropologist, but also by the fact that much of his case study material was gathered during the course of a commissioned, historical study. Bryan Bruns' chapter represents a distillation of his lifetime of contributions to common property theory (Bruns et al. 2005) and his extensive experience as a water management consultant. Unlike all the other studies in this volume, Issaka Osumanu's chapter on water inequalities in northern urban centers in Ghana is based mainly on survey research and statistical analysis. It fits well within this volume, however, because of his detailed analysis of the complex relationship between water, politics, and poverty in an urban setting in a developing nation.

According to the initial plan for the volume, chapters were to be organized in three sections: (1) water scarcity issues (depletion of water supplies, inequitable distribution, privatization, drought, climate change); (2) water quality issues (pollution, disease, endangered species); and (3) symbolic and spiritual issues not directly associated with scarcity or quality (ritual uses of water, protection of sacred water bodies, aesthetic values). As authors submitted their first drafts it became clear, however, that the lines dividing these three categories were blurred in almost every case. In the end, another more useful set of thematic categories emerged as the basis for organizing the volume, all of them critical to understanding the social life of water as it is constructed in the twenty-first century. These thematic categories are also not mutually exclusive and not all chapters fall neatly within just one theme, but grouping the chapters in this way creates a framework that highlights the central themes of the book while accommodating the diversity of chapters included.

Commodification

Water commodification, the organizing theme for part I of the book, is one of the most important and controversial issues in the world today and almost all chapters address it to some extent. Three chapters have been placed in this section. Chapters by Li and Strang describe processes of water commodification in the Andes and in Australia; the chapter by Mehta describes how the appropriation of water by the Indian state results in the commoditization of the economic life of displaced communities. This particular grouping creates the opportunity for readers with a special interest in commodification to focus on that theme alone. The introduction to part I indicates which other chapters in the volume make significant contributions to this theme as well.

Technology

Part II of the book is organized around the theme of water and technology. Given the global and ubiquitous presence of water technologies in our lives, in the form of piped water infrastructure, dams, irrigation equipment, and water treatment plants, to name just a few examples, it is surprising that no single publication that I am aware of is devoted to an analysis of the role technology plays in shaping our relationship to water. Brara takes an explicitly Latourian approach in her analysis of groundwater assemblages in Malerkotla, in the Punjab region of India. It is partly through the assembling and reassembling of irrigation technologies, she demonstrates, that social life itself, including kinship, ritual performance, and class structure, is assembled. Naguib describes the surprising outcomes of something as simple as the construction of a piped water system in a Palestinian village. Although women requested the piped water system and appreciated the other modern conveniences it made possible, older women remembered with fondness the time they formerly spent together at the village well, sharing stories and having a short respite from their dawn-to-dusk work lives as they waited their turn to draw water. Most chapters in this volume describe water technologies, but the chapters in part II are explicit in describing the social life of technology in detailed and sometimes surprising ways.

Urbanization

Part III focuses on water and urbanization. More than half the population of the world now lives in cities, and that percentage continues to increase everywhere. In some countries, persistent and deepening rural poverty is the most immediate cause of urbanization, and in many cases lack of access to water is a direct cause of rural poverty. But it is within urban environments that we face some of our most troubling challenges—in the form of water pollution, contaminated drinking water, and poor sanitation infrastructure. Osumanu, in the first chapter in part III, provides an excellent introduction to these issues, noting that the urban South as a whole is at risk, but that income and poverty statistics do not adequately account for the inequalities in water access that exist in places like northern Ghana, where urban development is generating new patterns of economic poverty that are directly correlated with water poverty. Leonard, by contrast, reminds us that cities in the global North are also very vulnerable to breakdowns in their ability to deliver safe drinking water to urban residents, and that economic downturns, such as those that we are experiencing today in Europe and North America, are likely to lead to more failures.

Governance

Part IV, the concluding section of the book, outlines some promising approaches to improved water governance and identifies some of the principles that should inform water governance systems everywhere, now and in the future. Wutich and colleagues address the issue of fairness of access to water through a comparative study of attitudes in four culturally and ecologically distinct settings. Sam and Armstrong point out the necessity of recognizing indigenous water rights as a human rights issue but conclude their article with the argument that state-level governance systems could benefit from the inclusion of indigenous principles. Bryan Bruns shares insights from his experience as a water management consultant in diverse institutional and community settings in Indonesia. Part IV concludes with John Donahue's description of how water users in Texas were able to come together to create a stakeholder process capable of resolving a particularly intransigent water conflict where the actors included whooping cranes, aquifer hydrology, and fluctuating rainfall patterns, as well as human beings.

There is no panacea for the world's multiple water crises and it is not our intention in this volume to suggest otherwise. The authors of this volume are committed, however, to the idea that a fuller understanding of the social life of water could help people from diverse backgrounds work together to accomplish more-equitable and sustainable relationships to water and to one another. The complex governance networks described by Bruns and Donahue may not be the sexiest of topics for social scientists to study, but they are the sites where the most important water wars of this century will be won or lost. This book therefore concludes with positive examples of what can be accomplished.

Notes

1. See, e.g., publications by Barlow (2007), Barlow and Clarke (2002), de Villiers (2001), Glennon (2002), Pearce (2007), Postel (1999), Sandford (2009), Shiva (2002), and documentary films by Bozzo (2008), Marshall (2011), Miller (2007), and Salina (2008).

References

Anderies, J.M., and B.K.A. Walker. 2006. Fifteen Weddings and a Funeral: Case Studies and Resilience-Based Management. *Ecology and Society* 11 (1): 21.
Appadurai, Arjun. 1986. Introduction: Commodities and the Politics of Value. In *The Social Life of Things*, edited by Arjun Appadurai, 3–63. Cambridge: Cambridge University Press.
———, ed. 1986. *The Social Life of Things: Commodities in Cultural Perspective.* Cambridge: Cambridge University Press.
Barlow, Maude. 2007. *Blue Covenant: the Global Water Crisis and the Coming Battle for the Right to Water.* Toronto: McClelland and Stewart.
Barlow, Maude, and Tony Clarke. 2002. *Blue Gold: the Battle against Corporate Theft of the World's Water.* Toronto: McClelland and Stewart.
Beck, Ulrich. 1992. *Risk Society: Towards a New Modernity.* London: SAGE.
———. 1995. *Ecological Politics in an Age of Risk.* Cambridge: Polity Press.
———. 1999. *World Risk Society.* Malden, MA: Polity Press.
Bennett, Jane. 2010. *Vibrant Matter: A Political Ecology of Things.* Durham, NC: Duke University Press.
Blaikie, Piers, and Harold Brookfield. 1987. *Land Degradation and Society.* London, New York: Methuen.
Bozzo, Sam, Director. 2008. *Blue Gold: World Water Wars.* Produced by Sam Bozzo. Narrated by Malcolm McDowell.
Bruns, Bryan Randolph, Claudia Ringler, and Ruth Meinzen-Dick. 2005. *Water Right Reform: Lessons for Institutional Design.* Washington, DC: International Food Policy Research Institute.
Button, Gregory. 2010. *Disaster Culture: Knowledge and Uncertainty in the Wake of Human and Environmental Catastrophe.* Walnut Creek, CA: Left Coast Press.
de Villiers, Marq. 2001. *Water: The Fate of Our Most Precious Resource.* New York: Mariner Books (Houghton Mifflin).
Folke, C., T. Hahn, P. Olsson, and J. Norberg. 2005. Adaptive Governance of Social-Ecological Systems. *Annual Review of Environment and Resources* 30: 441–473.
Franke, Richard, and Barbara H. Chaisin. 1980. *Seeds of Famine: Ecological Destruction and the Development Dilemma in the Western Sahel.* Montclair, NJ: Allanheld, Osmun.
Freitas, Robert A. Jr. 1999. *Nanomedicine*, vol. I, *Basic Capabilities.* http://www.nanomedicine.com/NMI.htm
Giddens. 2003. *Runaway World: How Globalization is Reshaping our World.* London: Routledge.
Glennon, Robert. 2002. *Water Follies: Groundwater Pumping and the Fate of America's Fresh Waters.* Washington, DC: Island Press.
Hastrup, Kirsten. 2009. Waterworlds: Framing the Question of Resilience. In *The Question of Resilience: Social Responses to Climate Change*, edited by Kirsten Hastrup, 11–30. Copenhagen: The Royal Danish Academy of Sciences.
Holling, C.S. 1973. Resilience and Stability of Ecological Systems. *Annual Review of Ecology and Systematics* 4: 1–23.
Kopytoff, Igor. 1986. The Cultural Biography of Things: Commoditization as Process. In *The Social Life of Things*, edited by Arjun Appadurai, 64–94. Cambridge: Cambridge University Press.
Latour, Bruno. 2004. *Politics of Nature: How to Bring the Sciences into Democracy.* Cambridge, MA: Harvard University Press.
———. 2005. *Reassembling the Social: an Introduction to Actor-Network Theory.* Oxford: Oxford University Press.
Marshall, Liz. 2011. *Water on the Table.* Documentary film. Produced in association with TVO, Toronto, ON. Distributed by Bullfrog Films and Kinosmith (in Canada).
Mauss, Marcel. 1990. *The Gift: the Form and Reason for Exchange in Archaic Societies.* London: Routledge.
Miller, Liz, Director. 2007. *The Water Front.* Montreal: Red Lizard Media.

Mythen, Gabe. 2004. *Ulrich Beck: A Critical Introduction to the Risk Society.* London: Pluto Press.

Orlove, Ben, and Steven C. Caton. 2010. Water Sustainability: Anthropological Approaches and Prospects. *Annual Review of Anthropology* 39: 401–415.

Pahl-Wostl, C., M. Craps, A. Dewulf, E. Mostert, D. Tàbara, and T. Taillieu. 2007. Social Learning and Water Resources Management. *Ecology and Society* 12 (2): 5.

Pearce, Fred. 2007. *When the Rivers Run Dry: Water, the Defining Crisis of the Twenty-First Century.* Boston: Beacon Press.

Postel, Sandra. 1999. *Pillar of Sand: Can the Irrigation Miracle Last?* New York: W.W. Norton and Company/Worldwatch Institute.

Prüss-Üstün, Annette, Robert Bos, Fiona Gore, and Jamie Bartram. 2008. *Safer Water, Better Health: Costs, Benefits and Sustainability of Interventions to Protect and Promote Health.* Geneva: World Health Organization.

Robinson, Harry, and Wendy Wickwire. 1989. *Write It on Your Heart: The Epic World of an Okanagan Storyteller.* Vancouver: Talonbooks/Theytus.

Salina, Irena, Director. 2008. *Flow: for the Love of Water.* Documentary film. Produced by Steven Starr for Oscilloscope Laboratories.

Sandford, Robert W. 2009. *Restoring the Flow: Confronting the World's Water Woes.* Surrey, BC: Rocky Mountain Books.

Schlüter, Maja, and Claudia Pahl-Wostl. 2007. Mechanisms of Resilience in Common-Pool Resource Management Systems: An Agent-Based Model of Water Use in a River Basin. *Ecology and Society* 12 (2).

Shiva, Vandana. 2002. *Water Wars: Privatization, Pollution and Profit.* Toronto: Between the Lines.

Swyngedouw, Eric. 1999. Modernity and Hybridity: Nature, Regeneracionismo, and the Production of the Spanish Waterscape. *Annals of the Association of American Geographers* 89 (3): 443–465.

———. 2004. *Social Power and the Urbanization of Water: Flows of Power.* Oxford: Oxford University Press.

UNESCO (United Nations Educational, Scientific and Cultural Organization). 2006. *Water, a Shared Responsibility. The United Nations World Water Development Report 2.* Paris and New York: UNESCO and Berghahn Books.

Wagner, John R. 2008. Landscape Aesthetics, Water, and Settler Colonialism in the Okanagan Valley of British Columbia. *Journal of Ecological Anthropology* 12: 22–38.

———. 2012. Water and the Commons Imaginary. *Current Anthropology* 53 (5): 617–641.

Wisner, Ben, Piers Blaikie, Terry Cannon, and Ian Davis. 2004. *At Risk: Natural Hazards, People's Vulnerability and Disasters.* London: Routledge.

Wolf, Eric. 1990. Facing Power: Old Insights, New Questions. *American Anthropologist* 92 (3): 586–596.

Part I

⬮⬮⬮

COMMODIFICATION

As outlined in the introduction to this volume, *The Social Life of Things* provides a useful starting point for theorizing about the agency of water within social systems (Appadurai 1986). Unlike Appadurai and his coauthors, however, we are not attempting here to generate new theory about commodification and exchange theory as universal categories of human behavior. Also unlike the Appadurai volume, most contributors to this volume see commodification as a fundamentally destructive process. Although some contributors to *The Social Life of Things* describe negative examples of commodification—most notably slavery—their overall emphasis is on the types of commodities that most people take for granted, such as food, cloth, and jewelry. It is worth noting that while very few people complain about the commodification of cell phones, cars, or other objects produced entirely for sale in a market, and few complain about the growing commodification of homes, food, clothes and other objects once produced mainly for personal use, many people do complain bitterly about the commodification of water. In previous publications I have criticized water activists for their tendency to demonize commodification without distinguishing one type of commodity from another (Wagner 2010, 2012), but in this volume I wish to agree with them that the commodification of water is, indeed, very often negative and even disastrous in its outcomes. Several contributions to this volume demonstrate how and why this is so.

Part I begins with a chapter by Fabiana Li in which she addresses the issue of equivalence as encountered by farmers and mining corporation engineers in two Andean settings. Many scholars have emphasized that, in order to be exchanged as commodities, unlike things must be assigned equivalent values. As Kopytoff (1986) has pointed out, the assigning of equivalent values can be the source of much conflict and has led in some cases to the creation of separate and limited spheres of exchange, as described by Bohannan (1959) for the Tiv of Nigeria. As summarized by Kopytoff, the Tiv traditionally had three spheres of exchange: one

that involved the exchange of subsistence items such as food and tools, one that involved the exchange of prestige items such as cattle and ritual objects, and one that
involved the exchange of rights-in-people, such as wives and offspring (1986: 71).
The globalized cash economy in which transnational mining corporations operate
demonstrates no comparable sensitivity to different systems of valuation, however,
and requires that all objects circulate within the same sphere of exchange and be
assigned a cash value. Reducing all values to cash values does violence to local systems of valuation and to the social relations through which valuation occurs.

This fundamental issue is addressed directly by Li when she describes the con-
trasting perceptions of canal water by Peruvian farmers and the engineers employed by the Yanacocha gold mine. Whereas the engineers perceive the water
diverted from irrigation canals to the mine as equivalent to the treated water that
is later returned to the canal, farmers perceive the latter as degraded and non-
equivalent. Engineers base their opinion on the fact that the original canal water is
classified by national regulations as suitable for irrigation but not for drinking, and
that the treated water also meets national irrigation standards. They ignore that
the original canal water was classified as irrigation water simply because it flowed
through an irrigation canal and was never tested for potability; they ignore that
for generations farming households have safely drunk that water; and they ignore
evidence presented to them about the obvious differences of smell and color and
the chemical residues that remain visible on the land following use of the treated
water for irrigation. The argument over equivalence thus signifies a deeply rooted
conflict that pits local agricultural communities against mining corporations and
against the state that licenses mining activities and water diversions.

Chapter 2, by Veronica Strang, provides a classic example of how water can be
commodified through a combination of neoliberal state water licensing strategies
and water markets, and how this form of commodification inevitably privileges
some groups at the expense of others. She begins her chapter by noting that Cubbie
Station, in Australia, is "the largest private irrigation scheme in the southern hemisphere," and then goes on to document the historical process through which a particular coalition of political and economic interests managed to capture the lion's
share of water from the upper Murray-Darling Basin, the most intensively farmed
watershed in the country. She also documents the environmental degradation it
has caused, and its negative social and economic impacts on downstream farming
communities. She does not romanticize the previous generations of European settler communities who began the process of transforming the Australian landscape
into a "miracle" of "productivity" (40), but rather demonstrates the ways in which
recent forms of commodification have contributed to processes of environmental
degradation and social fragmentation among farming communities.

In chapter 3, Lyla Mehta describes how a series of dams built on the Narmada
River in India have displaced *adivasi* (indigenous) communities and how relocation
policies have led to a reduction in their quality of life despite government claims to

the contrary. Mehta does not explicitly describe this as a process of commodification, but her chapter is included in part I for two reasons: first, she demonstrates how the government's relocation policy is grounded in a particular concept of modernity according to which the monetization of livelihoods is automatically perceived as an improvement in well-being; Mehta points out that the forcible relocation of adivasi communities and simultaneously the forced monetization of their economic lives, in fact has reduced their quality of life (61–63). Second, Mehta documents one of the most common ways in which water is used to support commodification of other goods and products. Even if Narmada water itself is not being sold as government-issued water licenses in a private water market, it is being used to generate goods produced exclusively for sale in commodity markets, such as electricity and a variety of industrial and agricultural products.

This volume does not contain case studies of the types of water commodification that have been most controversial over recent decades, most notably the privatization of public water supply systems and the issuing of water licenses to the bottled water industry in situations that lead to deprivation for other users or environmental degradation. In chapter 8, however, Osumanu does touch on the issue of privatization of water services in Ghana, a topic he has addressed in more detail in another publication (Osumanu 2008). De Burgos ends his historical study of water provisioning in Suchitoto, El Salvador (chapter 5), with an account of the human rights abuses committed by the national government following protests against an attempted privatization of the state water utility. Readers who are particularly interested in the issue of commodification will also find a valuable contribution to this topic in chapter 12, where Sam and Armstrong, writing from an indigenous North American perspective, point out the ways in which globalization recapitulates colonization, and how both are directed toward the wholesale commodification of natural resources, including water, and the dispossession of indigenous peoples.

References

Appadurai, Arjun, ed. 1986. *The Social Life of Things: Commodities in Cultural Perspective.* Cambridge: Cambridge University Press.

Bohannan, Paul. 1959. The Impact of Money on an African Subsistence Economy. *Journal of Economic History* 19: 491–503.

Kopytoff, Igor. 1986. The Cultural Biography of Things: Commoditization as Process. In *The Social Life of Things*, edited by Arjun Appadurai, 64–94. Cambridge: Cambridge University Press.

Osumanu, I.K. 2008. Private Sector Participation in Urban Water and Sanitation Provision in Ghana: Experiences from the Tamale Metropolitan Area (TMA). *Environmental Management* 42: 102–110.

Wagner, John R. 2010. Water Governance Today. *Anthropology News* 51 (1).

———. 2012. Water and the Commons Imaginary. *Current Anthropology* 53 (5617–641).

Chapter 1

CONTESTING EQUIVALENCES
Controversies over Water and Mining in Peru and Chile

Fabiana Li

∞

The aggressive expansion of mineral extraction in Latin America, made possible by an influx of transnational investment beginning in the 1990s, has generated a number of conflicts throughout the region. Although motivated by a diversity of issues and demands, one of the common features of these conflicts is the centrality of water in disputes over rights (see Bebbington and Williams 2008). One of the most highly publicized conflicts in recent years involved the Pascua-Lama gold mine, straddling the border between Chile and Argentina. The Pascua-Lama mining project attracted international attention when the company in charge of its development proposed to relocate three glaciers at the mine site. Activists opposing the project contended that the glaciers played a crucial role in regulating the water flow in this arid region, and helped sustain agricultural production in the valley downstream. In Peru, meanwhile, the Yanacocha mine (the world's second-largest gold mine) also sparked protests that focused on the effects of mining activity on the quantity and quality of water used by communities in the vicinity of the mine. In both cases, some campesinos (peasant farmers), agriculturalists, environmentalists, and other critics contended that mining operations compromise the availability of water for irrigation and domestic consumption, even as the mining companies and their supporters steadfastly denied these claims.

The consequences of large-scale mining operations raise important questions for an anthropological analysis of water. The issues that emerge relate to areas that long have interested anthropologists: the social organization of irrigation systems, rural politics, and conflicts over resources (see, e.g., Gelles 2000; Orlove 2002;

Trawick 2002). At the same time, controversies over mining bring to light new concerns and questions that challenge us to rethink ideas about the use and commoditization of resources, and about the specificities of water itself—its chemical properties, symbolic value, and everyday use. In this chapter, I focus on the Yanacocha and Pascua-Lama mines to examine how water became a contentious site of politics.[1] I use the concept of equivalence to explore the tensions that emerged as campesinos, mining engineers, and other actors formed their respective arguments about the availability, use, and control of water resources.

My interest in equivalence is informed by studies that variously examine how things are made comparable and exchangeable. Anthropologists have long been interested in theories of value and exchange, which have been foundational in the discipline and remain influential (e.g., Graeber 2001; Guyer 2004). More recently, scholars have moved discussions about equivalence and commensurability in new directions. For example, these concepts have been used in explorations of number as a site of anthropological enquiry (see Guyer et al. 2010; Verran 2001), the anthropology of money (Maurer 2006), and medical anthropology (e.g., Hayden 2006; Lakoff 2005). The concept of commensurability has been used to query how people interpret and negotiate radically incompatible knowledge practices and forms of ethical reasoning (Povinelli 2001), a question that is especially relevant for examining environmental politics.

Espeland (1998) has examined the role of commensuration in water conflicts. She defines commensuration as the comparison of different entities according to a common metric. Price, cost-benefit ratios, and other forms of quantification and standardization make different entities comparable (see, also, Espeland and Stevens 1998). In other words, commensurability is determined based on a common standard of value (e.g., money), unit of measurement (e.g., rate of flow), or system of classification (e.g., chemical composition). Like Espeland, I am also interested in exploring the various ways in which disparate entities are brought into relation. I use the term "equivalence" to capture two related processes: First, equivalence refers to the scientific and technical tools used to make things quantifiable and comparable. Second, I take equivalence to be a political relationship that involves constant negotiation over what counts as authoritative knowledge.

I argue that the environmental mitigation strategies of mining companies, including compensation agreements and water management projects, rest on a logic of equivalences that makes the consequences of mining activity seem commensurable with the mining companies' proposed mitigation plans. Equivalences shift the focus of discussions about water resources toward technical solutions and economic calculations that make different forms of water measurable and comparable. Thus, in the context of mining, a logic of equivalences make it possible to reconcile different forms of value, such that, for example, the potential consequences of mining development can be offset with monetary compensation, or a company's water usage can be measured against the needs of agriculturalists downstream from the mine. These ways of making equivalences rely on a series of assumptions. The first is that water from different sources (e.g., springwater, chemically treated water, and glacier melt water) are interchangeable. Another assumption is that water availability can be determined based on average flows, without taking into account the extreme seasonal variability that characterizes Andean ecosystems (e.g., droughts and floods). Finally, a logic of equivalences assumes that changes in water availability and quality can be compensated with monetary payments, employment opportunities, and development projects.

These assumptions fail to consider the different ways in which people determine water quality and quantity through their everyday interactions with the landscape. Their evaluation of water quality does not necessarily relate to the chemical properties of water, and may not coincide with scientific studies. On the one hand, hydrologists often conceptualize the water cycle as behaving in a consistent, uniform manner (Budds 2009), and mining companies require a regular flow of water in order to sustain their operations. On the other hand, farmers are acutely aware of the uncertainty, unpredictability, and irregularity of water flows. This awareness and their assessment of the changes taking place makes them reluctant to accept the terms of equivalence embedded in compensation and mitigation programs.

A logic of equivalences reinforces dominant forms of knowledge that rely on scientific, legal, and technical mechanisms to measure and manage water resources. Equivalences must be negotiated and are always open to contestation, however. While water makes some equivalences possible, it also makes possible relationships and connections among a wide range of actors that can challenge or disrupt the logic of equivalences. In Peru, for example, relationships built around an irrigation canal helped mobilize people to protest against the mining company, and complicated the company's water management and compensation programs. In Chile, the glaciers at Pascua-Lama helped to widen national and international networks of activism, and introduced an element of nonequivalence in discussions about the mine. The loss of the glaciers and related effects on water flows, activists argued, could not be compensated or mitigated through the company's environmental management plans. I will describe each of these cases in turn to show the ways in which equivalences are established, and how they come to be accepted or disputed in conflicts over mining.

The Yanacocha Mine: Expanding Relations

In 2005, while I was conducting fieldwork in a community neighboring the Yanacocha mine, people explained to me that their irrigation canal did not carry as much water as it once did, and that the water had changed. They could no longer drink it and it had a strange smell; some worried that it was harming their pastures and livestock. For the majority of campesinos living in this area, dairy farming, sheep herding, and small-scale agriculture are the principle sources of livelihood. All these activities depend on water, and campesinos were concerned that the mine was radically altering ground and surface water flows.

Minera Yanacocha, a joint venture between the United States–based Newmont Corporation, the Peruvian company Buenaventura, and the World Bank's International Finance Corporation, began operating in 1993. As the mine expanded its operations, the company diverted water from one of the streams that fed the Tupac Amaru canal, since acidic runoff from the mine's waste rock pile had made it unsafe for human use. In 2004, more than 200 registered canal users received $4,000 each, plus the equivalent of $6,000 per user in community development projects, as compensation for the damages (all dollars are given in U.S. currency). Furthermore, the company promised to pump chemically treated water from its treatment plant into the canal to make up for the reduced water flows.

A year later, however, people again began to notice a reduction in water flows, and canal users believed the water being returned to them by the mining company did not adequately compensate for the water they had lost—in quality or in quantity. Although canal users have numerous ways of detecting the mine's effects—the changing color and taste of the water, diminished flows, or the strange specks on the pasture after it has been irrigated—they had to substantiate this sensory knowledge with measurements, technical data, and expert witnesses. One way to do this was to request a *inspección ocular* (visual inspection) of the canal. I was invited to participate in one of these inspections, which involved walking the length of the canal with technicians from the Ministry of Agriculture and mining company representatives while measuring water flows at various intervals (see figure 1.1).

Canal users wanted to begin the inspection at the canal's point of origin, now located within the mine's property. While the group of campesinos felt entitled to inspect the full length of their canal, entering the mine constituted a security breach, and we were met by a large contingent of security guards. The canal users demanded to speak to the mine's representatives, and a group of engineers arrived to answer their questions.

"Why does the water look yellow?," asked one man, pointing to the water trickling down a rocky slope. An engineer replied that Minera Yanacocha operates within the legal framework, and that the water met the legal water quality standards. The engineer admitted that this water was not suitable for human consumption, but said that according to Peruvian law irrigation water did not need to meet the same standards as drinking water. Furthermore, he stated that none of the naturally occurring springs in the area were legally fit for human consumption.

During this tense confrontation, the canal users and the engineers seemed to be talking about the same thing—water quantity and quality—but from different sets of assumptions. From the point of view of the campesinos, the mining company was stealing water from the natural sources that once fed their canal, and the treated water being returned to them was not the same as the water they once had.

Figure I.I. Inspection of the Tupac Amaru Canal, August 2005

For the engineers, the mine's responsibility was to ensure that the water met legal standards for water quality, and that the irrigation canals were replenished with water from its treatment plant to make up for reduced water flows. The engineers' assumptions rest on a logic of equivalences, which maintains that (1) water from a natural source is interchangeable with water from a treatment plant, (2) water quality is acceptable if proven to meet the established legal standards, and (3) the mine's effects on the canal can be reversed by returning the same amount of water that was lost and compensating canal users with monetary payments, employment, and development projects.

The controversies surrounding the Tupac Amaru canal exemplify how modern mining irrevocably transforms landscapes and ways of life. New mining technologies, along with neoliberal reforms and increased foreign investment, enabled mining to expand into areas formerly used for agriculture and farming. At the Yanacocha mine, more than half a million tons of ore are moved each day, and cyanide leaching makes it profitable to extract microscopic traces of gold from the ore. To mine safely and prevent flooding, groundwater must be pumped from wells around the mining pit to lower the water table, altering ground and surface water flows. When the Yanacocha mine disrupted the irrigation infrastructure running through its property, the canals became the focal point of protests against the company. Minera Yanacocha's recognition of the damages to six canals helped make the effects of mining tangible, something that was not always possible given that the consequences of modern mining (such as the health and environmental risks of toxic chemicals) are not necessarily immediate or visible. The canal thus solidified the sometimes indeterminate impacts of mining activity while also solidifying relationships that revolved around it.

The canal binds people through kinship ties as water rights are passed from parents to children, and it is also a crucial element of community membership, since it requires users to fulfill certain obligations related to the maintenance and administration of the canal. The canal users' associations (which organize the distribution of water and canal maintenance) played a key role in mobilizing people in the countryside to protest against Minera Yanacocha, and their large member base helped ensure a good turnout at marches and rallies (figure I.2). The canals connected campesinos from various communities, and brought the involvement of other actors, including government agencies and NGOs. Though not always internally coherent or free of political infighting, canal users' associations provided campesinos with a unified front to present their demands to the company.

The canals also enabled campesinos to negotiate new forms of livelihood: compensation packages, employment at the mine, donations, and development projects. In these negotiations, the mining company sought to establish equivalence between the canal users' loss (as measured in terms of water flows and decreased

Figure I.2. March against the Pascua-Lama mining project, December 2009

economic productivity) and the benefits to be obtained from the compensation agreements. What the company did not consider, however, was that the canal was already the center of a vast network of relationships that could thwart their calculations. These various relationships—made through affect and kinship, antagonisms and necessity—were built around the canal during its construction and its daily maintenance, and continued to expand through disputes with the mining company. The canal connected people and landscapes through relations that encompassed but were not reducible to economic or utilitarian concerns, and that disrupted the equivalences underpinning Minera Yanacocha's compensation and mitigation plans.

The Tupac Amaru Canal

My understanding of the material and affective connections between people and landscapes inspired me to think about the Tupac Amaru canal as a site of entangled social and natural histories (Cruikshank 2005; Raffles 2002). If the canal and those who make use of it are conceived relationally, it becomes possible to see the ways in which landscapes are made through constant engagement and interactions between people, land, and other elements of the environment (Ingold 2000). Knowledge and skill are gained through experience and people's everyday involvement in these landscapes. The socionatural history of the canal reveals this process of engagement, and informs the way people talk about the changes brought about by the mine.

In the early 1980s, campesinos themselves worked to channel water to their communities, a task that required hard physical labor, resourcefulness, and an intimate knowledge of the landscape. Building the canal was a slow, frustrating process of trial and error; completion of the canal took more than five years. In popular accounts, the canal came to stand as a symbol of campesino ingenuity, sacrifice, hard work, and self-sufficiency. The construction of the canal was the first of a series of events that together transformed the landscape and economy. With the canal came a gradual shift from small-scale agriculture to dairy farming, since growing crops became unprofitable due to poor soils, decreasing yields, and low prices for agricultural goods. Those with access to irrigation water turned their fields into pastureland for grazing cattle. When the mining company arrived, a new road leading to its installations connected peasant communities to the city of Cajamarca. The road provided access for delivery trucks from a transnational dairy company, facilitating the transport of milk and leading to the intensification of dairy farming.

While wage labor and mine-related employment were usually temporary and unstable, the sale of milk provided most families with a small but steady income. Water became a crucial asset, yet the reliance on dairy farming posed the constant challenge of producing enough pasture to feed the cattle. In communities around the mine, there were few conversation topics as recurrent as the availability of pasture, the coming of the rain, and the amount of water in the canals. For people who relied on water for their animals and dairy production, the initial problems with the Tupac Amaru canal were immediately evident. "The water didn't taste good anymore," explained a woman living in one of the communities that used the canal. People used to drink from the canal when they took the animals out to pasture and were far from other sources of water, but they had to stop doing so.

The potability of canal water has been a key point of contention in recent mining controversies, since the mining company argues that by legal standards irrigation water is never fit for human consumption, and was not safe to drink even before the mine's arrival. According to Peru's general water law, water is classified into three types: Class I and II are for domestic consumption, and Class III is for irrigation and drinking water for animals. According to the law, canal water falls into Class III, and therefore is not required to be safe for drinking. These legal standards do not coincide with the campesinos' use of the water, however. In an area where water is scarce, drinking from the canals when one is far from natural water springs is part of life in the countryside—*costumbre*, people say, meaning force of habit. Before the mine's arrival, the Tupac Amaru canal was used not only for irrigation, but also for cooking, washing clothes, and other household activities. Regardless of its legal classification, campesinos have different criteria for determining if water is fit for human consumption: its taste, its coloration, its source, and its effects on animals and pasture.

According to a user of the Tupac Amaru canal that I interviewed, giving up their rights to the diverted stream did not amount to selling the water to the mining company. What they wanted was for the company to return the water that was taken. I asked, If Yanacocha were to give back the same amount of water (from its treatment plant) that it took away, why should they agree to pay them, in addition, thousands of dollars in cash? To this, the user replied, "The water is no longer the same. It's treated. It's no longer natural." He conceded that the water being pumped into the canal was clean, meaning that it had gone through a treatment process in the mine's facilities. This coincides with the arguments of mining engineers, for whom the water was clean because it had been treated to meet legal quality standards. Yet for this canal user, the fact that the water was clean did not mean it was the same water—the two were simply not equivalent—and this lack of equivalence required additional compensation.

Multiplying Connections

Faced with ongoing problems with their water, campesinos have tried to circumvent the logic of equivalences implicit in the company's water management and

compensation plans. One of their strategies has been to modify the registry of canal users to increase the number of people who might benefit from the company's
compensation schemes. The canal registry contains the names of all water users
and the number of irrigation hours that correspond to each. If we examine the registry from three different periods, an increase in the number of users is evident: In
1985, when the canal had just been constructed, 90 people were registered as users.
In 2002, when negotiations to compensate canal users began, there were 205 users.
In 2004, there were 460 users—with many more people wanting to be added.

The increase in users from 1985 to 2002 reflects a general pattern of inheritance in which parents tend to divide their irrigation hours among their children.
However, the fact that the number of canal users doubled between 2002 and 2004
was a result of the compensation agreements. Many canal users divided their irrigation hours among their family members, so it was not uncommon for an individual to have a mere half-hour of irrigation time, which they may or may not use
to irrigate their fields. Another way to get on the registry was to purchase hours
from a registered user, though the legality of these transactions was questionable.

The growing registry of users challenged the company's attempts to limit its
liability by restricting the number of relations it was willing to recognize. In a
study of the Ok Tedi mine in Papua New Guinea, Kirsch (2006) argues that compensation claims reveal competing assumptions about responsibility and liability.
In some contexts, property rights are established by cutting the network, or restricting the number of claimants to an object of value. For compensation claims,
however, strategies of ownership incorporated a wider range of claims on persons,
which had the cumulative effect of keeping the network in view. Similarly, in the
Yanacocha case, the registry provided a way for campesinos to keep in view the
many relations among people, water, and other elements of the landscape that
were concealed or excluded from Minera Yanacocha's efforts to create equivalences.
Compensation agreements rely on technical and legal frameworks to resolve the
disputes, but in the case of the ever-growing canal registry, the canal's effect of
multiplying relations also led to the multiplication of conflicts.

As I have tried to show, water is not simply a resource—it is also part of a
relationship that emerges from particular experiences and interactions with the
landscape. What the logic of equivalences conceals are the different ways of knowing and inhabiting the landscape that do not correspond to technical language or
scientific systems of measurement. Equivalences help produce a dominant form of
knowledge that disqualifies opposing viewpoints, making it possible to justify mining projects in the name of progress and development. In the section that follows, I
turn to a mining project in Chile to show how a logic of equivalences underpinned
the water management plans proposed for the Pascua-Lama project, and how these
equivalences were challenged as various actors produced contested knowledges of
water use and availability.

The Pascua-Lama Mine: Disputed Glaciers

In 2000, Canada's Barrick Gold Corporation solicited the approval of the Chilean and Argentinian governments to construct what it touted as the world's first binational mine, Pascua-Lama. The gold mining project, straddling the border between both countries, was made viable when the governments of Chile and Argentina ratified the Tratado de Integración Minera (Mining integration treaty) in 2000 (Luna, Padilla, and Alcayaga 2004).[2] The Chilean part of the Pascua-Lama mine would be located in the Atacama region, in the country's Norte Chico (near north). What made this new project particularly noteworthy in a country known for its long mining history, was the presence of glaciers in the area to be mined. Activists contended that the glaciers played a crucial role in regulating water flows in this arid region. The project's opponents argued that the mine would sit at the headwaters of a watershed that sustains agricultural production in the Huasco Valley, which covers an area of 9,850 square kilometers stretching from the Andes to the Pacific Ocean.

The development of the mine added another layer of complexity to existing tensions over land, access to water resources, and the rights of Diaguita Los Huascoaltinos, newly recognized by the Chilean government as an indigenous group in 2006. The initial phase of mining exploration followed a dramatic change in agricultural production in the valley. Beginning in the 1990s, agricultural land was transformed into vineyards dedicated almost exclusively to the production of table grapes for the international market. Small-scale farmers, who had previously grown vegetables, grains, and assorted fruits for the regional market and local consumption, switched to the more profitable production of grapes for export. A majority of the valley's farmers continue to cultivate small land parcels (less than one hectare to five hectares in size), but coexist with commercial growers, the most powerful of which control more than 1,800 hectares.[3]

This uneven land distribution influenced the allocation of water rights in the Huasco Valley. According to Chilean water law, water rights (in the form of shares) can be bought, traded, and sold in the market. Water rights are separate from land rights, meaning that ownership of a piece of property does not automatically give the owner rights over the water that flows through it (Bauer 1998). In the administration of water resources, access to water is a technical matter, governed by market forces; this approach contributes to the neoliberal project while erasing the differential power of water users (Budds 2009). For small farmers in the Huasco Valley, however, unequal access to water and their relative lack of power in the Junta de Usuarios (the association that represents canal users) were a constant preoccupation. Farmers perceived the introduction of large-scale mining projects such as Pascua-Lama as a new threat that exacerbated these long-existing tensions and competition for scarce water resources.

The Environmental Impact Assessment (EIA) that Barrick presented for approval in 2000 to the governments of Argentina and Chile contained a plan to relocate three glaciers at the mine site. The company proposed to use a hydraulic digger to remove almost twenty hectares of ice sitting on top of the mineral deposit and transport the blocks of ice to another glacier of similar geological and geomorphological characteristics (Barrick 2000), where they would fuse together over time. In this unusual attempt to make equivalences, the hydrological impacts of moving the ice were deemed to be minimal, since they would be moved to a site similar to their original location. The glaciers were not treated as part of a larger ecosystem and water cycle, but as *trozos de glaciar* (pieces of ice) to be displaced. In spite of the controversy generated by the glacier management plan, Chilean authorities approved Barrick's EIA in 2001.[4]

Over the project's long trajectory, the glaciers appeared in environmentalist campaigns as emblems of the destructive character of extractive industries. In response to public pressure, Barrick modified its proposed project, and promised to "only access the ore in a manner that does not remove, relocate, destroy or physically intervene [the glaciers]" (Barrick 2010a). While the previous glacier management plan was abandoned, the new design continued to conceive of the bodies of ice as distinct entities that could be isolated from the effects of the mining operations going on around them. Additionally, the company commissioned scientific studies to show that these bodies of ice were not true glaciers, since they did not exhibit the movement and basal sliding that characterizes glaciers (Barrick 2005). In newsletters and public presentations, the company downplayed their importance by calling them ice reservoirs, or glacierettes, that were insignificant in terms of their contribution to the hydrological balance of the watershed.

Amidst disagreement about the nature of the glaciers, water availability emerged as a key issue in arguments presented by the company and its critics. In spite of the common focus of water, however, each party involved had a different understanding of how water availability should be measured, managed, and safeguarded. From the company's perspective, acquiring the legal right to use the water it needed for its operations legitimated its claims to this resource. Small farmers who were already critical of an unequal water distribution system saw the company as an unwelcome competitor, however. According to Barrick, the water to be drawn from freshwater sources on the Chilean side was a small amount: a total flow of forty-three liters per second. Based on the company's calculations, forty-three liters per second was equivalent to 0.3 percent of the water in the Huasco River flowing into the reservoir that feeds the canals (Barrick 2010b). In an interview, a Barrick engineer emphasized that water availability was simply not a problem—there was enough water, and it was agriculturalists, rather than mining companies, that needed to learn to use it more efficiently by improving irrigation systems and using it responsibly. Furthermore, he argued that because of their small size, the bodies of ice around the mine site did not contribute significantly to surface water flow.

A logic of equivalences is implicit in the assumption that the mine's water use can be measured proportionally to the flow rate of the river, and that glacier snow-melt is the same as other kinds of water (which are much more abundant and thus more important for the water cycle). For small farmers in the valley, however, these arguments did not assuage fears about seasonal water shortages and long periods of drought. Furthermore, the company's calculations did not correspond to the farmers' experiences of seasonal variability and chronic water scarcity that made glacier snowmelt seem crucial during the dry summer months. While agriculturists must adapt to the irregularities of water flows, a mine requires a constant supply of water to sustain its operations. The company's water use was presented as insig-nificant in relation to average flows, but agriculturalists' experience of seasonal and interannual variability raised concern about a further reduction in water levels.

Glaciologists calculate the amount of water that a glacier holds based on its size and density, which gives the glacier's water equivalent (Azócar and Brenning 2008). Compared to other glaciers, the water equivalent of the Pascua-Lama glaciers is small. When I mentioned this to a local agriculturalist, a key player in the antimin-ing campaigns, he insisted that while the glaciers' water equivalent might be low, it was significant when seen proportionally to the very small amount of water avail-able. He argued that it was not possible to make generalizations about a glacier's water contribution, since different places have distinct geographical and climatic conditions. By emphasizing the particularities of place, he privileged the knowl-edge of local farmers over the generalizations of scientific and corporate data.

Agriculturalists are attuned to the cyclical pattern of drought and floods that characterize the valley's climate. For example, periods of heavy rains are marked by the blooming of the nearby desert. The *desierto florido* (blooming desert) is a phenomenon usually seen every ten years or so, and its delay was cause for concern. During my visit in early 2010, a section of the river had dried up, something that people had never experienced in their lifetimes. Some people attributed this change to the mine's operations, including prospecting, road building, and other con-struction activities they believed were already depleting the glaciers and disrupting surface and subterranean water flows. While these changes are not reflected in scientific studies or the company's environmental monitoring, people rely on visual indicators and sensory experiences of environmental change (Strang 2004): for example, the fluctuation of water levels in irrigation canals, the location of water springs, and the position of the snowline relative to other years in recent memory (cf. Budds 2009).

Equivalence through Compensation

For those who supported the Pascua-Lama mining project, the legitimacy of the company's actions rested on its abidance of legal regulations; after all, the EIA had gone through a lengthy process of evaluation. I would argue that the making of equivalences is a fundamental aspect of the EIA process, and its approval is in fact contingent on the acceptance of these equivalences. This is because in the EIA a company must demonstrate that it has developed a mitigation plan to counter every potential risk associated with its operations (Li 2009). Equivalence between a mine's potential consequences and the company's proposed solutions is necessary to establish a project's technical viability. If an environmental risk cannot be dealt with through technical means, however, monetary compensation can serve as a way to redress environmental and social risks that cannot be prevented. Compensation agreements allow companies to negotiate directly with communities, making it more difficult for activists and government agencies to block a project's approval. In this way, the equivalences established through compensation help support a company's claim that a project is technically and environmentally feasible. In the Pascua-Lama case, signing a compensation agreement with the association of canal users provided a way to circumvent local opposition and secure the EIA's approval.

As was the case in Peru, water in the Pascua-Lama controversy was both a central point of contention and a mobilizing force. People's concerns over the potential consequences of mining activity focused on the canals used by small farmers and medium- and large-scale agriculturalists, and the administrative units that grouped canals users provided the structure for people to organize. The Junta de Vigilancia del Río del Huasco, the administrative entity that represents canal users and manages water use in the valley, initially took a stance against the mining project. Barrick needed to show that it had the social license to operate, but since it was unable to win the approval of individual farmers, the company's strategy was to obtain the support of the Junta. During the EIA consultation process, Barrick entered into an agreement with the Junta's board of directors, offering $60 million in compensation for "direct or indirect potential damages to the canal users" (Flores 2005). The money would be payable to the Junta in $3 million yearly installments over the mine's twenty-year lifespan, and was to be invested in the maintenance, construction, and improvement of irrigation infrastructure.

In the media and its own publications, Barrick presented the compensation agreement as evidence of the company's commitment to corporate social responsibility. The agreement elicited much criticism from local residents, farmers, and environmentalists, particularly because it was negotiated without the full participation of water users and before the government had approved the project's EIA. A logic of equivalences enables companies to make the costs of mining activity appear commensurate with the benefits derived from compensation agreements, but it also does more than this. Efforts to make equivalences through compensation have the effect of limiting the number of participants involved in the negotiations and obscuring the vast networks that connect people, irrigation canals, water, and glaciers.

Compensation agreements are intended to help resolve conflicts between companies and communities, but as the case of the Tupac Amaru canal showed, they can sometimes contribute to the perpetuation of conflict. In the Pascua-Lama case, the compensation agreement exacerbated existing tensions over land and access to water resources. For small farmers living in the upper part of the watershed, the agreement was a reflection of a historic power imbalance within the Junta. The Junta represented more than two thousand canal users, who were divided into four sections of the watershed, stretching from the mountains to the coast. The eight directors on the Junta's board were elected by the canal users, and the number of directors for each section was proportionate to the number of water shares in that section. In practice, this meant that some sections were more powerful than others, and some shareholders had so many shares they essentially had the power to elect themselves as directors. Meanwhile, farmers with a single share or a fraction of a share felt they had no voice in the decisions taken by the Junta.

Only the board of directors of the Junta voted in the negotiations with Barrick, while the rest of the canal users were informed of the decision after the agreement had been signed. The agreement was ratified through a system that reflected a recent concentration of land and water rights in the hands of a few landowners, especially with the introduction of vineyards for export production over the past two decades. Meanwhile the majority of people living closest to the mine site saw themselves increasingly marginalized, and mining represented an additional threat to their water resources and way of life.

In the Pascua-Lama case, technical arguments, calculations of water availability, and compensation agreements were necessary for establishing equivalences. As Barrick's negotiations with the Junta illustrates, compensation agreements can sidestep legal frameworks and exploit existing power asymmetries. Companies face increasing public scrutiny, however, and must present their actions as legitimate and in line with their commitments to social responsibility. A logic of equivalence can help create an image of environmental responsibility and accountability to the public, while at the same time limiting the spaces for public participation. The company did not consider that water, and more specifically glaciers, have the effect of expanding relations and forging connections across difference in ways that can bring about unexpected consequences.

Glaciers and the Limits of Equivalence

When Barrick acquired the concession for Pascua-Lama, the area to be mined might have seemed no different from other mines in Chile or other parts of the world, in spite of its high altitude. Barrick soon found, however, that the mine site's particular characteristics inspired antimining campaigns powerful enough to put in question the future of the project. For the territory to be mined was not simply land that the company could buy or acquire rights to, nor was it water in the form of a lake or lagoon, which are sometimes engulfed by open-pit mines. That the mineral deposits lay beneath glaciers—a particularly symbolic form of water and earth—put Pascua-Lama in the international spotlight and at the center of debates around climate change, water scarcity, and indigenous rights.

Before the arrival of the mining company, the valley's local residents (including small farmers and those who self-identify as Diaguita Los Huascoaltinos) had not used the term "glaciers" to describe the ice bodies at Pascua-Lama. Although they are not visible from the communities in the valley, animal herders were familiar with the area, and most people talked in more general terms about the snow and ice on the cordillera. Some people said their parents and grandparents had talked about *hielos eternos* (eternal ice). All these ideas became incorporated into the campaigns against the mine as they converged a transnational imaginary shaped by scientific and environmentalist discourses about global climate change and environmental conservation. In addition to agriculturalists and other local residents, solidarity activists in the capital city of Santiago, NGOs, and international supporters were drawn to the Pascua-Lama struggle. The image of glaciers was crucial for disseminating their message through emails, petitions, and in the national and international media.

Glaciers have become (controversial) standards of measurement, useful to scientists as markers of climate change. Their value has also come to be understood not simply in relation to their economic importance (for tourism and recreation, for example), but also in terms of their aesthetic qualities and ability to elicit an affective response (Carey 2007; Orlove, Wiegant, and Luckman 2008). Glaciers bridge social and natural worlds (Cruikshank 2005), and as such they make it possible to bridge the interests of multiple groups of people, including canal users concerned with water scarcity, and environmentalists (based in Chile and beyond) wanting to protect fragile ecosystems.

The poignancy of the Pascua-Lama campaign rested on the principle of non-equivalence: for many of the project's local and international critics, glaciers were an incommensurable value. As Espeland (1998) has shown, defining something as incommensurable can serve as a political strategy, as when land is described by indigenous groups as something that is not simply a resource, but that is constitutive of their identity. In order to be effective in opposing the mining project, activists had to make the glaciers into something that exceeded their definition as water resources. At the most powerful points in the campaign, the glaciers were presented as non-negotiable entities: their value exceeded any utilitarian use, and environmental campaigns led by national and international NGOs focused attention on their aesthetic qualities and value to future generations.

Crucially, these representations of glaciers connected with local sentiments that focused on an attachment to place and the defense of an agricultural way of life. Oftentimes, the specific concerns of agriculturalists did not coincide with the more generalized antimining stance of their supporters (for example, student activists and environmentalists from Santiago). Yet even if local ways of defining the problems did not always merge with that of nonlocal supporters, the glaciers provided a point of convergence without the need for agreement. The Pascua-Lama case shows how glaciers brought an affective dimension to the controversy, and this was something that corporations could not control. This affective quality of glaciers—their ability to evoke sentiments that coincided with but also exceeded preoccupations about water scarcity—placed it outside the realm of calculations and beyond a scientific rationale that relies on a logic of equivalences.

Conclusion

The two cases discussed in this chapter show that disagreements over what is and is not equivalent are at the heart of many current struggles over water. In the context of mining conflicts, different ways of knowing and experiencing the landscape inform disagreements over water quality and availability, as well as disputes over how water resources should be used and managed. For small farmers and others living in the vicinity of a mine, extractive activity produces dramatic changes that are altering ways of life and relationships between people and elements of the landscape, including the irrigation canals on which farmers depend. Yet as I have tried to show, these relationships and the conflicts that they generate do not relate solely to the utilitarian value of water resources. Furthermore, people's way of assessing and responding to environmental changes are based on observations and experiences that do not necessarily correspond with scientific studies or legal norms that establish quality standards (in the Peruvian case) or rights of access (in the Chilean case).

In both cases, water creates a complex set of relationships and reveals tensions and power inequalities. These relationships have shaped the terms of equivalence as companies sought to use the canals as the basis of compensation and technical management plans. At the same time, these relationships challenged the terms of equivalence, both by incorporating new actors into the debates and by revealing the affective dimensions of water and glaciers that exceed technical and legal concerns, hydrological studies, and economic tradeoffs. The connections that water makes possible brought about two different responses: in the Peruvian case, the mining company negotiated an agreement with local communities that would have limited compensation payments, but campesinos sought to circumvent that agreement by expanding the canal registry. In the Chilean case, Barrick also attempted to use a compensation agreement to obtain the support of agriculturalists, but this led to increased tensions and opposition from small farmers who believed they were not represented by the Junta. In this case, the glaciers had the effect of expanding relations and putting the project in the international spotlight. Although activism was not able to stop the development of the Pascua-Lama mine, glaciers remained a central focus in ongoing disputes over water, land, and community rights.

Notes

I wish to thank Marisol de la Cadena, Joe Dumit, Kate Dunbar, Penny Harvey, Ben Orlove, and Suzana Sawyer for discussions that inspired the ideas presented here. My thanks to Ben Orlove and John Wagner for the opportunity to participate in the conference panel that preceded this book project and for helping me define the focus of this chapter. The research was made possible by grants from the Social Science Research Council, the University of California Pacific Rim Research Program, and the British Academy's Newton International Fellowship Scheme.

1. My analysis of the Yanacocha case is based on two years of doctoral research conducted in 2005 and 2006 primarily in the city of Cajamarca in the Northern Highlands of Peru. Research on the Pascua-Lama case was conducted over two months between November 2009 and January 2010 in the capital city of Santiago, Chile, and in communities in the Huasco Valley, Chile. In both Chile and Peru, I interviewed small farmers, NGO representatives, community leaders, engineers, and other actors involved in the conflicts.

2. The treaty opened up the border region to mining development by creating what some critics have called a virtual territory (Luna, Padilla, and Alcayaga 2004) intended to attract transnational investment and facilitate the use and movement of natural resources across the border.

3. A survey of eighty table-grape growers in the Huasco Valley (CORFO 2008) illustrates this disparity: fifteen large agriculturalists (more than twenty hectares) control 85 percent of cultivated land; fifteen medium agriculturalists (five to twenty hectares) occupy 10 percent of cultivated land; and fifty small farmers (less than five hectares) cultivate the remaining 5 percent.

4. The project was put on hold because of the low price of metals. Over the next few years, however, the company identified additional gold reserves, which increased the size of the reserves (from 2.3 million to 17.8 million ounces of gold) and the company's investment in the project.

References

Azócar, Guillermo, and Alexander Brenning. 2008. Interventions in Rock Glaciers in the Los Pelambres Mine, Coquimbo Region, Chile: Technical Report. Department of Geography and Environmental Management. Waterloo, Ontario, Canada: University of Waterloo.

Barrick. 2000. EIA Proyecto Pascua-Lama (EIA Pascua-Lama Project). Santiago, Chile: Barrick Gold Corporation.

———. 2005. EIA Modificaciones Proyecto Pascua-Lama. (EIA Pascua-Lama Project modifications). Adenda N° 2. Santiago: Barrick Gold Corporation.

———. 2010a. Editor's Reality Check: Myth versus Reality. http://barrickbeyondborders.com/2010/09/reality-check/

———. 2010b. Pascua-Lama 2010 Responsibility Report. http://www.barrick.com/Theme/Barrick/files/responsibility-reports/2010/Pascua-Lama.pdf.

Bauer, Carl. 1998. *Against the Current? Privatization, Water Markets and the State in Chile.* Boston: Kluwer.

Bebbington, Anthony, and Mark Williams. 2008. Water and Mining Conflicts in Peru. *Mountain Research and Development* 28: 190–195.

Budds, Jessica. 2009. Contested H2O: Science, Policy and Politics in Water Resources Management in Chile. *Geoforum* 40 (3): 418–430.

Carey, Mark. 2007. The History of Ice: How Glaciers Became an Endangered Species. *Environmental History* 12: 497–527.

CORFO. 2008. Estudio de Levantamiento de Necesidades de Capacitaciones de los Trabajadores Agricolas del Rubro Uva de Mesa del Valle del Huasco (Study of Training Needs of Agricultural Workers in the Area of Table Grapes in the Huasco Valley). Vallenar: Corporación de Fomento a la Producción (CORFO).

Cruikshank, Julie. 2005. *Do Glaciers Listen? Local Knowledge, Colonial Encounters, and Social Imagination.* Vancouver, BC: University of British Columbia Press.

Espeland, Wendy Nelson. 1998. *The Struggle for Water: Politics, Rationality and Identity in the American Southwest.* Chicago: University of Chicago Press.

Espeland, Wendy Nelson, and Mitchell L. Stevens. 1998. Commensuration as a Social Process. *Annual Review of Sociology* 24: 313–343.

Flores, F. 2005. Pascua Lama Anuncia: Estamos Dispuestos a Modificar el Proyecto (Pascua Lama announces: We are willing to modify the project). http://www.chilesustentable.net/web/2005/07/07/pascua-lama-anuncia-estamos-dispuestos-a-modificar-el-proyecto.pdf

Gelles, Paul. 2000. *Water and Power in Highland Peru: The Cultural Politics of Irrigation and Development.* New Brunswick, NJ: Rutgers University Press.

Graeber, David. 2001. *Toward an Anthropological Theory of Value: The False Coin of Our Own Dreams.* New York: Palgrave.

Guyer J. 2004. *Marginal Gains: Monetary Transactions in Atlantic Africa.* Chicago: University of Chicago Press.

Guyer, Jane, Naveeda Khan, Juan Obarrio, Caroline Bledsoe, Julie Chu Souleymane Bachir Diagne, Keith Hart, Paul Kockelman, Jean Lave, Caroline McLoughlin, Bill Maurer, Federico Neiburg, Diane Nelson, Charles Stafford, and Helen Verran. 2010. Introduction: Number as Inventive Frontier. *Anthropological Theory* 10: 36–61.

Hayden, Cori. 2006. Generic Medicines and the Question of the Similar. In *Cinvestav*, 50–60. Mexico City: Cinvestav.

Ingold, Tim. 2000. *The Perception of the Environment: Essays on Livelihood, Dwelling and Skill.* London: Routledge.

Kirsch, Stuart. 2006. *Reverse Anthropology: Indigenous Analysis of Social and Environmental Relations in New Guinea.* Stanford, CA: Stanford University Press.

Lakoff, Andrew. 2005. *Knowledge and Value in Global Psychiatry.* Cambridge: Cambridge University Press.

Li, Fabiana. 2009. Documenting Accountability: Environmental Impact Assessment in a Peruvian Mining Project. *Political and Legal Anthropology Review* 32 (2): 218–236.

Luna, Diego, César Padilla, and Julián Alcayaga. 2004. *El Exilo del Condor: Hegemonía Transnational en la Frontera.* (The exile of the condor: Transnational hegemony at the border). Santiago, Chile: Observatorio Latinoamericano de Conflictos Ambientales.

Maurer, Bill. 2006. The Anthropology of Money. *Annual Review of Anthropology* 35: 15–36.

Orlove, Ben. 2002. *Lines in the Water: Nature and Culture at Lake Titicaca.* Berkeley: University of California Press.

Orlove, Ben, Ellen Wiegant, and Brian H. Luckman, eds. 2008. *Darkening Peaks: Glacial Retreat in Scientific and Social Context.* Berkeley: University of California Press.

Povinelli, Elizabeth A. 2001. Radical Worlds: The Anthropology of Incommensurability and Inconceivability. *Annual Review of Anthropology* 30: 319–334.

Raffles, Hugh. 2002. Intimate Knowledge. UNESCO.

Strang, Veronica. 2004. *The Meaning of Water.* Oxford: Berg.

Trawick, Paul. 2002. *The Struggle for Water in Peru: Comedy and Tragedy in the Andean Commons.* Stanford, CA: Stanford University Press.

Verran, Helen. 2001. *Science and an African logic.* Chicago: University of Chicago Press.

Chapter 2

DAM NATION
Cubbie Station and the Waters of the Darling

Veronica Strang

∞

We are extremely proud of what we have achieved out there. We
have created a magnificent piece of irrigation infrastructure. It is at
world standard, if not the best in the world.
　　—Cubbie Station CEO John Grabbe, quoted in Uhlmann,
　　　"Government Tight-Lipped Over Cubbie Station Sale"

The darker it gets, the faster we're driving.
　　　　　　　　—Adams and Carwardine, *Last Chance to See*

Introduction

Australia's Cubbie Station is the largest private irrigation scheme in the southern
hemisphere. With ninety-six thousand hectares, stretching along twenty-eight kilometers of the Culgoa River floodplain in southern Queensland, it holds fifty-one
water licenses and has a water storage capacity of 537 gigaliters. It grows mainly
cotton, a crop that—though highly profitable—mines nutrients from the soil, depends heavily on pesticides, herbicides, and fertilizer, and—above all—requires
major quantities of water (figures 2.1, 2.2).

Cubbie is located in the upper Murray-Darling Basin (MDB), Australia's largest
and most intensively farmed river catchment. Since it diverts a significant proportion of the water that would otherwise flow across the NSW border into the
Darling River, the station is widely regarded as a major contributor to the Basin's
rapidly worsening ecological crisis. To many downstream farmers, environmental

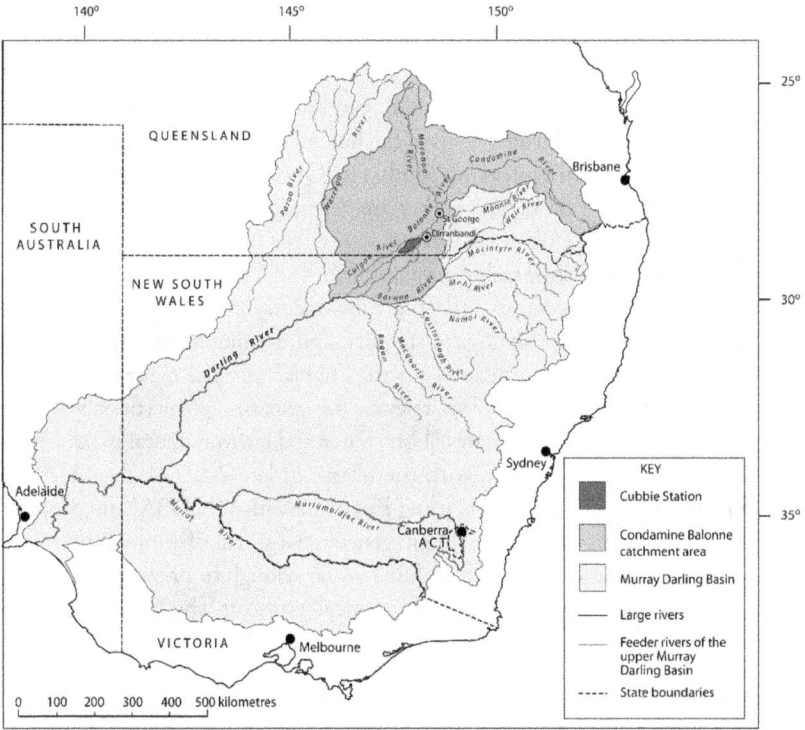

Figure 2.1. Cubbie Station and the Murray-Darling Basin

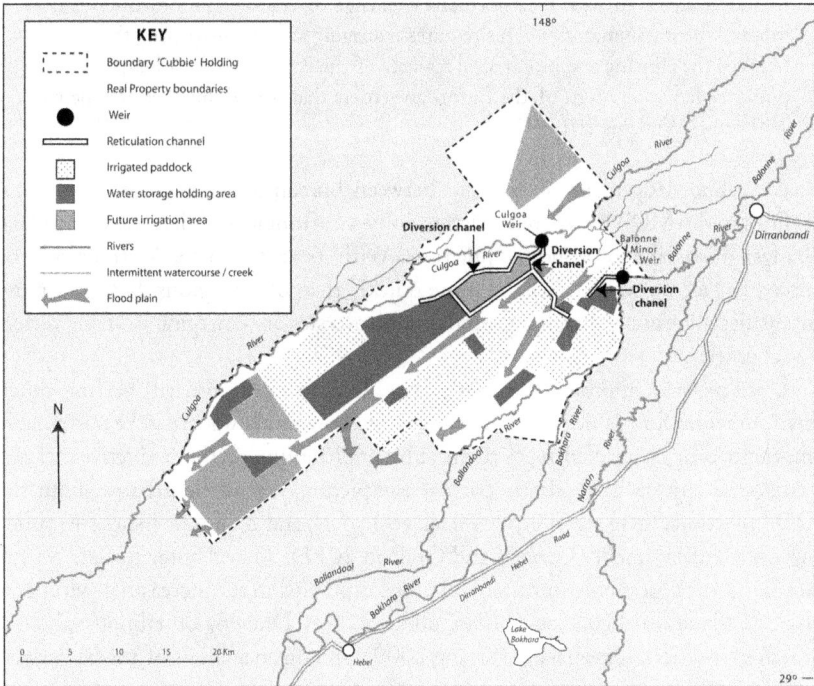

Figure 2.2. Cubbie Station diversion channels

groups and media commentators, it is the epitome of corrupt capitalism: a vampire sucking the lifeblood from the river catchment and its human and nonhuman inhabitants. But to the irrigators of south-east Queensland, the rural communities—Dirranbandi and St. George—that are dependent on them, and the owners and shareholders of the Cubbie Group, the station is a shining example of successful and (in their terms) sustainable development.

These alternate perspectives reflect a fundamental divide between concerns about long-term social and ecological stability, and dominant ideologies committed to competitive productivism and growth. Cubbie's size has made it a focus for controversy, but the issues it raises are relevant for irrigation properties across Australia and central to national debates about water and environmental management. These divisions were highlighted with the release in late 2010 of the *Guide to the Proposed Basin Plan* by the Murray-Darling Basin Authority (MDBA). Its proposed reform of water management in the Basin generated a storm of protests both from conservation organisations who felt it didn't go far enough to prevent further degradation of the aquatic ecosystem, and from local irrigators who felt that their way of life was being threatened.

> The competing tensions between water extraction for immediate human use and water essential to the long-term ecological function and sustainability of the rivers and groundwater systems in the Murray–Darling Basin (MDB) sit at the centre of public policy debate on water reform. Yet it is much more than this. The people of the Basin are faced with the enormous challenge of transforming themselves into more resilient communities. This requires managing and reconstructing the conflict between the climatic and biophysical realities of the Basin and the earlier private and public policy aspirations of the European settlers that have dominated for the past 150 years (Williams 2011: 1).

An acknowledgement of the gap between human aspirations and ecological realities in the MDB has therefore led to some confrontational demands for radical changes in social and economic practices. With its arid climate, Australia may be forced to face up to this demand more quickly than other regions, but as farming intensifies all around the globe, similar controversies are emerging in many different countries.

Until recently, approaches to water crises have focused primarily on how much environmental flow is needed to sustain rivers; how much water can be sustainably impounded or abstracted from them. The implications of this reductive techno-managerial approach are clearly part of the picture, but as the debates about the MDB illustrate, there are multiple social, ecological, and economic issues surrounding water management (Connell and Grafton 2011). This chapter focuses on the social, political, and environmental relations expressed in the increasingly vituperative representational contests about Cubbie Station. Drawing on ethnographic research in southern Queensland (Strang 2009), and on an analysis of media debates and political records, I consider how and why—at what appear to be enormous ecological and social costs—irrigation has become so central to Australia's agricultural economy. What can be done to avert the looming collapse in the MDB and in similarly overexploited ecosystems? What kinds of changes might enable more-sustainable modes of human-environmental engagement?

History and Heresy

Cubbie Station is the product of a particular history of place-making. Prior to European settlement, Aboriginal Australians maintained, for many millennia, a highly sustainable mode of environmental interaction in which small communities were inalienably tied to specific areas and, in particular, to their waterways, enjoying close affective relationships with a sentient ancestral landscape. Critical to this sustainability was a flexible and diverse use of resources, well designed to accommodate extreme variations in water availability (Holdaway, Wendrich, and Phillips 2009).

European settlers' engagements with the environment cast the new continent, its indigenous people, and its uncooperative "nature" in more adversarial terms. The arid landscape, with the vast void of *terra nullius* at its heart, was something to be forced into submission by the heroic agricultural efforts of the early battlers (Rose 1992; Schaffer 1988; Strang 1997). From the beginning, places with reliable water were the most highly prized and—being also the most important sites for the indigenous inhabitants—the most aggressively appropriated.[1]

Nevertheless, early settlers in Australia initially employed modes of engagement that had some features in common with Aboriginal practices. They established interdependent rural communities practically and symbolically joined by shared waterways, and early farming practices were sufficiently low-key to have (relatively) limited ecological impact. From this stable network of landowners, came a squattocracy that formed the backbone of Australian governance.

The process of civilizing the unruly land and waterscape entailed clearing and fencing spaces, building settlements, and, above all, capturing water. Dams, pipes, and water diversions transposed economic practices from a reliably wet and temperate environment to the most arid and variable climate on Earth. Irrigation schemes introduced in the late 1800s were seen as a great step forward, transforming dusty death into fertile life and providing farmers with a buffer against the vagaries of floods and droughts.

Accounts of this period describe with missionary zeal how, through the taming of water, a hellish and recalcitrant land was miraculously turned into a green, productive heaven. These representational tropes recur in contemporary discourses. The most central theme is one that appears cross-culturally, presenting water as a generative substance, and as a basis for wealth and health (Strang 2004). As elsewhere, the ownership and control of water provided social agency (Strang 2009, 2010). Thus, irrigation established the agency and power of the early settlers, giving men God-like directive powers over nature.[2] These ideas are amply illustrated in Ernestine Hill's classic *Water Into Gold: The Taming of the Mighty Murray River*, which describes the introduction of irrigation in the waterless wasteland of the MDB in the late 1800s:

> A great story, the transfiguration of a continent by irrigational science. This miracle of yesterday ... in the drought-stricken Murray River Valley ... inspired the stupendous national schemes for water preservation, conservation, allocation, that are changing the face of Australia today. (Hill 1958: v)

Chapters entitled Miracle of the Murray, Apostles of Irrigation, Utopia on the Murray, Acts of God, and Reining in the River describe how irrigation helped farmers to combat "Pharaoh's plagues" and "the fiery breath of a terrible drought" (1958: 38).

> Australia *Felix* was an arid waste, a hell of heat and flies, the old grey Murray ... pouring millions of gallons a day into the salt sea.... One man questioned the divine Creator's plan.... Irrigation was his cry ... to Hugh McColl was given the first vision of the Australia-to-be. (Hill 1958: 39)

This vision was embraced by the young Alfred Deakin who

> [s]aw the bare and blinding desert transmuted by industry and intelligence into orchards and fields of waving grain.... It was an age of big undertakings in a very young country; of valour rather than discretion; of venturing to win. The Victorian Government listened with interest to the youthful St Paul ... and set him to achieve the miracle. (Hill 1958: 40)

Thus, superimposed on a foundational view of the continent as a hostile adversary (which placed little value on its ecological needs) was a heavenly vision of irrigation and unlimited productivity. It expressed a desire to usurp an apparently less-than-divine plan and seize dominion of the material world, to the extent that any water leaking away to the sea was seen as wasted.

Within a few decades, the basin was transformed with dams, weirs, channels, pipes, and pumps, and by the mid-1900s an expanding population and an enlarging market were driving agricultural intensification across the continent, invariably at the expense of local ecosystems. Much of Australia's population—like those in other industrialized countries—shifted from rural to urban areas and acquired greater mobility. Homes become properties to be bought and sold with increasing frequency, and the localized communities that characterized earlier rural settlement began to dissolve, creating more fluid, ephemeral—and thus more individuated—social relations. In the late 1900s, a neoliberal emphasis on deregulation, competition, and growth encouraged the centralization of government and a concurrent decrease in local governance. Nevertheless, strengthened by their wealth and long-term social networks, the dynastic families of the rural squattocracy were able to extend their political agency upwards and outwards, retaining their central position in Australian politics.

The Murray Darling

Covering 14 percent of the continent, the MDB today contains 75 percent of Australia's irrigated agriculture, 25 percent of the country's cattle, 50 percent of its sheep flock, and 50 percent of its cropping land. It supplies drinking water for three million people. This transfigurative process has been repeated in river basins all around Australia. In addition to thousands of water bores tapping into aquifers: "[t]here are close to 450 dams over 10 metres in height, plus numerous regulatory structures … In Victoria alone, it is estimated that there are more than 300,000 farm dams" (Crabb 1997: 7). Irrigation now comprises the major form of water use in Australia, taking approximately 74 percent of the total.

Areas of southern Queensland previously devoted to cattle grazing have been transformed by the capture of overland flows and irrigation on a grand scale. To local irrigators, this is a positive outcome. A codirector of Cubbie Station described its designer, Des Stevenson, as "one of the first Australians to really understand river hydrology and the future value of water" (Paul Brimblecombe, as cited in Commonwealth of Australia 2006a: 17). To John Grabbe, Cubbie's CEO, the station is "a world leader in the efficient use of water for irrigation farming" (Commonwealth of Australia 2006a: 17).

But the MDB is also a world leader in ecological degradation. It has lost 90 percent of the wetland areas forming critical breeding areas for migrating birds (C. Kerr 2010: 1). A scientific audit described thirteen of its river valleys as in "very poor health" and a further seven as in "poor health" (Murray-Darling Basin Commission 2008: xii). "Both irrigation-induced and dryland salinization are present over large areas" (Murray-Darling Basin Commission 2010: 2) and there is now so little flow in the system that the Murray estuary requires constant dredging.[3]

Similar impacts of irrigation-based farming can be seen all around Australia, threatening both aquatic and terrestrial biodiversity:

> Some 450 species of plant, insect and bird life are under threat in … south-west Western Australia alone … [and] 2.5 million hectares of land are already affected by salinity, and there is the potential for this to increase to 15 million hectares. Much of this is Australia's most productive agricultural land. (Commonwealth Scientific and Industrial Research Organisation [CSIRO] 2010: 1)[4]

This degradation has been largely blamed on a prolonged drought that lasted from 2003 to 2009. These occur regularly in Australia: in the past century widespread droughts (and crippling stock losses) were recorded in1911–16, 1939–45, 1963–68, 1972–73, 1982–83, and 1991–95. Now, though only publicly admitted as a reality since the election of a Labor federal government in 2007, climate change is seen increasingly as a factor:

> Since the middle of the 20th century, Australian temperatures have, on average, risen by about 1°C ... Rainfall patterns have also changed—the northwest has seen an increase in rainfall over the last 50 years while much of eastern Australia and the far southwest have experienced a decline. (Australian Bureau of Meteorology 2010)

But a focus on drought and climate change—though clearly relevant—obscures a more basic issue. Irrigation in Australia has doubled in every decade since the 1960s. Many rivers, including those that compose the MDB, have been massively overallocated to divert as much water as possible into primary production. This has little ecological impact during extreme floods, but these are as intermittent as droughts and the constant curtailment of environmental flows is critical, particularly during drier periods. "The states have utterly mismanaged water entitlements in the Murray-Darling Basin, granting many more of them than the system can sustain" (Uhlmann 2009: 1).

The system is usually considered in ecological terms, but it is not just biodiversity that is at risk, but also cultural diversity and well-being. The reframing of land and water as commodities and the construction of a more brutally competitive environment has removed many local mechanisms of social and economic support, further undermining rural communities and the sense of belonging these provide. As ecosystems reach a point of crisis, farmers are increasingly in conflict with each other, with conservation groups, and with the urban populations of Australia. Depending on increasingly exploitative farming practices, they have gone out on a limb and placed too much weight on it. To understand why this bough seems about to break, it is necessary to consider the social and political patterns that have led to this alienation.

Alienating Water

Objects often move through stages of becoming commodities in "a shifting compromise between socially regulated paths and competitively inspired diversions" (Appadurai 1986: 17). Kopytoff notes that these progressions reveal "a moral economy that stands behind the objective economy of visible transactions" (1986: 64). Despite regular contests over public and private ownership in industrialized countries, water, until recently, has remained largely positioned as a common good. This term encapsulates the central meanings of water as the basis for collective well-being and reproductive potential, and acknowledges a moral imperative to maintain access for all. In essence, it asserts that water cannot (or should not) be treated wholly as a commodity.

However, a recent ideological trend toward narrow economic rationalism has resulted in less "socially regulated compromise" and more "competitive diversion" (Appadurai 1986: 17) to the extent that water in Australia has been radically repositioned as a resource, effectively separating its economic meanings from the broader, more-connective ideas previously encapsulated in notions of water as a common good. This conceptual diversion appears to have redirected water's social meanings into a side reservoir containing art, culture, and recreation, and reduced the mainstream to a focus on economic value.

While water commoditization has been presented in Australia as a positive step toward rational resource management, it requires the privatization of water and is fundamentally competitive. This chapter argues that competitive modes of economic engagement are intrinsically adversarial and so work against the cooperative relationships necessary for the collaborative conservation of hydrological flows, and against the valorization of noneconomic social and ecological needs. This is not news: consistently, in the history of water management, wherever privatization has occurred, social and ecological costs have tended to be externalized. Such externalization constitutes a form of alienation among people, and between people and place. In this way economic practices have become, in Polanyi's terms, disembedded from their social and material contexts (Polanyi 1944). What the case study highlights is how this process can be manipulated by elite groups so they accrue benefits at the expense of wider democratic and environmental interests.

In Australia, this disembedding made it possible to alienate water both from the land and from its inhabitants. By the 1980s, it was painfully clear that water resources were far from being the unlimited potential imagined by the irrigation apostles. Aided by a discourse of crisis, governments began to exert greater control over freshwater. While represented as an effort to protect local ecosystems and communities, moves toward quantification and volumetric measurement also enabled the reframing of water as an economic asset. Initially, Water Allocation Management Plans (WAMPs) regulated the amounts of water that farmers could abstract each year. There were moves to demand licenses and meters for new bores (and even for the thousands already drilled into aquifers), and to limit the amount of water that farmers could impound with dams and weirs. But, like land leases, water licenses explicitly mandated that these should be used for productive purposes, and neither state nor federal governments questioned the wisdom of further economic growth and intensification.

Advised by the influential Wentworth Group of Concerned Scientists, a right-wing think tank, the Howard government promoted market rule in which privatization and competition were presented as the basis for sustainable development. According to this economic logic, commoditization would lead to more-efficient water use, persuading farmers to irrigate only the most high-value crops. The Council of Australian Governments (COAG) initiated a scheme for water trading, reconstituting the allocations previously tied to land as private assets that could be traded on a speculative market. Caldecott describes this as privatization by stealth:

> Those who support water reform have been careful not to use the privatization word and accordingly the public have not been told the truth by the Council of Australian Governments (COAG) that their water reforms are about enhancing the market privatisation of the waters of the MDB. What was once a licence granted by state governments to extract a quantity of water by property owners ... has been unbundled into a marketable property right that can be owned by anyone.... It remains unclear to me how Australian governments achieved this monumental theft for markets without firstly gaining the direct approval of the Australian people ... this is truly an economic rationalist experiment on a grand scale. (Caldecott 2008: 10)

Some farmers were conscious of the social alienation implied by market rule: "It really worries me that some oil-rich Arab sheik might come in and decide he wants to buy half our water" (Graham Moon, fieldwork interview 2006). But the reframing of water allocations as tradable assets subsumed more-comprehensive visions of water as a connective social substance and as the basis of collective regenerative capacities, asserting instead a far more reductive economic view of water as gold. Public discourse became dominated by this metaphor. The Irrigation Association, meeting in Brisbane in 1998, echoed Hill's well-known text by taking Water Is Gold as its conference title. A butcher in Dirranbandi commented, when the 2010 floods finally came, "It doesn't rain money but this rain is almost as good" (Greg Stephens, as cited in Elks 2010: 2).

Assisted by emotive discourses about crisis and scarcity, water trading certainly increased water's economic value and thus its cost. To this extent it pushed farmers to target irrigation more precisely, leading the National Water Commission (NWC) to claim, "[W]ater trading is providing significant economic, social and environmental benefits across the southern Basin" (2010: 1). This positive assessment, however, was (and remains) based on the supposition that further agricultural intensification and growth are both sustainable and beneficial.

Overall, water trading appears to have further intensified resource use, awakening sleeper (unused) water allocation licenses and shifting attention to more profitable but ecologically exploitative crops such as cotton (Isaac 2002). It also has encouraged farmers to circumvent the need for costly abstraction licenses by constructing water harvesting impoundments to capture diffuse overland flows.[5] Until recently, as long as these were off-stream and below a certain height, hastily bulldozed ring tanks or turkey nest dams remained unlicensed, and unregulated.[6] In *The Dams that Drank a River*, Hodge describes how "a frenzy of dam building and land clearing ... has turned the Condamine-Balonne river system into a slave to cotton ... 70 percent of the water that once flowed into the Narran Lakes is now extracted" (2001: 1) (figures 2.3, 2.4, 2.5, 2.6).

Figure 2.3. Irrigation equipment typical of what is now used for introduced European crops

Figure 2.4. Wivenhoe dam, South-East Queensland

Figure 2.5. Farm dam, South Queensland

Figure 2.6. Irrigation channel, Queensland

The replacement of governance with market rule led to a competitive scramble to control freshwater resources, generating intense conflicts between upstream and downstream water users. It pushed Australia's trajectory of human-environmental engagement in an increasingly unsustainable direction. In the past two decades these social and ecological issues have become a source of deepening anxiety, gaining support for countermovements and for critiques of the externalization of production costs to ecosystems and communities. Australia therefore contains rising social and political tensions: contests over water allocations between powerful primary producers and increasingly angry smaller farmers; a widening gulf between cosmopolitan urban populations and conservative rural communities; and conflict (which is similarly socially and geographically located), between countermovements focused on ecological and social concerns, and right-wing elites that remain politically dominant and committed to further growth.

Cubbie Station encapsulates all of these issues. It is also both metaphorically and literally over the borderline: Queensland contains the top end of the MDB, and the Culgoa floodplain is just above the NSW border. Thus different state identities, interstate rivalries, and transboundary conflicts also play important parts in this story.

Over the Line

Australian federation was fiercely resisted by states determined to maintain their legislative independence. Queensland agreed to join in very reluctantly in 1901, and from 1968 to 1987 its cultural and ideological singularity was reinforced by Joh Bjelke-Petersen's far-right National Party government. With this government's "develop at all costs" approach, all forms of resource utilization were permitted—

indeed exhorted—to proceed with a free hand. Dams, bores, irrigation channels? The more the better, the bigger the better.

Recently, the federal government has made strenuous efforts to try to wrest the control of freshwater away from the states. John Howard established the NWC in 2004 and initiated regional catchment groups that, armed with federal funding, were partially successful in usurping state and local water governance. But the Queensland government continued to hand out water allocations with little regard for the consequences, most particularly those across the border in its major state rival—the wealthy, urbanized and effete New South Wales.

In the Queensland of the 1980s and early 1990s "bloody greenies" were scornfully dismissed as dangerous cranks—a fringe movement best ignored. Echoing the Great Divide between (male) Culture and (female) Nature established in the early colonial era, such terms often went hand in hand with views about "bloody communists" and "bloody pooftas." Though public discourses have become slightly less extreme, demographic analyses of activist groups and interviews with informants suggest that social and environmental concerns are still seen as largely feminine.[7]

This draws useful attention to the machismo that often attends competitive economic practices. In contemporary secular discourses, agriculture is less often described in patriarchal religious terms, though the tropes apparent in Hill's *Water Into Gold* (1958) recur in farmers' self-representations as paternal guardians of the land with a sacred duty to provide food (Strang 2009). Today, economic activity more often borrows imagery from armed conflict and sports, recounting battles, fights, captures, crashes, and sinkings; and the deeds of heroes, gamblers, winners, and losers. In this discursive context, Cubbie's aspirations to be a world leader and a St George resident's lack of empathy for "those whingeing buggers ... down there" (Lockyer 2010: 1) carry a strong whiff of testosterone.

Strategically positioned on the Culgoa floodplain, Cubbie impounds a significant percentage of the water that would otherwise flow into the Darling River. Its vast storage infrastructure is designed to take advantage of flood events. When major floods occur, water continues to flow into the Basin. But such overwhelming floods are infrequent, and with climate change are becoming even more so. Farmers and dryland graziers south of the border maintain that Cubbie and neighboring irrigation properties prevent regular medium and small floods from proceeding downstream, sacrificing the social and ecological health of the Murray-Darling to the profits of cotton farmers upstream. "The issue that we had all along is with the small and medium-sized flows, which can be impacted very severely by the extraction upstream ... it's a gross transfer of wealth from the floodplain down here to the irrigators upstream" (Rory Treweeke, as cited in Lockyer 2004: 1–2). Ian Douglas, national coordinator of the activist group, Fair Water Use, observed that Cubbie Station can store about a third of the 1,500 gigaliters needed to flush water through the Murray-Darling system: "The statistics are clear. We know the impact" (in Guest 2008: 1). Peter Black, the NSW member of parliament for the Murray-Darling, was similarly forthright:

> Cubbie Station is stealing water from the three States as I speak. Yesterday when I held up a picture of the lower Darling River with virtually nothing in it, the urgency of the matter was obvious. (Black, as cited in Parliament of New South Wales 2006: 2909)

Cubbie Group chairman Keith de Lacy countered that the station is unfairly accused of guzzling water, dismissing downstream concerns and requests for governments to act as "silly talk from … silly people" (in Guest 2008: 2). Cubbie's (then) joint managing director, John Grabbe, argued that it only abstracts between 0.8 and 1.0 percent of all the water abstracted from the MDB, and there are thus between 100 and 120 equivalents of Cubbie Station in the Basin as a whole (Commonwealth of Australia 2006a: 17).[8] The implication is that the "whingeing buggers" downstream are merely peeved that they have been beaten in a competition for water and profit. Or as another major irrigator on the northern Darling River said, "The people who complain would like no development at all upstream of them so they could have it all" (John McKillop, as cited in Roberts 2008: 1).

Water Power

Hydrological flows, of course, do not take place on a level playing field: Cubbie is advantaged in part by its location. But, more importantly, in the decision-making processes directing water management and distribution there are major inequalities in the power and agency of different groups. Cubbie's ability to capture licenses and water demonstrates the importance of understanding how social and political networks function, and how their genealogies influence events. The critical light cast on the station has illuminated some linkages that its opponents see as outright corruption. A little bit of political context helps to locate these. First, at a federal level, the right-wing Liberal Party government held power for over a decade under John Howard's leadership, in coalition with the far-right National Party. It was succeeded in 2007 by the Australian Labor Party (ALP), led by Kevin Rudd.[9] Like New Labor in the United Kingdom, the ALP is purportedly center-left, but espouses many neoliberal policies.

In Queensland, following a major exposure of political corruption, over thirty years of National Party rule ended in 1989 with the election of a Labor government under Wayne Goss' leadership. The Labor Party also prevailed in 1998 and, despite another scandal in 2000 (this time over electoral processes), has remained in power since, largely because of urban support in Brisbane. However, the Liberal National Party (as the right-wing coalition is now called in Queensland), remains dominant in rural areas, and its core is still composed of dynastic land-owning families whose national and international connections have been strengthened by participation in global markets.

The primary industries in Queensland often have provided leaders for the National Party at both state and federal levels. They also have powerful associations, including the Australian Irrigators Association, the Queensland Irrigators Council, the National Farmers' Federation, and the Queensland Farming Federation (until recently rather tellingly called Agforce). The Business Council of Australia brings together the CEOs of 100 of the country's leading corporations "to contribute to public policy debates" (Business Council of Australia 2010: 1). As this suggests, the major purpose of these associations is to represent farming and commercial interests to government bodies. There is considerable crossover, and commercial interests are well represented on influential government committees. For example, the federally appointed MDB Commission was formed in 1988. Its president was a former National Party leader and its CEO used to be the executive director of the National Farmers' Federation.

The owners and managers of Cubbie Station are similarly involved. Keith de Lacy held roles as treasurer and minister for regional development in the Goss Labor government, whose chief of staff at that time was Kevin Rudd. As well as being a former Water Resources Commission advisor and a member of the MDB Community Advisory Committee, John Grabbe has long been a major player in the National Party. Cubbie sponsored the election of Barnaby Joyce as the senator for Queensland in Canberra and as the leader of the Nationals in the Senate. Thus Cubbie "has its own Senator, Barnaby Joyce, a grateful recipient of donations from Cubbie and Grabbe and a dogged advocate of Cubbie's interests" (Keane 2010: 1).

In 2002, in response to increasingly vocal protests about Cubbie's activities, Queensland Premier Peter Beattie proposed that the state should buy the station and its water allocations with $160 million of federal financial assistance. This proposal met with public protests in Dirranbandi and fierce and successful political opposition. "The Queensland Nats will die in a ditch to protect Cubbie Station" (Sheehan 2005: 1).

The failure to rein Cubbie in, and the Queensland government's issuing of a tradable water allocation of 94,655 megaliters a year to the station shortly before COAG provided $350 million to buy back overallocated water, led to widespread accusations that due process had been circumvented. It is clear that there were longstanding weaknesses in the system, and John Grabbe, as an ex-Water resources officer, was very *au fait* with the rules, well connected, and thus particularly well placed to secure the water allocations for the station.

Downstream graziers share these allocation concerns: "We believe that the whole process in Queensland of allocation of overland flow in the Lower Balonne has been a corrupt process." The same spokesperson described "the flippant, nonchalant and at times dismissive attitude that successive Queensland natural resource ministers have shown to due process," and "the appointment of a chair [to the Water Resources Advisory Council] who will enjoy significant personal gain from the process" (Edward Fessey, as cited in Commonwealth of Australia 2006b: 4).[10] Similar remarks were made by one of Cubbie's major critics, Senator Bill Heffernan, who noted that the water policy advisory committee

> [g]ave the Queensland government advice.... There was a proposal to issue a licence for 469,000 megalitres for Cubbie Station. Funnily enough, on the licence, as part of the title in a "financial" commercial-in-confidence arrangement, is the downstream neighbour, who is the independent chair of the process. (in Dusevic 2009: I)

John Grabbe was untroubled by this relationship:

> I have absolutely no problem with it.... I cannot think of one member of that committee who does not have a financial interest in the district. Quite frankly I would be concerned if the make-up of the committee was not of people who have interests in the community. (Commonwealth of Australia 2006a: 24)

It is no secret in rural Queensland that the way to influence events is through involvement in local government and industry bodies. Primary producers describe this as keeping an eye on things, looking out for the farmers, and educating greenies about the economic realities (Strang 2009).

This has become increasingly problematic since elected bodies in Australia, as elsewhere, are now being replaced by community advisory committees, regional management groups, and similarly unelected quangos (quasi-autonomous nongovernmental organizations) composed of stakeholders. Such groups are represented as enabling grassroots involvement and, as Paley suggests, may illustrate a more fluid notion of democracy (2008). But as this case study shows, much depends on the groups' demographic composition and genealogical connections. It is clear that such bodies also have considerable potential to override collective interests, opening up decision-making processes (and thus resources) to capture by powerful and well-established commercial elites. (See also Davies 2007; and Fraser 1997.)

In addition to populating public committees, farmers have their own advocacy groups. Under the umbrella of the Balonne Community Advancement Committee (BCAC), Cubbie and other local irrigators formed the Dirranbandi Irrigators' Association and the St George Water Harvesters. A further joint initiative, SmartRivers, was set up to challenge the science underpinning the state government's 2000 Draft Water Allocation Management Plan. Assisted by funding from Cubbie and advice from its public relations firm, SmartRivers spent $200,000 hiring its own scientists and making selective use of the Cullen report on the MDB (Cooperative Research Centre for Freshwater Ecology 2000).[11] It argued that there was no degradation evident in the rivers, and that its findings "brought into question a lot of the science that is done by government for government" (Todd 2003).

But Rory Treweeke, a grazier from the heavily impacted Narran River flood-plain in New South Wales, argued that SmartRivers science was a travesty:

Professor Cullen has stated to me that he believes that system to be at least 50 per cent over allocated.... [But] from my personal experience of involvement for some 20-odd years, any point of view that has been contrary to that which was pro-development to the ultimate extreme was derided, ignored or refused to be accepted. (Commonwealth of Australia 2006b: 2–3)

Treweeke chairs the Lower Balonne Floodplain Graziers Association, established to confront the irrigators upstream. This is linked with a larger organization, Fair Water Use, that has begged the government to exert some control over the cotton industry and to buy back water allocations. The downstream graziers feel that their communities and their livelihoods are being destroyed:

For the last decade massive surface-water and river diversions on Cubbie and other large irrigation properties have been impeding the natural flow of water across the flood plains and headwaters of the Darling, disconnecting this water from the river and, as importantly, from its groundwater storages. Natural underground reserves are the blood-bank of the river, sustaining the Basin and its communities during times of drought, which is, and always will be a natural occurrence in this region. However, the vast tourniquets constructed by the likes of the Cubbie Group are now devastating the hydrology of the Darling Basin, dramatically worsening the severity of the current drought and crippling those involved in appropriate and sustainable agricultural activities downstream. (Fair Water Use 2010: 1)

This issue has caused immense grief to our family and immense grief to about 74 other families on the flood plain in New South Wales.... Basically the unsustainable and irresponsible over allocation of water in the Lower Balonne has had a profound effect. (Edward Fessey in Commonwealth of Australia 2006b: 4)

Though these groups do not challenge the framing of water as a commodity, their plea implies the reinstatement of a moral community in which social equity is ensured through access to a common good. Their discourse valorizes social and ecological connection, and there is a nod toward the meanings of water as a social substance, albeit expressed in economic terms as a blood-bank rather than the more usual lifeblood. This collective flow is dammed by the enclosing tourniquet upstream. Concerns about loss of community are underpinned by ideas about cultural heritage, identity, and the value of rural settlements, still seen in the national imagination (though urban living has long been the norm) as the pioneering heartland of the country.

But Cubbie Station and other major irrigators also present themselves as integral to the survival of rural communities. Although some research has suggested that the upstream irrigators in the MDB have a more brutal focus on economic values than the more collectively oriented graziers farther south (Baldwin 2008), Cubbie and its neighbors claim that in damming the rivers they are also stemming the flow of people from rural areas and protecting "the right to live and work in rural Australia" (David Carson, as cited in Dirranbandi Irrigators' Association and the St George Water Harvesters 2000).

Cubbie employs about fifty people directly, and a further 120 or so indirectly. Though state and federal governments recently produced some funding to buy back water allocations, the Balonne Shire Mayor, Donna Stewart, suggested that any raking back of those handed out to Cubbie would be "catastrophic ... Everything we go to in this Shire is supported by Cubbie" (in Cawood 2008: 1). A Queensland farmer, Ian Cole, argued that an appropriation of irrigation rights would immediately drop "highly developed land back to dryland grazing values— an impossible proposition" (in Cawood 2008: 1). Greg Grainger, Cubbie's code-signer, said that it would be "a tragedy if Cubbie's water was taken away.... Not just for us but [for] the number of people who work on this place and draw their incomes from this place and the amount of money that this place can produce" (in Roocke and Douglas 2010: 1).

Those committed to competitive development see moves to reduce water allocations as regressive. When the NSW government bought an irrigation station called Toorale (for $23.7 million) and—to return its allocations to the Murray-Darling—turned it into a national park, its former CEO described this as a "great act of bastardry" (Boyd 2010: 1). A Liberal National Party representative claimed that the closure would cost 100 jobs and take $4 million from the local economy (Ray Hopper in Commonwealth of Australia 2008b: 3395). Another, clearly envisioning Nature slipping the leash, said

> Toorale ... will become a haven for the wide-scale breeding of feral pigs and other vermin that will be absolutely destructive to agricultural industries regardless of where they are. The same situation will be applicable to the upper reaches of the Darling River and the Balonne River system in Queensland if this voluntary buy-back system is undertaken. (Vaughan Johnson as cited in Commonwealth of Australia 2008b: 3393)

A Water Crisis

The purchase of Toorale, and other proposals for state and federal governments to buy back water allocations emerged from concerns about the freshwater "crisis." But irrigators argued that groups sufficiently concerned about ecological issues should simply purchase allocations, "for the environment."

In 2010, this adherence to market rule produced the bizarre spectacle of the Australian Conservation Foundation asking the inhabitants of the Mildura area to contribute $15 each so it could buy 200 megaliters to save the Hattah Lakes.

But the positioning of the environment as a charity case is hardly likely to produce sustainable outcomes and, even with successful fundraising, most environmental organizations do not have sufficient resources to compete with commercial groups for increasingly expensive water "assets."

Labor governments in Australia are (slightly) less enamored of privatization and more open to the prospect of buying or even taking back overallocated water resources for the national or collective interest. Australia's change of federal government in 2007 had several important effects. The new government had a less confrontational relationship with the Green Party, and although right-wing MPs continue to deny the anthropogenic reality of climate change—"nothing but a good old drought" according to an Labor National Party representative (Vaughan Johnson in Commonwealth of Australia 2008b: 3391)—one of Kevin Rudd's first moves as premier was to sign the previously shunned Kyoto Protocol.

In 2008 he consolidated federal control of freshwater by establishing the MDB Authority with powers across several states. As conservation groups pleaded with state and federal governments to buy back Cubbie's water, the Inland Rivers Network also called for "a major and immediate targeted water purchase in the Murray and Darling systems … to avert the ecological and social crisis" (Amy Hankinson in Commonwealth of Australia 2008a: 15). But a Queensland Liberal National Party representative described the federal legislation as

> [m]oving away from a long history of collaborative management … between the states and the Commonwealth … to a more centralized approach.… It is unfortunate that ill-informed southern commentators and ill-informed southern academics concentrate on Queensland in respect of the significant difficulties facing the Murray-Darling.… [They] seem intent on demonizing Queensland. (Andrew Cripps in Commonwealth of Australia 2008b: 3374)

In March 2010 the federal government put $100 million toward buying back overallocated water for the MDB, but Penny Wong, the water minister, was careful to say that the government would only take water from willing sellers. At the time it seemed there might be some of these. A twist in this tale is that, generating much *schadenfreude* around Australia, Cubbie Station went into voluntary financial administration in October 2009, owing its creditors in excess of $300 million. Though critics suggested this was because its activities were fundamentally unsustainable, Cubbie maintained it was merely a result of the prolonged drought. Arguments raged about whether the government should buy the station, or some of its water allocations: some said this was an opportunity, others said it would make no hydrological difference to the Basin, and would simply be a huge (and corrupt) waste of public money.

Reprieved by heavy rains in 2010, Cubbie hoped to "recapitalize the balance sheet" with new investors and owners, and fulfill its aim to create profits of $50 million a year. "Western Gulf Advisory's bid for the giant Cubbie Station would

see founders the Stevenson family return as minority shareholders in a recapital-
ized business" (Condon 2010: 1). In July 2010, the station managed to borrow
sufficient funding to plant a cotton crop (Rego 2010: 1); pricing Cubbie's market
value at $450 million, John Grabbe suggested it might be willing to sell about a
third of its water, but must keep sufficient allocations to remain "sustainable." As
Keith de Lacy put it: "[W]hat we want is for Cubbie to continue, effectively in its
current form" (ABC Rural 2010: 1). With a profitable crop in play, the station was
able to start negotiating with its bankers with a view to coming out of voluntary
administration, thus closing the door to any cheap government buyback of its
water allocations.

   In April 2011, the station harvested a crop calculated to be worth $150 mil-
lion (Douglas 2011). The administrators then fired John Grabbe (allegedly for
appearing in a media report and commenting on the financial situation of the
station) and, within a week, were reportedly seeking new purchasers in China for
the station.

   The prospect of losing Cubbie's water allocations to private investors overseas
reignited a number of concerns, not just about sustainability, but also about national security, highlighting the key relationship between water and political power.
Concerns were raised by the federal government, which at one point made a bid
of $40 million to try to regain some of the station's allocations. As an unnamed
government source put it,

> There will be defence considerations, food and crop sustainability considerations,
> whether we can afford to practically kiss goodbye to the water—private investors
> will have no reason to sell the water off.... And then there's the simple issue of
> allowing foreign investment in what is a huge whack of ... land in Queensland. It's
> not a headache for us, it's a migraine. (Quoted in Wright 2011)

As ever, the decision rested on which ideologies prevailed: those directed by
market forces and notions of unlimited growth, or those oriented toward more
sustainable human-environmental relations and a vision of water as a common
good. Despite all of these concerns, in October 2012 Cubbie Station was sold
to a Chinese-led consortium with 80 percent of the shares held by the Shandong
RuYi Group.

Sustaining Water Values

For the primary producers quarrelling about water allocations, sustainability seems
to be largely about whether their activities remain financially viable, and, in a secondary sense, whether their local communities can maintain sufficient critical mass
to sustain themselves in place. The farther downstream farmers are located, the
more they also have to confront issues of ecological sustainability, and consider

how to achieve what Fair Water Use calls "appropriate and sustainable agricultural activities" (2010: 1). But for most this simply means ameliorating impacts on local ecosystems, and making better use of scarce water resources, rather than instituting more-fundamental changes. If only water can be shared equitably, they say (eliding the internal contradictions implied), they can continue with business as usual.

Australia's urban population, though critical of industrial farming practices and anxious about ecological health, is unwilling to decrease its levels of consumption. The designed obsolescence of the fashion industry helps to keep cotton profitable; ongoing demand for SUVs fuels climate change. The prospects for real change seem small: "Past behaviour does not fill us with much confidence" (Sandeman 2008: 728). There are some alliances between "light green" conservation groups and farmers purportedly sympathetic to their concerns: both hope to maintain an only slightly altered way of life. But "deeper green" environmental organizations argue that radical changes are needed if the ecology of the Murray-Darling Basin is not to collapse, bringing farming and other resource-based activities down with it.

> The MDB has all the hallmarks of a developing catastrophe.... To remedy the problem would require a complete reorganisation of farming, its methods and crop types across the whole area. This will mean massive disruption to the present farmers and townsfolk now living in the area, as well as to the food supplies of the nation. (Sandeman 2008: 727)

And this viewpoint is no longer merely the province of the deep greens. Emerging research on the MDB underlines the point that the well-being of ecosystems and their human inhabitants are inextricably intertwined:

> Business as usual is not working in the Basin. Contrary to what some people think, these problems have most certainly not been resolved by the floods of 2010–11, just as they were not created simply by the Millennium Drought. Decisions must be made, and made soon, and these cannot be half-measures or we risk irreparably damaging the future of all those who live, work and care about our basin, our home. (Connell and Grafton 2011: 1)

Nevertheless, it is easier to contemplate collapse at some ill-defined time in the future than it is to make major preemptive changes. At present such anxieties are trumped by fears of a more immediate crisis in Australia's rural heartlands, and a related economic crisis at a national level. Although the federal government's willingness to divert some (limited) funds toward water buyback may seem progressive, this has really served only to maintain the status quo.

> It is a strategy that has so far minimised any political damage to the Government. But it has also meant Rudd and his senior ministers ... have been unable to deliver the kind of sweeping reform necessary ... buying back water rights piecemeal simply won't address the scale of these issues. (Eltham 2009: 2)

While it is tempting to criticize the government's current water policies as fiddling while Australia burns, in reality decades of neoliberalism have shifted so much political and economic power into the hands of private commercial interests and transnational corporations that politicians and the law now have little control of events (see Tan 2001). History also has considerable momentum: a commitment to irrigation and visions of growth and development as the basis of wealth have become deeply ingrained, and the notion of a steady-state economy—one that limits expansion and internalizes the real costs of production—is commensurately alarming.

What would enable Australia to move from mere performative mitigation to a more sustainable adaptive trajectory? Looking at the conceptual, epistemological, social, and political alienations that over time have fragmented social and environmental relationships at every level, it is difficult to see how these can be reconciled into a coherent national shift in direction. A major ecological collapse may force real changes, but by then the MDB and other catchment areas may be degraded beyond recovery. Possibly the localized collapses already appearing will increase in number, frequency, and intensity until a tipping point is reached, but even then real change will require real consensus.

Thus, although the arguments over Cubbie continue to rage across the Queensland border, the real borderline lies between the groups committed to protecting business as usual, and those willing to initiate change. Australia's countermovements are growing in strength, and there is some potential for these to demand a government that is more directed by collective social and ecological interests, rather than protecting the interests of longstanding elites and abdicating its responsibilities to the market.

As Wittfogel noted half a century ago, political leadership depends on the control of water (1957). To have sufficient power to lead change the Australian government would have to reject what Caldecott calls its "tired privatisation agenda" (2010: 2). Partial collapses elsewhere have begun to open this door: for example, with the renationalization of banking in Europe and America, and in the return of British Rail to public ownership. In Australia, the reacquisition of water licenses by the government "for the environment" is potentially the thin end of an important wedge, and the fear and rage that this has sparked in far-right groups suggests that this is precisely how it is perceived. As Kerr points out (M. Kerr 2004), there is a related need (that will no doubt generate similar angst) for a thorough reform of corporate law, requiring primary producers to take responsibility for the social and environmental impacts of their activities.

A collective reclaiming of resources implies a reopening of flows between disparate groups within Australian society; between divided levels and areas of government; and between people and their material environments. Such reclaiming demands conceptual and practical integration, and a reconciliation between Culture and Nature. Perhaps most critically, it requires people to reconsider what is meant by wealth and health: to repudiate reductive economic visions of "water as gold," reestablishing it fully as a meaningful substance that connects people to each other and to the ecosystems they inhabit.

Notes

I am grateful to the many water users and managers in southern Queensland who participated in this project; all were immensely generous with their time and their input. I would also like to thank my colleagues in Auckland for their thoughtful comments on a first draft of this chapter, and most particularly my research assistant, Mira Taitz, who sought out further material about Cubbie Station with unflagging enthusiasm and efficiency. This research was funded primarily by the Australian Research Council, with some additional support from the University of Auckland.

1. Key sacred sites for Aboriginal people are frequently located at reliable water sources. This made them equally attractive to colonial settlers.
2. I use the gendered term deliberately. As has been noted elsewhere, irrigation schemes in Australia have always been a male enterprise (Lahiri-Dutt 2006; Shiel 2000).
3. Salination occurs when groundwater tables are artificially raised by irrigation, or by the replacement of native vegetation with shallow-rooted crops. The salt deep in the soil substrata is drawn up into the surface topsoils.
4. The area damaged by salinity to date represents about 4.5 percent of present cultivated land, and estimated current costs include $130 million annually in lost agricultural production, $100 million annually in damage to infrastructure, and at least $40 million in loss of environmental assets. Salinity affects regions in all parts of Australia (CSIRO 2010: 1). All dollars are in U.S. currency here and throughout the chapter.
5. It is worth noting that "abstraction license" is a relatively recent term, clearly meant to get away from the connotations of damming, impounding, and diverting.
6. A recent report by the National Water Commission reports, "[T]he total volume of water unaccounted for as a result of land use activities outside our current water entitlement regimes and planning frameworks equates to almost one quarter of all the entitled water on issue in Australia" (National Water Commission 2010: 1).
7. There are some clear gender differences in the groups involved in these issues. Commercial lobby groups are often almost exclusively male, while those devoted to social or ecological concerns tend to be largely female in composition. Informants similarly affirmed the persistence of these gender roles in community-based groups (Strang 2009).
8. It should be noted, though, that this often-repeated claim only refers to Cubbie's licenses for direct abstraction, and not to its massive overland water harvesting activities.
9. Rudd was replaced by Julia Gillard in 2010.
10. Leith Boully is the chair of the Lower Balonne Ministerial Water Resources Advisory Council, which advises the Department of Environment and Resource Management on issues relating to water policy in this area. Her property sits alongside Cubbie Station and she has a reportedly highly profitable (confidential) commercial arrangement with the Cubbie Group.
11. This was a major scientific report by a freshwater ecologist on the state of the MDB. It was supposed to be definitive and to resolve divisions about water allocations.

References

ABC Rural. 2010. *Middle Eastern Buyer Could Save Cubbie Station.* Report dated June 3. http://www .efarming.com.au/News/agricultural/02/06/2010/100501/middle-eastern-buyer-could-save-cubbie-station.html

Adams, D., and M. Carwadine. 1990. *Last Chance to See.* New York: Ballantine Books.

Appadurai, A. 1986. *The Social Life of Things: Commodities in Cultural Perspective.* Cambridge: Cambridge University Press.

Australian Bureau of Meteorology. 2010. *Climate Change.* http://www.bom.gov.au/climate/change/

Baldwin, C. 2008. Integrating Values and Interests in Water Planning: using a consensus-building approach, PhD thesis, University of Queensland. School of Natural and Rural Systems Management.

Boyd, David. 2010. Toorale Station, Bourke. http://davidboydsblog.blogspot.com/2010/03/toorale-station-bourke.html

Business Council of Australia. 2010. http://www.bca.com.au/

Caldecott, J. 2008. National Water Market: Privatisation on the Murray Darling. *Australian Options Magazine* 54: 9–13.

Cawood, M. 2008. Darling Water Buyback Causes Ripples in Queensland. In *The Land.* http:// theland.farmonline.com.au/news/nationalrural/agribusiness-and-general/general/darling-water-buyback-causes-ripples-in-qld/1244115.aspx

Commonwealth of Australia. 2006a. Water Policy Initiatives, Senate Rural and Regional Affairs and Transport References Committee. In *Hansard,* August 2.

———. 2006b. Water Policy Initiatives, Senate Rural and Regional Affairs and Transport References Committee. In *Hansard,* August 16.

———. 2008a. Management of Murray-Darling River Basin System, Senate Standing Committee on Rural and Regional Affairs. In *Hansard,* August 18.

———. 2008b. Water (Commonwealth Powers) Bill, Second reading. In *Hansard,* November 11.

Commonwealth Scientific and Industrial Research Organization (CSIRO). 2010. *Salinity: How Big Is the Problem?* http://www.clw.csiro.au/issues/salinity/

Condon, T. 2010. Business As Usual at Cubbie: Western Gulf Advisory. In *The Australian,* June 10. http://www.google.co.nz/search?hl=en&rlz=1T4RNWN_enNZ269NZ272&&sa=X&ei=V mJXTL2bN4uWsgP-vLXaAg&ved=0CBMQBSgA&q=Western+Gulf+Advisory+bid+for+t he+giant+Cubbie+Station+would+see+founders+the+Stevenson+family+return+as+minori ty+shareholders+in+a+recapitalised+business&spell=1

Connell, D., and Grafton, R.Q. (eds) 2011. *Basin Futures: Water Reform in the Murray-Darling Basin,* Canberra: ANU e-press. http://epress.anu.edu.au/apps/bookworm/view/Basin+Futures+Water +reform+in+the+Murray-Darling+Basin/5971/intro.xhtml

Cooperative Research Centre for Freshwater Ecology. 2000. *Annual Report of the Cooperative Research Centre for Freshwater Ecology 1999–2000.* Canberra: CRCFE.

Crabb, P. 1997. *Impacts of Anthropogenic Activities, Water Use and Consumption on Water Resources and Flooding, Australia:* State of the Environment Technical Paper Series (Inland Waters). Canberra: Department of the Environment.

Davies, C. 2007. Grounding Governance in Dialogue? Discourse, Practice and the Potential for a New Public Sector Organizational Form in Britain. *Public Administration* 85: 47–66.

Dickie, Phil, and Susan Brown. 2007. *The Rise and Rise of Cubbie Station.* http://www.melaleucamedia .com.au/01_cms/details.asp?ID=257

Dirranbandi Irrigators' Association and the St George Water Harvesters. 2000. *Farmers Take to the Highway and the Superhighway in Wamp Debate.* Media Release. December 8th. http://www.ozcotton .net/news/dirranbandimr.html

Douglas, A. 2011. *Cubbie Group Boss John Grabbe Fired by Administrators.* ABC News, April 15. http://www.abc.net.au/rural/qld/content/2011/04/s3192866.htm

Dusevic, T, 2009. Cubbie Gets a Basting from Bill. *Australian Financial Review,* August 22. http://www.billheffernan.com.au/news/default.asp?action=article&ID=142

Elks, S. 2010. Singing in the Rain at Cubbie Station. *The Australian,* Mar 6, 2010. http://www.theaustralian.com.au/news/nation/singing-in-the-rain-at-cubbie-station/story-e6frg6nf-1225837533053

Eltham, B. 2009. Don't Buy the Farm, Penny. *Water Policy,* Aug 18, 2009. http://newmatilda.com/2009/08/18/dont-buy-farm-penny

Fair Water Use. 2010. Why Canberra Must Move on Cubbie and Murray-Darling Control. http://www.fairwateruse.com.au/content/view/170/

Fraser, N. 1997. *Justice Interruptus: Critical Reflections on the "Postsocialist" Condition,* New York: Routledge.

Guest, A. 2008. Expansion Plans for Cubbie Station. *ABC Report.* http://www.abc.net.au/am/content/2008/s2288587.htm

Hill, E. 1958 [1937]. *Water Into Gold: The Taming of the Mighty Murray River.* London, Sydney: Angus and Robertson.

Hodge, A. 2001. *The Dams that Drank a River.* ABC Report, http://www.fairwateruse.com.au/index2.php?option=com_content&do_pdf=1&id=173

Holdaway, S., Wendrich, W. and R. Phillipps. 2009. Identifying Low-level Food Producers: Detecting Mobility from Lithics. *Antiquity* 84: 185–194.

Isaac, M. 2002. *To Market, To Market: Why Dogma Hasn't Worked with Water.* Brisbane: The Brisbane Institute.

Keane, B. 2010 [2008]. How Cubbie (and Labor) Consumed the Murray Darling. http://www.crikey.com.au/2008/08/21/how-cubbie-and-labor-consumed-the-murray-darling/

Kerr, C. 2010. ACF Buys Water to Repopulate Lake Species. *The Australian,* March 16, 2010. http://www.theaustralian.com.au/news/nation/acf-buys-water-to-repopulate-lake-species/story-e6frg6nf-1225841121176

Kerr, M. 2004. Greening our Corporate Law: A Prerequisite for Achieving Sustainable Development. Masters thesis, Faculty of Law, University of Sydney.

Kopytoff, I. 1986. The Cultural Biography of Things: Commoditization as Process. In *The Social Life of Things: Commodities in Cultural Perspective,* edited by A. Appadurai, 64–91. Cambridge: Cambridge University Press.

Lahiri-Dutt, K., ed. 2006. *Fluid Bonds: Views on Gender and Water.* Kolkata: Stree.

Legislative Assembly of New South Wales. 2006. Consideration of Urgent Motions, Water Management. In *Hansard,* Oct. 18.

Lockyer, P. 2004. Bitter Water Feud Grows in Queensland, NSW. In *The 7.30 Report,* Australian Broadcasting Commission, February 24. http://www.abc.net.au/7.30/content/2004/s1052459.htm

———. 2010. Queensland Floodwaters Move South. In *The 7.30 Report,* Australian Broadcasting Commission, March 16. http://www.abc.net.au/7.30/content/2010/s2847702.htm

Murray-Darling Basin Commission. 2008. *Sustainable Rivers Audit-SRA Report 1: A Report on the Ecological Health of Rivers in the Murray-Darling Basin, 2004–2007.* Canberra: Murray-Darling Basin Commission.

———. 2010. Water and Land Salinity. http://www2.mdbc.gov.au/salinity/land_and_water_salinity.html

National Water Commission. 2010. *Impacts of Water Trading in the Southern Murray-Darling Basin: An Economic, Social and Environmental Assessment.* Canberra: Australian Government.

Paley, J., ed. 2008. *Democracy: Anthropological Approaches.* Santa Fe: School for Advanced Research Press.

Polanyi, K. 1944. *The Great Transformation.* New York: Farrar and Reinhart.

Rego, F. 2010. Creditors Give Go-ahead for Massive Cubbie Cotton Crop. ABC News, July 13th 2010. http://www.abc.net.au/news/stories/2010/07/13/2951777.htm

Roberts, G. 2008. Farmers in Fight Over Floodwaters. *The Australian,* January 24. http://www.theaus tralian.com.au/news/farmers-in-fight-over-floodwaters/story-e6frg6oo-1111115385726

Roocke, N., and A. Douglas. 2010. Cubbie Water Take Misunderstood: Designer. *ABC Rural,* March 15. http://www.abc.net.au/rural/qld/content/2010/03/s2846346.htm

Rose, D. 1992. Nature and Gender in Outback Australia. *History and Anthropology* 5 (3–4): 403–425.

Sandeman, J. 2008. The Water Crisis Facing Australia. *International Journal of Environmental Studies* 65 (6): 721–729.

Schaffer, K. 1988. *Women and the Bush: Forces of Desire in the Australian Cultural Tradition.* Cambridge, Melbourne: Cambridge University Press.

Sheehan, P. 2005. A National Party That Is Anything But. *Sydney Morning Herald,* August 29. http://www.smh.com.au/news/paul-sheehan/a-national-party-that-is-anything-but/2005/08/28/1125167548103.html

Sheil, C. 2000. *Water's Fall: Running the Risks with Economic Rationalism.* Annadale: Pluto Press.

Strang, V. 1997. *Uncommon Ground: Cultural Landscapes and Environmental Values.* Oxford, New York: Berg Publishers.

———. 2004. *The Meaning of Water.* Oxford, New York: Berg Publishers.

———. 2009. *Gardening the World: Agency, Identity and the Ownership of Water.* Oxford, New York: Berghahn Books.

———. 2010. Fluid Forms: Owning Water in Australia. In *Ownership and Appropriation,* edited by V. Strang and M. Busse, an ASA Monograph. Oxford, New York: Berg Publishers.

Tan, P. 2001. Irrigators Come First: Conversion of Existing Allocations to Bulk Entitlements in the Goulburn and Murray Catchments, *Victoria. Environmental and Planning Law Journal* 18: 154–187.

Todd, Ian. 2003. ABC Television interview, Feb. 16, 2003.

Uhlmann, C. 2009. Government Tight-Lipped Over Cubbie Station Sale. *Agricultural News,* August 17. http://www.abc.net.au/7.30/content/2009/s2658646.htm

Williams, J. 2011. Understanding the Basin and its Dynamics. In *Basin Futures: Water Reform in the Murray-Darling Basin,* edited by D. Connell and R.Q. Grafton. Canberra: ANU e-press. http://epress.anu.edu.au/apps/bookworm/view/Basin+Futures+Water+reform+in+the+Murray-Darling+Basin/3491/intro.xhtml#toc-anchor

Wittfogel, K. 1957. *Oriental Despotism: A Comparative Study of Total Power.* New Haven: Yale University Press.

Wright, J. 2011. *China May Buy Up Cubbie Station,* ABC News Report, April 4. http://www.smh.com.au/environment/water-issues/cubbie-station-puts-up-for-sale-sign-in-china-20110423-1ds5c.html

Young, I.M. 2000. *Inclusion and Democracy.* Oxford: Oxford University Press.

Chapter 3

WATER AND ILL-BEING
Displaced People and Dam-Based Development in India

Lyla Mehta

∞

Displacement Politics in India

The history of large dams parallels the history of development. In the 1950s and 1960s, when the modernization paradigm reigned supreme, development tended to be project-focused and was considered a unilinear way to progress. The large dam, executed in a top-down way, epitomized the development and the project of modernity. In the 1950s, projects such as large dams generating water and power were supposed to help India to "catch up with the West" and promote modernity (Fernandes and Thukral 1989; Mehta 2009). It was unquestioned then that such megaprojects would require the displacement of large population numbers. Forced uprooting was considered to be the cost of development due to overarching national interest. By drawing on the notion of eminent domain, which is linked to the colonial Land Acquisition Act of 1894, the government has the power to appropriate private property which is justified due to so-called national purpose. Dam-based development policy and planning have largely followed the utilitarian and Benthamnian logic of 'the greatest happiness for the greatest number' (Rayner 2003). This has allowed for millions to be displaced in the interest of the so-called common good (cf. Roy 1999). In recent decades, postindependent India has witnessed the emergence of new social movements questioning the logic of dam-based development. While large dams might have made some parts of the desert bloom and led to full granaries and enhanced food security, they have not been without high social and environmental costs (cf. Goldsmith and Hildyard

1992; McCully 1996). Also, as James Scott (1998) notes, this high modernism and technical progress often leads to hegemonic planning that has excluded diverse perspectives and alternative paths to development as well as the agency of local people.

For the people adversely affected by large dams, they have meant displacement and homelessness. The rivers, expected to be transformed by dams and reservoirs to harness power and water, have become rivers of sorrow for the displaced people living on their banks (Thukral 1992). Unfortunately, there is no consensus on how many people have been displaced by large projects such as dams since India's independence. Estimates range from twenty-one million to fifty million (see Hemadri, Mander, and Nagaraj 2000). Largely, the planning and implementation of resettlement and rehabilitation has varied from state to state and has proceeded on a very ad hoc and incremental basis (cf. Dreze, Samson, and Singh 1997; Fernandes and Thukral 1989; Thukral 1992). Little wonder, then, that many of the oustees of dams such as Ukai in Gujarat, Hirakud in Orissa, and Bhakra Nangal in the Punjab have joined the ranks of urban slum dwellers or migrant workers, or have fallen into the cycle of debt bondage. Out of the millions displaced, only 25 percent have been rehabilitated (Parasuraman 1997). As many studies have documented, the rest have experienced a significant decline in their economic, sociocultural, and nutritional status (see Fernandes and Thukral 1989; Mehta 2009; Morse and Berger 1992; Thukral 1992).

Thus, not surprisingly, resettlement schemes have led to impoverishment (Cernea 1997) and immiseration, not only due to their top-down style of decision making and the suppression of the ousted, but also due to the inability of resettlement schemes to rebuild lives and livelihoods. They have also often led to a decline in the standard of living of the displaced. While relocation and resettlement are largely physical and economic initiatives, rehabilitation is more protracted and difficult, as it involves restoring a community's and individual's livelihood, income, dignity, well-being, and the capacity to interact in the new environment as an equal (Asif 2000). But as extensive research in the India context has documented, rehabilitation rarely takes place (Asif 2000).

The dams on India's famous Narmada River stand out as some of the most controversial dams in the world. The Narmada Project comprises 2 megadams, 30 large dams, 135 medium dams, and 300 small reservoirs and dams. All these projects, if realized, will most certainly totally transform the Narmada River, India's holy and last free-flowing river. One of the megadams, the Sardar Sarovar Project is supposed to bring water to some thirty million people and irrigate 1.8 million hectares of land with a capacity of 1,450 megawatts of power (Raj 1991). The 135-meter-high dam, if completed, will submerge 37,000 hectares of forest and prime agricultural land. Apart from the various disputes about its purported benefits and environmental impacts, it has been criticized due to its deleterious human consequences. The project will negatively affect the homes, lands, and livelihoods of about a million people. About 250,000 people (largely tribal, adivasi people) will be directly impacted and lose their homes due to reservoir submergence in Gujarat, Maharashtra, and Madhya Pradesh. The adivasi groups relevant for this chapter are the Tadvis and the Vasava, who fall under the generic category of Bhil. For reasons of space, it is not possible to discuss differences between Tadvis and Vasavas or, indeed, all the many debates around tribe and caste in India. For purposes of this chapter, however, it suffices to say that the Tadvis have always been more exposed to the outside world, consider themselves to be superior to the Vasavas, and had already seen themselves as Hindu, even in the submergence village. By contrast, the Vasavas had less exposure to the outside world. Their dialect, dress patterns, and customs have changed dramatically over the past decade since resettlement and induction into the market-based economy and Hindu caste society has had more profound impacts on the Vasavas than it has had on the Tadvis (cf. Hakim 1997).

This chapter examines the changing water worlds of project-affected people of the Sardar Sarovar Project. The main focus is on communities from their forest village on the banks of the Narmada River to a small resettlement site in the plains of Central Gujarat twenty years ago. It also briefly examines the situation of those who have refused to leave and continue to live along the dammed river. I argue that dominant models of water and well-being tend to mold official discourses concerning displacement that ignore the multifaceted dimensions of both water and well-being. It is these dominant models that serve to legitimize both forced displacement processes and controversial water projects.

Water, Well-Being, and Displacement Processes

Well-being, in the traditional approach, is often defined as physical needs deprivation due to private consumption shortfalls (largely with respect to food) (Schaffer 1996: 24). By contrast, more participatory and qualitative approaches focus on a much broader conception of ill-being or deprivation, including "physical, social, economic, political and psychological/spiritual elements" (Chambers 1995:vi). Thus, sources of both well-being and ill-being include income and nonincome sources of entitlements, social relations of consumption and production, and the more qualitative aspects of security, autonomy, self-respect, and dignity.

The latter more holistic concept of well-being is at the core of the work of authors such as Amartya Sen (1993, 1999). For example, Sen argues that even though it is common to "use incomes and commodities as the material basis of our well-being ... what use we can respectively make of a given level of income, depends crucially on a number of contingent circumstances, both personal and social." (1999: 70). Hence, well-being is firmly anchored in a particular social and personal context. This is why Sen advocates—for evaluative purposes in particular—the capability approach as a means to measure well-being. This approach focuses on "substantive freedoms—the capabilities—to choose a life one has reason to value" (Sen 1999: 74; see also Sen 1985, 1993). At the heart of this approach one must look at the freedoms that an individual can enjoy. Thus development, according to Sen, is a process of expanding the real freedoms that people enjoy (1999: 3)

In this broader sense, well-being should increasingly be understood as a multidimensional phenomenon ranging from income to the public provision of goods and services, access to common property resources, and other intangible dimensions such as clean air, water, dignity, self-respect, and autonomy (Razavi 1999). Unfortunately, conventional approaches to poverty, well-being, and ill-being still focus on the consumption of traded goods or incomes. They ignore natural resources and the consumption of nonmonetary goods and services (e.g., Baulch 1996; Razavi 1999), along with the sociocultural values that are placed on them by individuals. Due to the close links between infrastructure projects and dominant discourses of development, this trend of measuring well-being through tangible and material gains and losses is also mirrored in many resettlement policies and programs.

That displaced people face increasing ill-being and a decline in their standard of living, of course, has been well documented in the extensive literature on displacement and resettlement processes (e.g., Cernea 1997; Grabska and Mehta 2008; Indra 1999; MacDowell 1997; Scudder 1995). Many studies have discussed how vulnerable communities tend to be impacted by dams in ways that require an evaluation that goes beyond the monetary loss of land (e.g., Fernandes 2009). In this chapter, I view water as a life-giving resource having material, symbolic, and cultural values used by different social actors for different social, political, and economic purposes (see Mehta 2005). I argue that it is important to take into account the multidimensional aspects of water and its relationship to well-being or ill-being. Conventional ways of evaluating water and well-being focus on aspects such as regular provision, distance and adequate quality. A more multidimensional approach, by contrast, would also need to focus on issues such as autonomy, links with identity, and the freedom to choose. I explore a range of representations of the water/ill-being nexus in the Narmada context to argue that it is necessary to focus on the broader capabilities approach if well-being is to be measured in a fair manner. As traditional riverbed communities move from river basins to settlements in plains, and as rivers are transformed to dams and reservoirs, dramatic changes occur in water quality and quantity. These changes have both tangible and intangible implications for a project affecting people's livelihood options, health, sociocultural identity, daily routine, and social relations. However, bureaucrats and policy makers focusing on conventional understandings of water and well-being neglect or even willfully ignore displaced people's subjective sense of ill-being. I demonstrate these issues in two situations. The first focuses on Malu, a resettlement village in Vadodara District, Gujarat, where the absence of the river and the poor water situation is one of the main causes of ill-being. I also briefly focus on activist communities who have refused to move from their homes and now live by a river that is steadily becoming unable to sustain lives and livelihoods.

The Changing Water World of Displaced People in Malu

The contrast between Gadher, the tribal village on the banks of the Narmada and Malu, the resettlement site, is striking. Gadher was a sprawling village spread out over the river valley with fields and houses scattered over the hills and forest. By contrast, Malu, even twenty years later, is still a *vasahat* (resettlement site) with little or no tree cover and half-complete houses situated close to each other in grim unaesthetic lines. Indeed, one could argue that this simple physical arrangement of housing by the government is a method of control. In Gadher, livelihoods were far more diversified with the forest, land, river, and livestock playing important roles in the subsistence-oriented economy with its few market-based linkages. In Malu, most people make their livelihoods through agriculture. Forest-based work, fishing, and extensive grazing, so prevalent in Gadher, are largely absent. The monetization of goods and services has led to dramatic changes; most displaced people still complain that money is always short.[1]

Changing Access to and Control over Water

In Gadher, the main source of water for the village was the Narmada River. Women would spend between one to four hours a day in a number of trips. Trips were made as and when necessary and the distance walked could be as far as three kilometers in each direction. Water was collected largely for domestic purposes. Tasks such as the washing of clothes, bathing, and the watering of livestock were usually performed at the river. Hamlets far away from the river also had access to a hand pump, streams, and various wells. These provided a good supply of water due to abundant groundwater levels in the hills.

In Malu, there are many more sources of water, though the supply is not as reliable. Bureaucrats claim that the water sources are close by to the homes. Indeed, there are about twelve standpoints near the houses, four hand pumps, and water from a government tanker. In 2000 and 2007, many of these were not working, however. A village pond is shared with the host villagers, and can be used to water livestock in the monsoon. A hand pump in the host village of Malu is located about two kilometers away from the resettlement site and is used when none of the local sources has water.

The quality of water is defined in three categories: *Meetu pani* (sweet water, which is of the best quality), *Moru pani* (bland water, but drinkable) and *Kharu pani* (salty water, which is undrinkable and does not quench one's thirst). Not one of the sources in the resettlement site is considered sweet. The only accessible sweet source is the hand pump in the host village but resettled families are hesitant to use it because it causes friction between them and the host village. By contrast, Narmada water, as they remember it, was always sweet.[2] Flowing water is available sporadically for periods of about twenty minutes a day, other than between 6:30 a.m. to 10:00 a.m., when it is available continuously. Displaced people have little control over the operation or maintenance of these sources or indeed the quantity of water available daily. The advent of water is the high point of the day, with many women crowded around a tiny trickle of water for a short while. Obviously, during these tense moments many conflicts can ensue. The water situation in the Vasava quarter is even more precarious than that in the Tadvi quarter; there is even less regularity and Vasava women are often forced to go to the Tadvi quarter or to the host village for water.

The largely erratic nature of the supply has led to water, once an element taken for granted in the lives of the settlers, to becoming one of their largest problems. A free-flowing river that gave them twenty-four hour access has been replaced by a variety of unreliable sources that provide water for very short periods. The autonomy that women enjoyed in collecting water whenever they wanted has been lost. Instead, they are dependent on the government, host villagers, and other people for their daily supply.

NGO workers, health officials, and officials of the Nigam believe that women have benefited tremendously because they no longer have to walk long distances to the Narmada River to access water for domestic use. Had the Malu water sources been functioning and providing safe and adequate water, this might have been true. But, as Vasava women have pointed out, they also sometimes need to walk long distances to the host village and engage regularly in battles with the host community over scarce water. To counter official views of drudgery, women have expressed the sentiment that they prefer the so-called drudgery of their submergence village to the situation in the resettlement site. This is because of the daily uncertainty around water, the taste and quality of which are questionable. Thus, even after two decades, they remain nostalgic for the river. This nostalgia can be seen in the practice of older villagers who keep jars of Narmada water in their homes. Another way of interpreting the women's views of drudgery would be that women would prefer less drudgery and more facilities in their ancestral homes rather than having to fight for them daily in the settlement that, for many, has still not become home.

Settlers are convinced that the resettlement authorities are indifferent to their needs vis-à-vis water. This belief is reinforced by the fact that Narmada water, released in February 2001 via pumps into the canal system, is still bypassing them, supposedly on its way to the drought-affected areas of Kutch and Saurashtra. As one educated resettler put it,

> We can go out to work to buy food and clothes. But without water, we cannot live. We will die waiting for this water. Even now, the water of our mother Narmada is bypassing us. We need to move (Tadvi male, sixty-five years of age).

On my last visit in 2007, Vasavas still complained bitterly about the water situation. They told me that they needed to walk to the main village to fetch water. By contrast, the situation of the Tadvis had improved: three months prior to my visit a new pipeline had been installed. Before that, Tadvi women used to walk to Malu or men brought water on their bicycles.

Changing Water Relations

Due to the importance that water plays in shaping people's daily schedules, there have been many changes in social relations amongst women, between men and women, between Vasavas and Tadvis, and between the host villagers and displaced people. The doxa (that which is taken for granted in any particular society; cf. Bourdieu 1977) around the gender-based division of labor around water in Gadher dictated that women were responsible for the collection of water and many other household chores. In Malu, in some cases this doxa seems to have become more relaxed. Although it is still primarily women who collect the water, there are many more instances in Malu where men help out in this task because of the limited time available to collect water. In some extreme cases, even fathers-in-law help their daughters-in-law collect water. This helpfulness is not necessarily due to the fact that men feel more compassion toward women's workloads, or indeed because there have been significant changes in the way they see the gendered division of labor. Given the limited time available for the collection of potable water, men often help out in water collection in order to secure an adequate daily supply. Thus the scarcity of water has relaxed gender relations vis-à-vis water. This is because chores need to be done collectively in the limited time available. Men realize that if they do not help out, there will just not be enough water for the household. This is one example of a positive change in gender relations, with men engaging in hitherto traditional "female" tasks.

However social relations around water have also been rather conflictual in large households. In one of the largest households, there are complaints of fights between the five sisters-in-law over water. Although they live under one roof, they do not collect water together, because each identifies herself as a member of an autonomous family. This is in keeping with their life in Gadher when, after marriage, the son traditionally moved into a separate house. Such arrangements are impossible in Malu where there is no forest to generously provide teakwood to build new houses and where space is scarce. Besides, water shortages, impoverishment, and low and poor agricultural yields have made the struggle for survival intense in Malu. Consequently, the collective pooling of resources is no longer possible and the nuclearization of families has become more pronounced. However, this atomization takes place amongst enhanced physical proximity. Several generations and families live under one roof, though they operate as separate economic units. Here water becomes a marker of difference: separate collection makes a statement about a family's separate identity in the now mixed household. Thus, the mere supply and collection of water encroaches on a person's time and space, giving rise to social and family tensions that were absent in the past.

The relationship between the host village and the settlement has improved significantly since the early years when the adivasis (especially the Vasavas) were ridiculed for their clothes, dialect and language. Much of this took place at water points, which are still the main sites around which interaction between the hosts and displaced people takes place. Now, both adivasis and the host villagers agree that the adivasis have gone through a process of *sudhar* (reform), where, at least, superficially, they have adopted the dress patterns and customs of the host community.[3] This view of reform is based very much on a modernist view of sudhar. By contrast, in a 1905 dictionary sudhar was defined as going back to one's roots and becoming more independent (David Hardiman, personal communication). An elderly Vasava woman, whilst rushing off to grab some water from the tanker, took this view of sudhar: "How can they call this sudhar? We barely have our own food and water to survive. In Gadher there was enough for everybody. Even our land holdings have decreased ... we had about fifteen acres in Gadher, here we have received only five."[4] Largely, displaced people and the host villagers agree that the adivasis have undergone a process of reform. Thus, much of the early animosity toward the displaced people has lifted. In 2007, a Vasava man, who in twenty years has emerged as quite a power-broker in the community, was elected the headman of the whole village, including the old village, which is non-adivasi. He is an exception to the rule, however. Most other Vasavas in Malu continue to live a here-and-there existence, because they find it impossible to make ends meet in Malu. Many have returned to Gadher and have moved high in the catchment where they can still find enough grass for their animals and cultivate so-called wasteland or forest land.

Both Tadvis and Vasavas feel that they have become unnecessarily dependent on the government to supply their most vital and needed resource, water. They never experienced such a created dependence at any time before in their ancestral lands. In Gadher, they had the choice to fill when they wanted, how much they wanted, and with whom. From their point of view, water-related tasks have given rise to social relations that are occasionally conflict-ridden and that lock them in relations of dependence with their hosts and the authorities. This has led to a decline in autonomy for the displaced people, leading to a decline in freedoms and consequently in well-being.

Water and Health

The resettlement literature acknowledges that the general health of a population can decline due to displacement (Cernea 1997). Causes include social stress, trauma, malnourishment, unsafe water supply, and so on. These trends were observed during fieldwork in Malu. People believe that the poor quality of water has led to chronic diarrhea, dysentery, colds, nausea, and so on. It has also led to an increase in mortality. The majority of households within Malu have lost family members in the village, especially children. It would be foolish to deny problems concerning low life expectancy and child mortality in the submergence villages, but research conducted by monitoring agencies such as the Tata Institute of Social Studies (TISS) reports that the health status in the submerging villages was significantly better than in the settlements (TISS 1997). While child mortality had taken place before, people believe its incidence has increased. For example, my host family lost one son while living in Gadher, but since moving to Malu the family has lost one teenage daughter, one teenage son, and one infant granddaughter. The last two of these deaths were due to water-related diseases: jaundice with septicemia and diarrhea, respectively.

The men, in particular, complain about the negative impact water has had on their immune system and the general fitness of their families. Displaced people say that due to the different types of water they consume, their bodies cannot become used to one type and thus fail to build up a strong immune system. During several interviews, people reminisced about how strong and healthy the children were in Gadher and that the boys would grow up to be strong, quick, and agile. "There we could climb trees, chop wood and carry loads of up to 100 kilograms for about ten to fifteen kilometers; today we cannot even run one kilometer without becoming exhausted" (Tadvi male, aged thirty, in conversation, whilst his child was in hospital). People also use the analogy of land and water to compare their health in Malu and Gadher. "There our kids were *masboot* (strong), just like our land. Here our kids and the land are both weak" (Tadvi male, aged sixty).

The authorities have tried to improve the water quality by supplying chlorine tablets to the households in the settlement so they can disinfect drinking water. Doctors complain that the people do not use the tablets. This is true in some households because the taste of chlorinated water is very strong. No one has explained to the villagers, however, the importance of dissolving the tablets into the water. No one has shown them that the tablets are most effective when crushed and mixed into the water. If simply added to a vessel, the tablet sinks to the bottom without dissolving or disinfecting. Whilst talking to a scientist in the public health office in Baroda, it was bought to our attention that the tablets should be sealed when supplied to prevent the evaporation of chlorine. However, the tablets that are supplied to the displaced people are unsealed, and at the most are wrapped in newspaper, reducing their efficacy. This is a good example to show that the interventions undertaken by the authorities, though well meaning, are pretty useless unless they are accompanied by consultation, follow-up, and feedback processes.

Many of the grassroots health workers believe that water is the cause underlying the increasing illnesses. They have filed reports about these problems, but their advice has gone unheeded. Following villagers' complaints and the paralysis of grassroots workers, my colleague and I collected water samples and had them tested by the Gujarat Pollution Control Board. All samples taken from all the set-tlement sources were found to be bacteriologically unfit for drinking. The official sources were also found to be chemically unfit for drinking due to the overpresence of fluoride and alkalinity. Our tests confirmed what the villagers have been saying all along, but still this was not enough evidence for the Nigam. The head of the Nigam, Vinod Babbar, informed us in 2000 that "water is universally bad all over Gujarat and that the displaced people should stop complaining" (Babbar 2000). To placate us, a health team was sent to Malu, which according to the displaced people, would not have visited the village had we not been around. Official water tests were conducted. The results confirmed the poor water quality.

The doctors who visit the village insisted that the Health Cell (operational since 1999) has done wonders for the sites. They argue that the nurse now pays regular home visits to displaced people; health workers register every oustee's health status on a form, hand out chlorine tablets weekly, and immunize children regularly. From the displaced people's point of view none of these interventions has led to an improvement in their health status. Their water problems have largely been ignored because health workers believe that water falls under the jurisdiction of the engineering department; the health workers and doctors are largely ignorant about adivasi life and culture and the former nutritional statues of the displaced people. No health profiles exist about the original villagers, in this case those in Gadher. Thus, the extension workers have assumptions about adivasis health that are based on stereotypical notions (e.g., all adivasis drink and smoke and that is why they are ill) rather than on context-specific knowledge of the changing health status of these people.

The Nigam Health cell completely ignores adivasi notions of healing; it bases its knowledge on Western clinical medicine and allopathic concepts of illness and diseases. Adivasi beliefs however, "do not exist in isolation; rather [they are] ... a part of their entire socio-cultural religious system" (Swain 1990: 17). Healing takes place through knowledge "which is gathered/learnt through traditional ex-perience which they have learnt through trial and error" (Swain 1990: 17). There has been no attempt by the health authorities to try and understand, complement, and strengthen the traditional systems of health care that were used in Gadher in order to make health care more accessible to displaced people both physically and psychologically.

If anything, the government, through its health program, is achieving even more control over the settlers' lives: "In Gadher my Grandmother lived to 100 years without ever taking a single tablet. Today children are given injections and tablets from the moment they are born, and the more they take them the more they need them" (Tadvi male, thirty years of age). Here we are reminded of Foucault's no-tion of biopower (1978). Indeed, it appears as though the resettlement authority, Nigam, uses its power and authority to assert control over people's bodies through medical discourses and programs that denigrate adivasi medicine and healing systems (cf. Foucault 1980: 170–171). Through biopower, Nigam established a hegemony of its own dominant discourse of water and health, which suppresses displaced people's problems and perceptions. It also provides legitimacy and a self-sustaining character to its ever-increasing programs and schemes.

Water and Sociocultural Identity

According to local cosmology, the Narmada is *mata* (mother) and even holier than the River Ganga. Legend has it that the mere sight of the river can absolve an individual of his or her sins.[5] Hence the damming of the river was initially inconceivable for many of the area's inhabitants. "How can they [the state] dam a mother?," was a common question I heard when I first went to the Narmada Valley in 1991. The river was useful for fishing, washing, bathing, and riverbed cultivation in the summer. It also served as an artery for communication with relatives across the bank; it was the home to many holy temples and facilitated the transport of goods downstream (e.g., logs). The river was a kind of map that served as a point of reference for the people and connected its inhabitants with each other. It was also life giving and the origin of creation in some adivasi knowledge systems (cf. Baviskar 1995). Their reverence for this river suffused their daily lives (cf. Baviskar 1995).

The loss of the river and its ecology has had a tremendous impact on the way people live in Malu. This chapter has discussed many of the physical changes. Many of the activities undertaken by the river now have to be done in the house. For women, this has taken away the times that they could leave the house, talk, and spend time with their friends whilst working by the riverbed or in the forest, away from their menfolk. Now the farthest many women go without their men is to the nearest source of water or fodder.

For the older generation, there is a greater attachment to the Narmada than for their children and their grandchildren. Whilst older members want to return to Gadher and are still eloquent about the river, their children and grandchildren are likely to stay in Malu and make the best of what they have been given. The importance of the river to the elders is manifest in their daily prayers to jars filled with Narmada water. This is almost to remind them of their daily ablutions by the river. Also every time they take a trip to the Gadher area, they bring back memories of their homeland through traditional grains or herbal medicines. Their children and grandchildren, however, do not keep Narmada water at home and do not have the same attachments. As an older Tadvi woman put it, "We feel happy near the river; our friends and relatives are also close by. In Malu we are far away from both our kinship circle and the river." Thus, the river and its ecology was crucial in determining both the adivasis' sense of space and of identity. Kibreab (1999) recognizes the importance of place as vital for a person's well-being. He explores the importance of territorial-based identity as being critical to human well-being. By trying to transform the displaced people from 'wild' adivasis to mainstream

Gujaratis, the authorities have totally ignored the links a person has with his or her surroundings. The authorities also have not understood how the absence of the river and the forest and the significant changes in water and land uses have led to increasing ill-being for the displaced people.

It would be wrong, however, to portray the displaced people as mere passive victims of displacement. Those who have moved have now adjusted to the re-settlement situation. They have "turned stony land into gold," as one resident of old Malu puts it.[6] Much of this, however, has been with little or no outside help, notwithstanding the inadequate compensation package.

The adivasis in Malu never actively resisted the project in the early years. Unlike their cousins on the opposite bank in Manibeli and elsewhere, they sadly but re-luctantly left their ancestral homes for Gujarat, some filled with hopes for a better future influenced by the promises of Gujarati NGOs such as the Arch Vahini. In Malu, at least, that dream went sour. Instead, for the first ten to fifteen years, they experienced impoverishment, poor health, and a diminishing sense of well-being. Some returned to Gadher with their livestock and now live in a half-empty village and makeshift homes that they feel is a better option to life in the plains. Those who remain in Gadher offer covert forms of resistance. I am tempted to view their complaining and dissatisfaction as their everyday weapons (cf. Scott 1985). They offer resistance by refusing to forget the river and the life of Gadher. These tiny acts of resistance serve as a reminder to the callous state that their removal from Gadher took place without consultation, fully informed and prior consent, and often through gross human rights violations.

The Situation of Those Who Have Refused to Move

Let me now briefly examine the situation of those who overtly refused to move and have since 1991 participated in the struggle of the Narmada Bachao Andolan (or Save the Narmada Movement, henceforth the Andolan). The dams on the Nar-mada River also stand out with regards to their high social costs, and due to the dynamic protest movement. The Andolan has successfully highlighted and made clear to millions all over the globe the plight of the displaced peoples affected by the Narmada dams and the dark sides of top-down projects such as large dams. It also has inspired several social and environmental struggles on the Indian subconti-nent and raised questions important for India's future such as sustainable develop-ment, participation, the rights of indigenous peoples, the viability or nonviability of large top-down centralist projects, and the mobilization of protest. Over the years, the Andolan has adopted a strategy of noncooperation, mass mobilization, and nonviolent forms of protest including rallies, picketing, sit-ins, fasts, and the more extreme case of *jal samapan* (save or drown actions). By following the slogan, "We will drown, but not move," activist villagers have refused to vacate their ancestral homes. As a result they have resisted and faced police atrocities and repressive tactics including mass arrests, harassment, the molesting of women, and the clear-felling of their forests.

On a visit back to the Narmada Valley in 2007, the dam was at 122 meters. The valley was completely transformed. It looked beautiful but rather sinister. Semisubmerged trees and hillocks were visible everywhere. A journey across the river that had taken thirty minutes by boat in the 1990s now took about five hours by motor launch. The river resembled a sea, and what were once small streams looked like swollen rivers. In the last monsoon, there had been massive submergence. Even though resettlement had begun almost twenty years ago, many of the villages close to the dam still had about sixty to one hundred families who had actively participated in the protest movement and still refused to move. Life had become very difficult for them, however. Many complained about the changing nature of their river, now a still reservoir. Their agricultural lands were either submerged or very saline and hence not very productive. In Dunel village in Maharashtra, I reunited with a family after ten years. Galiben, the mother, told me that they had still not got land to their satisfaction so they were not going to leave. The forest still provided them with wood, vegetables, and work, but agricultural land was scarce so they relied on grain from outside. She said:

> The government has ruined our Mother. The Narmada is not Narmada anymore. It is not a river anymore. It is dammed and doesn't flow. It is dirty and we suffer from scabies and other skin problems. It doesn't flow anymore. It is difficult to grow food now because most of our lands have been submerged. There is no forest work anymore. We are supposed to get development, but this is only on paper. In reality, we *Adivasis* are going through a slow death.

Even though many resisting oustees are yet to receive compensation and rehabilitation, and even though they complain of increasing livelihood insecurity on the banks of the Narmada, they are still proud that they have been a part of such a dynamic movement. This helped them gain a new awareness as citizens, both of India and of the globe. As Noorjibhai, a villager from Mokhdi, Maharashtra, told me:

> If there had been no protest movement, nobody would have got anything. At least now, many have received some land and compensation. I still refuse to leave my ancestral home. The government is incapable of providing us with just compensation. We are now aware of our rights as citizens. We have waged battles in the streets of all the major cities and our struggle has been taken to several countries of the world and Washington. We will continue to fight for our rights.

Not all those who have remained still have this fighting spirit. The few families who still remain in Manibeli, the closest village to the dam and once the stronghold of protest, told me they were fed up. They were waiting for decent land, but the government was increasingly callous. The villagers had refused to accept land twenty-odd years ago and wanted productive land (ideally with irrigation facilities), but the government was offering them nothing decent. In the heyday of protest, these families had stood in the rising waters of the dam, had risked their lives, and had refused to move. They had courted arrest and been beaten up by the police. But twenty-five years on, resistance fatigue had set in. This is largely due to the impossible situation of life on the banks of the Narmada. A free-flowing river that once sustained them is now an inhospitable reservoir. The silt makes it difficult to fetch water and the daily trudge for water has become increasingly dangerous and difficult due to the slippery mud. Moreover, the water is a breeding ground for disease (e.g., malaria, skin rashes, scabies, etc.). Many therefore unsurprisingly said that if they received decent land they would be prepared to move to the plains.

Conclusions

This chapter has explored the water/ill-being nexus. It demonstrated that the dramatic transformation of the water worlds of different groups of displaced people led to different articulations of ill-being. Resettled people encounter increased control over their lives by government, livelihood insecurity, and impoverishment; they also suffer due to poor water quality. Families who still live along the banks of the river have faced the rising waters despite police repression. Their protest has led to a greater awareness of their rights and also to improvements in nationwide resettlement programs. But many of these protesters have yet to see any benefits, and now live by a dammed river that is no longer able to sustain them and their needs.

Oddly enough, even though the sociocultural aspects of this project have been extensively studied, most officials, health workers, and NGO workers do not understand why the adivasis articulate such ill-being. Why? There are many narratives around the displacement of adivasis in Gujarat. Widespread simplistic notions abound about their backwardness and wildness, and it is argued that resettlement processes ultimately help them reap the benefits of modernity. "Resettlement is development," is a popular slogan in Gujarat. These narratives give health workers, officials, and high-level bureaucrats the illusion that the pains of displacement can actually rapidly become gains during the resettlement and rehabilitation processes. Adherence by government officials and pro-dam proponents to economistic and reductionist models of well-being also fail to help them comprehend the water/well-being nexus.[7] Similarly, the narrative that dams are necessary because otherwise all riverwater would flow waste into the sea, justifies technicoengineering world views and discounts the ecological perception that the river is the lifeblood of an ecosystem.

These, however, are charitable interpretations. The more-harsh interpretation would contend that there is a certain instrumentality and intentionality in bureaucratic indifference toward adivasis' ill-being. There is ample evidence to show that dam-based development, as represented by the Sardar Sarovar Project, can lead to an unjust spread of pains and gains and the colonization of rivers and their people. The political economy of dams and large-scale irrigation is well known. Clearly, very powerful interests and lobbies are being served all over Gujarat (cf. Mehta 2005). Unholy institutional alliances have emerged between politicians, bureaucrats, social workers, academics across the state to give rise to one dominant discourse of water, namely that there is no alternative to the Sardar Sarovar Project (TINA) (Mehta 2005). The displacement of adivasis is seen as a small price to pay for the promise of water to drought-prone Kutch and Saurashtra.[8] Consequently, debates concerning alternative broader ways to view water and rivers and the water/well-being links are being suppressed. In these ways, dams-based development has emerged as a powerful discourse of development with a totalizing effect on water management debates in the region.

The hegemony of dams-based development has been challenged in many ways, especially by those activists who have refused to move. Others, such as those in Malu, despite initial acquiescence have displayed covert resistance. The interplay of covert and overt resistance on the part of the oustees will hopefully allow their perceptions of ill-being/well-being/water to be taken seriously. For their struggle over water is not only one of access, but also one of meaning.

Notes

This chapter draws on research and experiences in the Narmada Valley since the early 1990s. Research in 2000 and 2002 was made possible with a DFID/ESCOR–funded project on "Gender, Displacement and Resistance: Drawing Lessons from the Narmada Experience." Funding from the Development Research Centre on Migration, Globalisation and Poverty allowed me to make a further trip in 2007. I am very grateful to the women and men of the Narmada Valley and the activists of the Narmada Bachao Andolan ("Save the Narmada Movement") for their warmth, inspiration, and friendship. Some sections of this chapter draw on Mehta and Punja (2006); I thank Anand Punja who worked with me on the research in 2000. All responsibility for any errors rests with me.

 1. Malu is located in Baroda District and has a population of 596 people, out of which 371 are Vasavas, 220 are Tadvis, and 5 are Rohits.
 2. Now, however, given the damming of the river, the water quality is no longer so good. Residents along its bank complain of silt, worms, and illnesses.
 3. Discussions of the various contestations around sudhar and what this means for adivasi women and men's identity are beyond the scope of this chapter and will be discussed in forthcoming work.
 4. Here she is referring to forest land or wasteland that in the eyes of the state was deemed to be encroached. Thus the so-called liberal package of five acres of land in the eyes of many displaced people does not compensate them for access to forest land, common property resources, and land along the river-bed.
 5. By contrast, one must bathe in the Ganges for one's sins to be absolved.

6. Indeed, their hard work on the land has led to its value increasing fivefold in just a decade (study survey).
7. For example, government officials and pro-dam proponents argue that the consumption patterns of displaced people have changed and there is a greater use of goods such as electric fans, motorcycles, televisions, and so on. This may be true, but it also needs to be said that displaced people do not have money to visit relatives around the Narmada, money to buy milk, ghee, and fresh vegetables, which they obtained virtually free of cost in their forest environment. The lack of the latter has led to their isolation and poor health.
8. For detailed analyses of how these dominant discourses of water have essentialized and naturalized the phenomenon of water scarcity in Gujarat, see Mehta 2005.

References

Asif, M. 2000. Why Displaced Persons Reject Project Resettlement Colonies. *Economic and Political Weekly* 10 (June): 2006–2008.

Babbar, Vinod. 2000. Interview by author in Baroda, September.

Baulch, Bob. 1996. Editorial. The New Poverty Agenda. A Disputed Consensus. *IDS Bulletin* 27 (1): 1–9. IDS, Brighton, UK.

Baviskar, A. 1995. *In the Belly of the River. Tribal Conflicts over Development in the Narmada Valley.* Delhi: Oxford University Press.

Bourdieu, Pierre. 1977. *Outline of a Theory of Practice.* (Translated from the French by Richard Nice). Cambridge: Cambridge University Press.

Cernea, Michael. 1997. The Risks and Reconstruction Model for Resettling Displaced Populations. *World Development* 25 (10): 1569–1587.

Chambers, Robert. 1995. Poverty and Livelihoods: Whose Reality Counts? *Environment and Urbanisation* 7 (1): 173–204.

Dreze, J, M. Samson, and S. Singh, eds. 1997. *The Dam and the Nation: Displacement and Resettlement in the Narmada Valley.* Delhi: Oxford University Press.

Fernandes, W., and E.G. Thukral, eds. 1989. *Development, Displacement, and Rehabilitation.* New Delhi: Indian Social Institute.

Foucault. 1978. *The History of Sexuality, Volume 1: The Will to Knowledge.* Translated from the French by Robert Hurley. New York: Pantheon Books.

Foucault, Michel. 1980. *Power/Knowledge. Selected Interviews and Other Writings 1972–77*, edited by Colin Gordon. Brighton: Harvester Press.

Goldsmith, E. and N. Hildyard, eds. 1992. *The Social and Environmental Effects of Large Dams: Volume III, A Review of the Literature.* Cornwall, UK: Wadebridge Ecological Centre.

Grabska, Katarzyna and Lyla Mehta. 2008. Introduction. In *Forced Displacement: Why Rights Matter*, edited by K. Grabska and L. Mehta, 1-25. Houndmills, UK: Palgrave Macmillan.

Hakim, R. 1997. Resettlement and Rehabilitation in the Context of "Vasava" Culture. In *The Dam and the Nation: Displacement and Resettlement in the Narmada Valley*, edited by J. Dreze, M. Samson, and S. Singh. Delhi: Oxford University Press.

Hemadri, R., H. Mander, and V. Nagaraj. 2000. Dams, Displacement, Policy and Law in India. Prepared for *Thematic Review 1.3: Displacement, Resettlement, Rehabilitation, Reparation and Development.* World Commission on Dams. http://unpan1.un.org/intradoc/groups/public/documents/APCITY/UNPAN021311.pdf

Indra, D., ed. 1999. *Engendering Forced Migration: Theory and Practice.* Oxford: Refugee Studies Program.

Kibreab, G. 1999. Revisiting the Debate on People, Place, Identity and Displacement. *Journal of Refugee Studies* 12 (4): 384–410.

MacDowell, C., ed. 1997. *Understanding Impoverishment. The Consequences of Development-Induced Displacement.* Oxford: Berghahn Books.

McCully, P. 1996. *Silenced Rivers. The Ecology and Politics of Large Dams.* London: Zed Books.

Mehta, L. 2005. *The Politics and Poetics of Water: Naturalising Scarcity in Western India.* Orient Longman: New Delhi.

————. 2009. Contexts and Constructions of Water Scarcity. In *State of Justice in India: Issues of Social Justice, Volume IV, Key Texts on Social Justice in India,* edited by S. Roohi and R. Samaddar. New Delhi: SAGE.

Mehta, L., and Grabska, K. (ed) 2008. *Forced Displacement: Why Rights Matter.* London: Palgrave Macmillan.

Mehta, L., and A. Punja. 2006. Water and Well-being: Explaining the Gap between Official and Displaced People's Perceptions of Water. In *Waterscapes: The Cultural Politics of a Resource,* edited by A. Baviskar. Delhi: Permanent Black.

Morse, B., and T. Berger. 1992. *Sardar Sarovar: Report of the Independent Team.* Ottawa: Research Futures International.

Parasuraman, P. 1997. The Anti-Dam Movement and Rehabilitation Policy. In *The Dam and the Nation: Displacement and Resettlement in the Narmada Valley,* edited by J. Dreze, M. Samson, and S. Singh. Delhi: Oxford University Press.

Raj, P. 1991. *Facts: Sardar Sarovar Projects.* Gandhinagar: Narmada Nigam Limited.

Razavi, Shahra. 1999. Gendered Poverty and Well-Being. *Development and Change* 30 (3): 409–433.

Roy, A. 1999. The Greater Common Good. *Frontline* 16 (11). www.flonnet.com/fl1611/16110040.htm

Schaffer, P. 1996. Beneath the Poverty Debate: Some Issues. *IDS Bulletin* 27 (1): 23–36. IDS, Brighton, UK.

Scott, James C. 1985. *Weapons of the Weak: Everyday Form of Peasant Resistance.* New Haven: Yale University Press.

————. 1998. *Seeing Like a State: How Certain Schemes to Improve the Human Condition Have Failed.* Yale: Agrarian.

Scudder, T. 1995. Resettlement. In *Handbook of Water Resources and Environment,* edited by A. Biswas. New York: McGraw-Hill.

Sen, A. 1985. *Commodities and Capabilities.* Amsterdam: North-Holland Press.

————. 1993. Capability and Well-Being. In *The Quality of Life,* edited by M. Naussbaum and A. Sen. Oxford: Clarendon Press.

————. 1999. *Development as Freedom.* Delhi: Oxford University Press.

Swain, S. 1990. Health, Disease and Seeking Behaviour of Tribal People of India. In *Tribal health in India,* edited by S. Basu. Delhi: Manak Publications.

Tata Institute of Social Studies (TISS). 1997. Experiences with Resettlement and Rehabilitation in Maharashtra. In *The Dam and the Nation: Displacement and Resettlement in the Narmada Valley,* edited by J. Dreze, M. Samson, and S. Singh, 184–214. Delhi: Oxford University Press.

Thukral, E. 1992. *Big Dams, Displaced Peoples: Rivers of Sorrow, Rivers of Joy.* Delhi: SAGE.

Part II

❦

WATER AND TECHNOLOGY

The hydrological cycle today is best understood as a social and technological, as well as an ecological, system. Huge volumes of water flow through elaborate networks of underground pipes, delivering freshwater and removing wastewater from our homes and workplaces. Irrigation infrastructure, including pipes, canals, wells, pumps, and all the industrial products needed to sustain them, carries 70 percent of all the freshwater appropriated by human beings to agricultural uses (UNESCO 2006). Very few large rivers anywhere in the world flow uninterrupted from their headwaters to the sea; most of those that do, carry with them the waste products of industrial and agricultural technologies. In the Okanagan Valley of British Columbia, where I live and conduct research, upland water is diverted into storage reservoirs during spring freshet; dams control the water level of large lakes in the valley bottom; water flows not just to our homes but to our lawns and gardens through underground sprinklers; even lake recreational activities are increasingly dominated by boats, jet skis, docks, and swimming paraphernalia. In rural Papua New Guinea, where I also conduct research, the role of technology is limited but still clearly present. A gravity-fed piped water system brings water from a nearby creek to a series of village standpipes; intervillage transportation is entirely by water via outrigger canoe and outboard motor dinghies; upstream logging activities have worsened seasonal flooding in an important agricultural area.

Given that the human relationship to water is mediated at every turn by technology, it is surprising that so few publications focus on the ways in which technology transforms our relation to water in sensual, emotional, and symbolic terms. Almost every chapter in this collection describes one or more water technologies in some detail, but most do not focus on the ways in which these technologies generate, support, or undermine emotive, embodied relationships to water. Technologies are usually positioned in our accounts as political or economic instruments that we use to accomplish certain goals, but that themselves are devoid of intent, agency, or social significance. Paar (2010), by contrast, has eloquently described the ways in which human bodies must be retuned at the sensory level when they adapt to technological and ecological change. In her account of the Walkerton crisis in Canada, for instance, when several people died from drinking municipal water contaminated with E. coli bacteria, she describes how residents learned to distrust their own ability to judge water quality on the basis of taste and smell, and how they also lost trust in one another as they struggled to understand the nature of the human error that had led to the tragedy.

The first two chapters in part 2 are exceptional for the attention they pay to the aesthetic, emotive, and symbolic significance of water and associated technologies. In the Palestinian village described by Nefissa Naguib in chapter 4, older women reflect on the changes brought about by the introduction of a piped water system. Having water in their own homes saves them a great deal of time and effort since formerly they had to carry water home from a nearby village spring or, during the dry summer months, from much-more-distant sources of water. Relationships among women changed dramatically as a result of the loss of sociality associated with the village spring, a site where women would formerly meet, share stories, and discuss the issues of the day. The reddish water from the village spring also possessed special aesthetic and spiritual qualities that piped water does not. Through their memories of the color, taste, and smell of springwater, older women reimagine the social fabric of a previous life and regret what has been lost as well as acknowledge what has been gained.

In chapter 5, Hugo De Burgos recounts the history of *La Pila de San Juan,* a community water tank, in Suchitoto, El Salvador. Unlike the village described by Naguib, the residents of Suchitoto did not have a source of communal water until the water tank was built in 1840. With its construction water comes into play as a public symbol for the first time, a metaphor for prosperity, the medium for new forms of sociality, and the source of new types of dispute and class distinction. When piped water comes to Suchitoto, displacing the use of communal water tanks, it introduces another series of transformations in the meaning of water and leads eventually to violent disputes over the right of the state to privatize the water supply.

In chapter 6, Rita Brara describes the technological innovations that are contributing to the depletion of aquifers in the Punjab region of India. But her focus is not on the political ecology of water depletion, or on the environmental costs of this particular pattern of agricultural intensification. She focuses instead, following Latour, on describing the "sociotechnological assemblages" being brought into play in this setting (this volume, p. 135). "Wells and energized pumping sets for groundwater," she writes, "appear in this narrative as dynamic assemblages connecting the lives of farmers, technicians, policymakers, and ancestors" (this volume, p. 135)

Chapter 7, by Swathi Veeravalli, is less focused on technology than the preceding three chapters but is included in part 2 because of the role water storage tanks play in the construction of water inequality in a rural setting in Kenya. Water possesses social agency in this setting, she argues, because of the way seasonal fluctuations in precipitation affect water access. Periods of low precipitation tend to reduce inequalities since under those circumstances everyone has relatively equal access to what little water is available. During periods of high to medium precipitation, however, better-off families capture water in storage tanks; it is at these times that significant degrees of inequality do occur. The social agency of water in this setting is thus mediated by the cost, size, placement, and functionality of water tanks.

References

Parr, Joy. 2010. *Sensing Changes: Technologies, Environments, and the Everyday, 1953–2003.* Vancouver, BC: UBC Press.

UNESCO (United Nations Educational, Scientific and Cultural Organization). 2006. *Water, a Shared Responsibility. The United Nations World Water Development Report 2.* Paris and New York: UNESCO and Berghahn Books.

Chapter 4

AESTHETICS OF A RELATIONSHIP
Women and Water

Nefissa Naguib

∞

We are both the mirror and the face it shows.
We taste the moment and savour all eternity.
We are the pain and of that pain the cause.
We are the sweet cold water, and the jar that pours.
—Rumi, Jalāl ad-Dīn, *Enlightening Poems*

Preamble

"Women fetching water! Why do a study on women gossiping by the spring?" A scornful remark by a young Palestinian archaeologist seemed to sum up the initial responses when I first presented my topic more than a decade ago. Why women and water? Because water makes and unmakes human life and bonds, sensory experiences, and relationships with cosmological forces.

Introduction

I have argued elsewhere that anthropology that is attentive to human, primary, sensory experience has much to tell us about the evocative workings of water memory (Naguib 2009). My point of departure is my experience of unexpected responses to my questions to women about their water chores. When I began my research on gender and women's access to water I thought I had set myself well-defined and regulated tasks; I believed that by piecing together the events surrounding modernization of water I would end up with a picture that would not only be complete in itself, but that also might make some contribution toward an analysis of development, gender, and water. As with many such tasks, however, this one turned out to be more complicated than I had initially thought.

Having assembled what reports I could find about centralization of water, I interviewed women in the village who had lived through the time of fetching water from the spring to getting it piped into their houses.[1] My conversations with them were during a particularly critical period of violent confrontations between Israeli forces and the Palestinian population. Their experiences and expressions about water from the spring to the tap were difficult to accommodate to the archival evidence I had collected about the modernization of water access. What weight, for instance, should I give to information about a girl's bearing as she walks to the spring, the sacredness of water, or the sensation of reddish water on the skin? I decided that such representations by women whose lives were chronically disrupted by long-drawn-out aggressions were an integral part of the story of the centralization of water in the village. I discovered I could not collect the material on water and on women's histories separately. Women, water, aesthetics, and global moments were so complexly intertwined that they could not be disentangled. I obtained more introspective, intimate, and personal accounts of women's lives when I asked them to "tell me about fetching water," and traced the impact of modernized water access and violent political events in interviews that I had intended to focus on water.

In this chapter, water tells the stories of women's lives. Water is analyzed from the perspective of transcending sheer emotional and physical memory and sensory experiences. For this reason, my arguments are an extension of my original research on water and historical consciousness, the sensed relevance of the past, and ruptures in the present among elderly Palestinian women. To this end, "Aesthetics of a Relationship" is about the relationship between water, aesthetics, and memory within the context of critical moments in a Palestinian village. The ethnography of water is an attempt to bring together themes of ruptures, synaesthesia, and memory around fetching and using water. I propose that such reflections on water, aesthetics, and memory will elaborate on various theoretical approaches, ranging from critical events and fabrics of social life, to aesthetic undercurrents and water consumption.

From the very first time I visited the village, clues abounded on the connection that existed between the aesthetics of water, struggle, and sensory recollections of the past. As I listened to the women's stories about the past, as part of my research project, which focused on how women perceived the centralization of water, I began to understand the extent to which water chores permeated their senses and aesthetic engagement. This was evident in sixty-year-old tales of water "sparkling like a rooster's eye," "reddish water trickles beautifully," "springwater between the fingers," wild hills where animals and women could move freely. Did they really remember such far-off details of their lives as sparkling water, the trickle of reddish water, laughing and crying, and plump figs? If not, then what are sparkling water, trickles of water, figs, and wild hills doing in their stories? In the context of this chapter, I suggest that these stories say something more about how water memories open up a world of interrelated activities, of systems of meanings, and about how water is the substance that ties lives together.

There is more. Recollections of water are recounted against the backdrop of immense human suffering in the village. On the everyday level, I have observed old Palestinian women not only making contingency plans in case their home is hit or demolished, but also having to deal with the constant agony that accompanies violent casualties and injured human dignity (Naguib 2005). Recollections in this chapter illustrate how history lies in the stories that people tell, and in their practices (Cole 2001; Comaroff and Comaroff 1992; Rosado 1980; Stoller 1995). As fragile as it is pervasive, water illustrates powerful human sensations and experiences, as well as how people adapt, resist, and transform their lives.

While much scholarship on ruptures points back to critical global events, and work on memory and the senses highlights primary human emotions and aesthetics, the stories that women tell bring these themes together. To bring the conjunction of the intimate and larger disruptive events more sharply into focus, I turn to Veena Das' (2007) writing on critical events that dominate local social imaginaries, events that change the shape of the lives of those who are caught up in them. She has characterized such moments as terrible instants when worlds are disrupted and destroyed. These events bring into being new modes of action; women in this village learn to live in the world in new or different ways. This chapter is about a village in a region where water is scarce and contested but water is not only a physical resource. Once again, there is more. Water is also a medium and a metaphor for how these women perceive the world. The concept of the critical event is evidenced by the existence of multiple and often muted voices that express the suffering visited on them. Das argues for an anthropology that does not search for the meaning of these events—they cannot be accounted for in any simple way. I agree with her that in constructing metanarratives of such events, certain kinds of institutions—including the state—appropriate the experience of individuals for their own ends. For this and other reasons let us go to the village, the women, faith, and the color of water.

A Muslim Village

The village is relatively small, and in many ways it still conforms to the external images of a traditional Palestinian village community. It is located in the Ramallah Hills and overlooks other hilltops on all sides. The village has a population of close to 250 inhabitants and is situated on the top of an elevated ridge approximately 720 meters above sea level. Ramallah is the closest large city, and the road from Ramallah is the easiest route to take to the village (figure 4.1).

The summers are hot and dry, with the seasonal hot winds coming during spring and early summer. This part of the West Bank has been under cultivation for millennia, with the characteristic terracing still common in the Ramallah Hills. Today, as in ancient times, olive and fruit trees grow on the terraces. Most of the fruit trees are found in the wadis, which are covered with wild flowers in the spring. Fruits such as figs, almonds, plums, peaches, and apricots are picked in spring and summer. Chickpeas and lentils are gathered during the early summer months; barley, tomatoes, and cucumbers are harvested in summer. Yet it is the olive trees that physically and emotionally dominate the landscape. Olives are picked from November to late January.

Figure 4.1. Entrance to village

The general impression is of an arid and barren environment. The rocky hill-sides are occasionally dotted with olive trees, virtually the only visible vegetation (figure 4.2). Between the hills and in the valleys there are patches of green bushes.

Today's village is built above the ruins of what is usually called the old village. The physical arrangements are remarkably different from the pattern of the old village. The homes are scattered in the landscape and are on top of the hill, which means that the village today is without a well-defined center. In Palestine such villages are often referred to as newer types of villages: a village with a main street that cuts through it.

One of the first families to move out of the old village and to the new village was Abu Ali's family. They moved before the installation of water. Other families followed once it became clear to everyone that the village was going to have piped water. Within a couple of years, a Jewish settlement was built on the highest point above the village, so Palestinian women fetching water in jars faced the risks of confronting settlers who patrol the areas close to the water springs. Irrigation is a different matter; it is still much easier to use the overflow of rainwater during the short winter and spring to water the olive trees and vegetables that a couple of families still maintain.

Figure 4.2. Olive grove

Lack of household water has always been a major problem. Before the arrival of piped water, households fetched it from two springs. The closest source, referred to as the village spring, was in a field just below the village. During the summer months, water was scarce in the village spring and the women had to walk for several hours to another source. Household water was not only used for human and animal consumption; the women also had to make sure that some was left for the small vegetable gardens in the backyard of each dwelling. The women also collected rainwater during the winter; in the village this water is known as winter water.

The village elders had made several attempts to solve the water problem, and finally in 1985 the village got its much longed-for piped water. The initial reaction to piped water in the village was relief. Finally, after years of stagnation, the village market would grow, children would be educated, the men would find good jobs, and life would be prosperous for all. Mothers and grandmothers were tired of fetching water. Their daughters and daughters-in-law no longer helped them. Younger women were busy getting a higher education or trying to enter the job market, and they were not always available to help the older women. For the first couple of years, piped water was identified among the women and the men in the village as "a good." Water was going to be available for them to use—at any time and all the time. It was assumed that the resource would be evenly and fairly distributed. Piped water was not only for household chores, but also for the watering of small gardens and the irrigation of fields. The dry toilet sheds were locked up, and flush toilets were built inside houses. Young mothers hoped that with modern sanitation their workloads would be lighter and that more infants would survive. With piped water, the older generation of women imagined that, finally, they might enjoy some of the blessings of old age: ease and comfort.

Piped water did not come alone. It was accompanied by connection to the electricity network, and in the two first years following the piped water and installation of electrical services, televisions, refrigerators, and washing machines also came to the village, brought home by sons, daughters, and husbands working abroad. Everything seemed to go according to expectations: children went to school, men had jobs, and more young women joined the labor market. There was water inside the houses, but the washing machines were prominently placed outside for all to see. The television set, outside or inside the house, was constantly switched on.

Following the first Palestinian intifada in 1987 the situation in the village changed, and villagers could no longer rely on the services from Mekorot, Israel's national water company.[2] The Jewish settlements were given priority and the village was left without tap water most of the summer. The springs were no longer accessible because settlers roamed the area and threatened the women. Rainwater cisterns became the main source of drinking water for the villagers. Following the peace process that started with the Oslo Agreement in 1993, water was considered an interim issue, and the Palestinian Water Authority assumed responsibility for local government. Again, Palestinians in the village looked forward to a more reliable water supply, yet Israel continued to control the flow and volume of water allocated to the Palestinian areas, and villagers still relied mainly on rainwater storage.

The civil unrest that followed the first and second intifadas led to economic sanctions from Israel, resulting in an increase in migration among the younger generation. Everyday life demands more expenses from the villagers, mainly because tap water is expensive. The cost of water makes the maintenance of a small vegetable garden difficult, so a growing number of villagers have become increasingly dependent on charity from religious institutions, NGOs, better-off neighbors, or kin.

Unlike the general representation of Palestinian villages as patriarchal, this village is dominated by older women, with some young women and children also living there. Many older women are household heads while others live alone; most are widows, or have been abandoned by husbands who have been gone for most of their married lives. Some women "know" that the men will come back to their family, and they still receive the occasional money order; others realize that they have been abandoned.

Sons, brothers, and young husbands are abroad: some are studying, some are working (also in Israel), and some are in Israeli prison cells. Several young families have moved to larger villages, to regional towns, or abroad. The initial delight tied to the new water supply has turned to anxiety, loneliness, poverty and deprivation.

The main source of household water today is rainwater collected in cisterns. Thus, although the women agree that tapwater would be easier, they consider the cistern water, or winter water, to be more reliable.

Women and Water

"You want stories about the stones that grind the corn (*Hajar al Tahoun*)? That is what you what us to tell you about?" asked my landlady, Um Qays, when I told her of the complicated things I wanted to know about the changes that came about when springwater was replaced by piped water. They have lots of stories to tell about water, she said. When I first met her, she was standing trimming a bush outside her house. I was struck by the straightness of her back and the strength in her movements. Her voice was soft, but somehow I felt she could have led a regiment; it did not surprise me later to hear that she has a history of fierce quarrelling. She took me on a round of calls to introduce me to the other women.

The spring was where women spent most of their time: walking to it, waiting in line, filling jars or cans with water, and walking home with them (figure 4.3). At the spring, women exchanged news, gossiped, planned the marriage of their children; "We washed our body, slept a little and laughed and cried together." Some women were born in the village, others came from neighboring villages. Some were financially comfortable, others were poor. Some were in their eighties and some were in their sixties. But to all of them, fetching water is about being like "the stones that grind the corn." These are stones that never stop turning and pounding. More than one woman told me softly, "We know our water like our body knows our heart."

Um Qays was born Wagiha Abdel Magid Ali in Musharafah in the beginning of the Mandate years, and she has always lived in the village. She had two older brothers and five younger sisters; "some lived, some died." She had a happy childhood. Her father was a good and pious man; everyone came to him for advice. Her paternal grandfather was a sheikh and taught all his children, including the girls, to read. For various reasons, unfortunately, Um Qays, did not have the same possibility. But her father kept a record of the birth of all his children, the girls included. She was born in the olive season in 1920. Her father held his daughters in high regard: "Just like the prophet, my father, too, cared for his daughters."

Figure 4.3. The village spring

Her father and her mother were God-fearing peasants who believed in hard work and "never took anything from anybody. God is my witness." In her mind they were true Palestinian peasants who preserved the Palestinian values of generosity, hard work, and honesty. Her father worked in the field for the local landowner. Unlike others, she spoke favorably of the landowner's family, which treated her family well and never offended them. Her father died an old man, but her mother died when she was still young, just after Wagiha was married. She remembers her mother as a woman who never sat down to rest, who was always doing chores. "I never saw my mother put food in her mouth. God bless her soul." Her reserve and restraint was passed on to her daughter, so it was very easy to find a husband; rumors about hard-working girls travel fast.

Wagiha was married at the age of twelve to her paternal cousin. After the wedding, she stayed inside her husband's house for one week. Then she was taken in a beautiful procession to the spring:

It was a wonderful day, and I was like a young gazelle, walking with my head held high, so that all saw me coming out of my husband's home. They all sang around me. I had on my beautiful gold bracelets, and they made lots of sound, a beautiful sound. My own family was large, and I was protected. We all carried water jars; I was very small, only twelve years old, so I carried a tin painted with red shapes, and I had placed herbs on it. I had with me sweets for the spirit of the spring, so that the evil spirit would not ruin my home and my mother-in-law would be satisfied with my work. Then I went out to the spring to drink water. The water felt good and was a little red. The taste was bitter to give my husband a boy. I filled a jug and carried water back to the house. This was a good marriage.

During the first weeks, Wagiha was happy and the center of everybody's attention. Then things changed. There were no more festivities. She was now under the command of her mother-in-law. With tears in her eyes she told me how home is always the domain of the oldest woman. Men come home only to sleep and eat; otherwise they are always outside. When the men come home from work, they expect everything to be in order; they must never experience that things are not as they should be. Men, Wagiha explained, do not care where the water comes from, as long as they get their tea and food. They do not want to see a tired woman making an effort; everything has to look easy, and a woman has to look as if she enjoys everything she is put to do.

> The man goes to the field and he feels he is great. And I had to find water and feed for his horse. Everything had to be ready. I wanted to take the horse down to fetch water. But he did not let me. I had to find the water. My mind went mad when I could not collect enough drops of water. His only concern was whether the horse had enough to drink. I had to find enough water for the horse and for the household. When it was hot, in the eighth and ninth months of the year, all night I was out looking for water, and I did not sleep. My head went around looking for water and making my husband and my mother-in-law happy.

Life with her mother-in-law improved slightly after the birth of her first son, "the springwater was bitter and God heard my prayers, I gave my husband a healthy son." She has several sons. Four daughters "also came ... springwater was sometimes a little sweet."

Um Qays used to wake up her daughters in the middle of the night to fetch water and feed the animals, "but here we cannot ask the man to do that." Maybe, she says, the spring was a curse that made the young girls want to leave the village, but today she knows that she is blessed with a son who sends money to pay for the water bills and electricity bills and the maintenance of the cistern "so I can wash myself and make a cup of tea with tasty water."

Now that Um Qays and the other women have electricity and water taps in the house, they no longer go out to do their chores. Um Qays misses going to the spring to fetch water and is upset that the spring is not being maintained. It is true that fetching water and fuel for the oven was very hard work, but bread baked in the traditional oven is best, and water fetched from the spring is "real water." Water springs are located in the landscape of "the land of God"; walking back and forth is good for the "body and mind."

Her cousin Um Khaled insists always on washing her granddaughter's hair with wellwater. Now that the spring is dry, she has the winter water to use on the little girl. The water is fetched up in a bucket; she cannot afford an electric pump. When the water is hauled up, she collects it in a big basin. The water is cool, yellowish, frothy, and to me has a distinctly stagnant smell. The basin is left outside in the sun to "soak up the warmth of the sun. It has blessings." Flies circle around the basin, which has a green rim on its inside. Every time she is going to bathe her granddaughter, there is a shouting match between Um Khaled and the girl's mother, Samia. The grandmother calls it winter water and Samia calls it "sewage water."

Samia is actually her ex-daughter-in-law, who lives next door with her widowed mother. Um Khaled is also a widow, living in a one-room house. Her son recently remarried, to a young woman from Ramallah. "The new bride refuses to move up to live in the middle of nowhere," laughed Um Khaled. At one point, Samia whispered to me "lucky her." Khaled and his new bride moved to Ramallah where both have good jobs. Samia is now Um Khaled's closest neighbor. When I asked why the child was not in Samia's care, Um Khaled was shocked and scolded me: "Are you out of your mind? I am his mother. This is the way we do it here. The mother of the boy is best." Looking crossly at me, she repeated "She is always best." She ignored me for the rest of that evening.

The next day during breakfast, Um Khaled handed me a glass of tea, stroked me on the head, and said, "I decide when Samia can spend time with the child. But just the other day, Samia started to talk about her rights." She laughed loudly at what she saw as a ridiculous idea. "Who has rights? Nobody! We are peasants; not even men have rights. But Samia works in Bir Zeit and watches all these films from Egypt. They put ideas into her head, like they put ideas into your head too." Um Khaled had spent long hours telling me horror stories about her mother-in-law, so I reminded her and asked whether she also would have liked to have more rights. Would she have liked the possibility to say what she wanted to her demanding mother-in-law? Would she have liked to refuse to go mad searching for water? Wouldn't she have liked to rest when she felt like it, rather than just stealing some moments of sleep while she was waiting by the spring? Instead of getting angry again, as I feared, she gave me a big smile. "Before, we had no time to think about rights and lack of rights. We thought of nothing. I had no time to think of what my mother-in-law did to me. We were between the spring, the oven, and the olives. Between being pregnant, nursing, pregnant again, and so on. And then the children, they also have to grow. The head has no time to think."

Um Khaled lost a child: "I was young, and they made me work all the time. I was tired walking back from the spring; I fell backwards. As I fell I pushed the two walking behind me, and they joined in the fall. But no one was angry with me. We were all young girls and friends. But it was too late; I lost the child in my stomach." I said something about tap water making it easier for girls today; they do not fall on the way to the spring and miscarry their baby. "You do not understand. I am tired in my head today. Like the young people today; they are tired in the head. Boys do the work of girls, and girls do the work of boys; it is not the way God made us. Before, I was only tired in my body. It is better to be tired in the body; then we do not think about the blood all around us. Now I sit and think a lot, and sometimes the tears start rolling."

Her older neighbor and closest friend, Um Muhammad, was the village healer and midwife. A midwife and healer is a woman with blessed powers. There is a place of pilgrimage on a hill just east of the old spring, the shrine of a village woman who won religious recognition because she "cured the unhealable" and was a washer of bodies. It was also said that she was close to the angels. Her shrine was often visited on the way to fetch water; especially at times of great sorrow, "nobody noticed if we took the long way to the spring."

Like all the other women in the village Um Muhammad fetched water and took her laundry to the spring. She points to her neck:

> You see how beautiful my neck is. It is because I was the best at carrying water on my head. A woman who could do that was clever. Not like today; they moan when they walk from the chair to the bed. Today women, when they give birth, they stay in bed. We never said anything. We always went together and joked and sang. I went at night between midnight and 2:00 to the spring and slept at the source. I even gave birth to one of my children at the spring. Wiped him, wrapped him up, fetched the water, and walked back. They were happy when I came back with a boy. We walked with our water on our head straight, proud, aware of our strength. A woman showed she was clever, carried the jug or can of water with pride. The ones who carry on their heads have more beautiful necks than the women of today.

"Look at Estehar," and she pointed to her daughter-in-law; "she has a neck like a man. But look at my neck," and she pushed her head shawl to the side to show me her long well-toned neck (figure 4.4).

Today she lives in the house she moved into when she moved out of the old village in 1987. She had been a widow for most of her life and was tired. She moved in with her youngest son and his family. Her son, she said, had built the home "all on his own." But they did not have much money, so could not afford to construct a well, and it is more expensive to build a well after the house is finished. "I am sorry that we could not afford a well when we built this house. It is cheaper with a well, and in summer the water is cut off for fifteen days," said Estehar. Um Muhammad is hoping that her oldest son, who is working in Jordan, will be able to send home enough money for them to dig a well and install an electric pump. "I know that it will not feel like springwater, but at least it will be water with coolness, color, and taste."

Being a healer she is concerned about the village diet: "We cannot grow vegetables without worrying about the water bill. Today we cannot use tap water from the water company to water the garden, so we go to the market. But in the market the vegetables and fruit are not always good and they cost money. In years when there is little water in the tap, the vegetables are too expensive to buy, and I feel sick from lack of vegetables. You know vegetables and fruits watered with water from the hills are healthy—they are best for you." Giving me cup of tap water, she asked me to smell it. "It smells like medicine. You know, like when you visit the doctor in Ramallah."

If it were not for the spring, women could never have stopped to rest. "The spring was the only place where young mothers could allow themselves to sit or lie down. We rested without losing our integrity, without showing any sign of weakness and being accused of being idle or bad workers. And because we are peasants we like to be out in nature. It makes us feel strong. Outside water is close to God. And it is so beautiful."

Figure 4.4. Um Muhammad

Faith in Water

We send down water from the sky according to due measure, and we cause it to soak in the soil; and we certainly are able to drain it off with ease.

—Qur'an Surah 23: 18).

To even attempt to understand the story of water in the village, "you have to know God." Water is not only a limited natural resource, but is also a gift from God and therefore divine. Women, young and old, make a clear distinction between water that comes from God, which should be free for everyone, and water that is chemically treated, controlled, and in pipes, and that therefore does not have the same virtues.

If a person drinks from a well that is not his or hers, the owner cannot claim the water back. The owner of the well may suggest, however, that the person go to an alternative watering place where water is located in public space and therefore not part of a property. If the water is not close by, no one can refuse another person the right to quench his or her thirst. There is also the water that is stored in jugs or vases: This is very different water, because it may even be sold by water vendors, who walk to water points and carry water in vases to sell. But if that water is stolen, the thief must not be punished. As to water quality and environmental conscious-ness, the women explained that if a peasant is irrigating his land, he must not harm his neighbor downstream, and animals must not pollute a spring.

God rules over creation. "He expresses His will directly to us" and everyone "knows" that water is life and therefore holy. This resource is not simply life; giv-ing it is endowed with legendary qualities and is a metaphor for divine interven-tion in their world. Water in their perception is part of creation, the good of this earth that is in the hand of God, and that God has bestowed on humans and can therefore also take away.

With Islamic water traditions as their frame of reference, the village women say water is found in four locations: in a river basin: "like the sweet water of the Nile"; in the sea: "they have that in Gaza"; and in canals in villages: "like in the villages in Gaza and Jericho." Then there is "water like the rooster's eye" from the spring and "yellow, cool, bitter, and frothy winter water" in containers or wells. While well water is acceptable, the best is the water that overflows from the spring: "and we all shared that water." In the mountains, peasants are close to God; to refuse another person water, is to sin against God. Even animals are taken care of, and they are not allowed to die of thirst. Several women talked about animals being watered before they quenched their own thirst. It is clear that when a peasant has invested in a well or cistern by his home it is private. Still, the right to slake one's thirst cannot be refused, so the women living alone with no well or cistern expect their neighbors to help when the tap runs dry or they cannot afford to use tap water.

Rain is the main source of water for small gardens and olive trees. The rains in the mountain are believed to have the quality of *nabiyya*, a word used by the women to describe water that is from heaven; water that flows with the purpose of bringing goodness. Reflections about water are always reflections about faith, grati-tude, humility, and submissiveness to customs and traditions endowed by God. "From water we made all living things," they would quote from the Qur'an. The process and scope of water is thus always "in the hands of God." Water is given also with regard to cycle of seasons; the rhythm of village life depends on the time spent fetching water. In the hot summer months, more time was spent "walking in the valley to find water. Walking here and there." And then when they eventu-ally got to the water there was the time spent waiting for the drops of water to fill the jar. What the women did and still do when they fetch, store, or use winter or spring water is to know that the water is from God. Rain is the sign of God's blessing on the lives of women.

Tasting Redder Water

Stories about faith and water show that water is both intensely intimate and intensely social. Every woman had her own story of water. With each story about fetching water, each woman is talking about how she has lived and continues to live, and about the challenges involved in being an honorable daughter, a good wife, obedient daughter-in-law, responsible mother, and respectable mother-in-law. Stories are about faith, about how the storytellers are torn between tradition and the modern world, between female companionship and patriarchal demands, between youthful duties, adult responsibility, painful old age, and the pain of occupation and conflicts. When a woman scoops up water to drink, pours it into my hand so I get a sense of the texture, or asks me to study its color or smell, it spurs personal memory and concentration. Scooping up water is also the first collective exchange about the texture and fragrances of "redder," "more yellow," "more bitter," "cooler," "more frothy" water. One particularly expressive example is when a women scoops up water from her cistern and boils water for tea, prepares the tea, and offers a glass to her neighbor, saying to her, "Here, like the old true times."

The experience of true times brings me back to where I began this chapter: Why water and memory? I have argued that we can understand the evocative power of water by looking more carefully at the expressions it evokes, and the aesthetics that can be carefully elaborated to different extents and in different ways. Discussing the sensibilities of water, I use Arnold Berleant's argument that aesthetics is "what is perceived by the senses" (2010: 4). Water is considered aesthetically—that is, as a construction of sensory and experiential human practice. Angela Hobart and Bruce Kapferer once said, referring to aesthetics in rituals, that aesthetics "is what ties art (and all other human endeavors) to life" (2007: 5; parentheses in original). This tie is captured by the women's universal approach to their water as discussed in this chapter. Water effectively dissolves preconceived distinctions between nature and culture, fetching and consumption, individual and community, body and mind. At the same time, the stories tell us that water is a profound medium for individual responses on different levels: to the divine, the body, society, and critical events. It is precisely through this capacity to make connections that water becomes a highly charged sensory resource.

To say that water is "sensory" is another way of drawing attention to the fact that many life directions constituted by and through the medium of water are also dynamic, and ought to be analyzed as such. They are processes of embodied practices that are representations of the complexities of human life. As an anchor of intimate and embodied experience, water is at the core of the village women's collective or individual acts of remembrance.

David Sutton (2001) provides further clues from his work on food and memory. He has explored these dimensions by drawing attention to the potential of food to evoke memory. Maintaining Connerton's (1989) argument on commemoration and Csordas' (1994) discussion of embodiment, Sutton shows how worlds of experience and interpretation are contained in food. In this chapter, water emerges as a material link that confirms and establishes, in a very sensual manner, the connections felt by the women to each other and to their village. This is important, and it is what, as Sutton (2001) says, makes Connerton's *How Societies Remember* (1989) such a powerful reference in the study of memory. In asking "how" rather than "why" societies remember, Connerton draws our attention to memory as sedimented in practices such as washing the body and hair with reddish water.

The issue of food and memory also points to the significance of human responses to the shortage of food. Megan Vaughan (1988: 1) once wrote, "Famines gather history around them" and they will continue to do so. Hunger is a memory stimulant; when people are hungry they remember food. But it is not quite as simple as that. Memory of food can also be a form of resilience. In *In Memory's Kitchen*, edited by Cara De Silva (1996), remembering food in Theresienstadt (Terezín) concentration camp was making a stand against dehumanization. Recipes giving careful instructions were based on daily conversations about the right way to prepare their beloved dishes, conversations among the women who were being starved to death in the Czechoslovakian camp. "Cooking with the mouth," was a stance that "strengthened their resolution to survive, if only because it made more vivid … not what they sought to escape from, but what they sought to return to" (De Silva 1996: xxviii).

The case of Terezín is a particularly extreme example of how women survive through their memories when their sense of normalcy is viciously destroyed. In recovering true good time, the village women held long—and for me gripping—conversations and occasionally, arguments, about the colors, textures, tastes, and smells of spring water and winter water. By their very nature, memories of color, taste, and smell of water tend to the peculiar and the randomly associative. That they have such a great capacity for the women to associate with episodic recognition helps to explain why such attributes are useful for making meaning of apparently random recollections of events. When aesthetic perceptions merge into mundane everyday activities, we recognize just how sensuous human practices can be. In the village, water is refined synaesthetically and emotionally, so the spring becomes a particularly strongly demarcated cultural site for the reimagining of social fabrics that have been displaced by dreadful violence in time and space.

Why women and water? Because few things in human life are more primary than sensory recollections and satisfactions.

Notes

At various stages and to various degrees this chapter is molded by the women from what I call the village. Their help and generosity is invaluable. This chapter is part of the "Aesthetics and Devotion" project, for which I thank the donor, the Research Council of Norway, as well as my colleagues in this project. I am deeply grateful to John Wagner for inviting me to contribute, and for his intelligent suggestions, and editorial patience. A number of other people have read and commented on various presentations and drafts of this piece. Thanks go to Ingvild Flaskerud, Nancy Frank, Randi Haaland, and Catharina Raudevere. Special thanks go to Bjørn and our daughters.

1. Village is the fictive name used in this chapter.
2. The first intifada erupted on December 8, 1987. It continues to be a sustained uprising in the West Bank and Gaza Strip. This Palestinian struggle has taken a mostly nonviolent form of manifesting resistance through labor strikes, waving the Palestinian flag, commercial shutdowns, and resignations of Palestinian employees in Israel.

References

Berleant, A. 2010. *Sensibility and Sense: The Aesthetic Transformation of the Human World*. Exeter: Imprint Academic.

Cole, J. 2001. *Forget Colonialism? Sacrifice and the Art of Memory in Madagascar*. London: University of California Press.

Comaroff, J.L., and J. Comaroff. 1992. *Ethnography and the Historical Imagination*. Boulder, CO: Westview Press.

Connerton, P. 1989. *How Societies Remember*. Cambridge: Cambridge University Press.

Csordas, T. 1994. *Embodiment and Experience. The Existential Ground of Culture and Self*. Cambridge: Cambridge University Press.

Das, V. 2007 *Life and Words. Violence and the Descent into the Ordinary*. Berkeley: University of California Press.

De Silva, C., ed. 1996. *In Memory's Kitchen: A Legacy from Women of Terezín*. Translated by Bianca Steiner Brown. Northvale, NJ: Jason Aronson.

Hobart, A., and B. Kapferer. 2007. *Aesthetics in Performance: Formations of Symbolic Construction and Experience*. New York: Berghahn Books.

Naguib, N. 2005. Stones and Stories: Engaging with Complex Emergencies. In *Gender, Religion and Change in the Middle East: Two Hundred Years of History*, edited by I.M. Okkenhaug and I. Flaskerud. Oxford: Berg Publishers.

———. 2009 *Water, Women and Memory: Recasting Lives in Palestine*. Leiden: Brill.

Rosado 1980. *Ilongot Headhunting: 1883-1974: A Study in Society and History*. Stanford: Stanford University Press.

Stoller 1995. *Race and the Education of Desire: Foucault's History of Sexuality and the Colonial Order of Things*. Durham, NC: Duke University Press.

Sutton, D.E. 2001. *Remembrance of Repasts: An Anthropology of Food and Memory*. Oxford: Berg Publishers.

Vaughan, M. 1988. *The Story of an African Famine: Gender and Famine in Twentieth-Century Malawi*. New York: Cambridge University Press.

Chapter 5

LA PILA DE SAN JUAN
Historic Transformations of Water as a Public Symbol in Suchitoto, El Salvador

Hugo De Burgos

∞

Introduction

Researchers at the Fraunhofer Institute for Interfacial Engineering and Biotechnology in Stuttgart have developed new technology aiming at solving the water crisis problem. With this new knowledge, they are able to extract water from air humidity. This novel way of obtaining water will also undoubtedly bring new social engagements, meanings, relations, and understanding of the vital liquid. A similar experience happened in Suchitoto, El Salvador, in 1840, when, for the first time in that community, a water tank was built on the banks of the San Juan River to harness and publicly store potable water. This pragmatic change of technology, however simple, created new social relations and public symbols, and an innovative water engagement.

In this chapter, I analyze and interpret some aspects of the social life of water in Suchitoto. The study explores the historical trajectory of water in this city from the introduction of a potable water supply in the early nineteenth century to the public and bloody demonstrations in 2007 against its privatization. I examine water as both a natural object and as a contested public symbol, and argue that since water in Suchitoto was stored in municipally owned water tanks for the first time, a new understanding for the engagement with water as a public symbol emerged in this community. An incipient system of *pilas* (water tanks) built by municipal authorities in the 1840s became the primary vehicle for the new social transformation of water in Suchitoto. Conceptually and historically, for many Suchitotenses (people from Suchitoto), water went from being a wild, naturally found, and free, life-sustaining substance to becoming an object of public dispute, a metaphor for social prosperity, an alienated commodity, a political and economic issue, and a matter of death.

The Setting

The city of Suchitoto derives its name from a Nahua word meaning flower bird, or *pájaro flor* in Spanish (Lardé y Larín 1957). It was originally an indigenous village occupied by invading Spaniards in the sixteenth century. Suchitoto is located around forty-eight kilometers northeast of San Salvador. Approximately 6,000 people live in its urban core. Greater Suchitoto, however, includes twenty-eight cantons, and seventy-seven rural communities, with a total population of about 20,000. During the 1980–92 civil war in El Salvador, Suchitoto was the scene of heavy fighting between government troops and guerrilla forces from the Farabundo Martí National Liberation Front, now the party in power.

In 1991, a year before the Peace Accords were signed in Mexico to end a twelve-year-old civil war, world-renowned community leader Alejandro Coto organized and instituted the Festival Permanente de Arte y Cultura (the Permanent Festival of Art and Culture), transforming Suchitoto through this initiative into what many Salvadoreans refer to as the touristic cultural capital of the country. In 1997, by legislative decree, the National Legislative Assembly declared Suchitoto Conjunto Histórico of cultural interest for El Salvador. The term *conjunto histórico* is a legal designation applied to buildings in a given locality. It is typically used to protect entire villages, towns, or cities that have historical and cultural value. Suchitoto is geographically close to Ciudad Vieja (Old City), the archaeological site where the capital city of San Salvador was originally founded, before it was permanently moved to its current location (Barberena 1977). Because Suchitoto has been able to preserve an architectural facade that mimics the colonial period, in 1998 this city also figured among the 100 most-endangered sites promoted by World Monuments Watch, an NGO based in New York. Today, Suchitoto is a thriving community deliberately maintaining some of its colonial architectural heritage and successfully selling it as a form of nostalgic tourism.

Methodology

In 1998–99, I conducted archival and ethnographic research, coupled with participant-observation in Suchitoto. Subsequently, I published a book on its urban development history (De Burgos 1999). Although during the archival research I read hundreds of pages from municipal minutes, resolutions, correspondence, and other documents, only manuscripts with complete information were part of the 1999 publication. Much of the incomplete or partial information I found, however, also caught my attention and fed my curiosity. Ideas of potable water as signs of prosperity, modernity, decency, and prestige were disseminated throughout several municipal documents dating roughly from 1840 to 1905. Notwithstanding the data was fractional and lacked solid context, I was able to glimpse the social life of water in Suchitoto of the nineteenth century. I shared this information with a few octogenarian cultural consultants, from a total of twenty people who participated in the research. In at least five informal and two semistructured interviews, six of the oldest participants related to me some of the stories their grandparents had told them about water in earlier times in Suchitoto. Most of their narratives coincided with what I had read in the archival documents. In addition to this oral tradition, I gathered some other data from an archaeological site that I found in Suchitoto in 1999 because it was described in old municipal written records.

Water and Prosperity in the Nineteenth Century in Suchitoto

Although as part of their nostalgic tourism many Suchitotenses zealously sell the idea of a thriving colonial past, and claim it is indelibly inscribed in the city's present urban appearance, historic Suchitoto could not be farther removed from this fictionalized golden era. Based on the historical evidence and documented events of those years, the colonial period in Suchitoto did not have the historical relevance and prosperous impetuosity that the first postcolonial decades had (Barberena 1777; Browning 1971; Castro 1978; De Burgos 1999; Goméz 1987; personal communication in 1998 with Enríque Kunymena, a Salvadorean historian; Lardé y Larín 1952, 1957). These perceived prosperous times included the social transformation of water.

El Salvador became an independent state after the Central American federation was dissolved in 1839. During its early history as an independent state, El Salvador was shaken by frequent revolutions, including an urban and infrastructural revolution that left permanent landmarks in its architectural urban face, particularly in Suchitoto. In that city the monumental Santa Lucía church, among other buildings, attests to the urban impetuosity of that period. Thus, contrary to what has been intentionally promoted by locals as a tourism strategy, Suchitoto's current urban face was mostly built during the republican, and not the colonial, period (De Burgos 1999: 30).

By executive decree on February 14, 1835, head of state Nicolás Espinoza created the department of Cuscatlán, and designated Suchitoto as the departmental capital city (Lardé y Larín 1957: 136). A year later, by legislative decree, on March 22, 1836, the village of Suchitoto was promoted to a higher rank and became a town. Twenty-two years later, on July 15, 1858, Suchitoto was finally granted city status (*Gazeta de El Salvador* 1858). The perceived symbolic prosperity of Suchitoto in this historical period was proportionally accompanied by rapid urban growth, a thriving indigo industry, the proliferation of haciendas, the building of important infrastructure, and the creation of valued social institutions—schools, churches, key administrative public buildings, and public water services (De Burgos 1999).

Demographic growth in Suchitoto was, in part, the result of the migratory waves precipitated by the indigo trade boom in the middle of the nineteenth century. Although 1625 was, according to Rubio Sánchez (1976: 17), the year when indigo became the main crop in Central America, it was not until 1854, after a powerful earthquake destroyed the city of San Salvador, that Suchitoto became an important indigo producer and market in the region (Browning 1971). Also during this time an important distribution of urban lots took place. Urban lines were demarcated. Old streets were paved with stones and new streets were built. Bylaws were decreed to maintain social order and guarantee an organized urban development. Culturally oriented projects were also initiated during these years of perceived prosperity. In 1849, municipal trustee Francisco Revelo deplored what he called educational negligence, and proposed the founding of the first school for girls in Suchitoto. Ahead of his time, through writing and community activism, Francisco Revelo created an important local discourse to promote urban development, community prosperity, gender equality, and the national identity of a nascent nation (De Burgos 1999). His discourse of urban prosperity had been part of the context in which new social configurations around water emerged in Suchitoto.

Since the 1840s, municipal authorities have expressed ideas of how potable water has a distinguishing feature between animals and persons, implying that animals relate to water instinctually, but humans morally. Accessing and administering potable water was seen as a sign of prosperity and human decorum. Water was central to all material, social and ideational forms of development. Urban planning and population growth in Suchitoto could not have taken place without water. As Vladimir Baiza, a key cultural consultant, poet, and resident of Suchitoto, told me in June 2010,

> Water conditions human settlements in Suchitoto and everywhere. Water defines the number of people who can live on the land, and the number of animals one can afford to have. Water influences the wisdom of a population. But above all, water also determines the quality of life, education and human development of a population.

Central to this hasty urban planning of the nineteenth century in Suchitoto was the creation of a rustic but initially effective system of *pilas de agua* (water tanks). They were equipped with *chorros públicos* (public faucets) to provide the population with plentiful and potable water without charge. The water system started with a single, but highly celebrated, water tank. Eventually, as public demand for this new water technology increased in Suchitoto, more water tanks were built. By 1880, forty years after the first public tank had been built, the system of *pilas* had evolved into a complex network of water dispensers all over town. With these *pilas* a new understanding of, and engagement with, water emerged in Suchitoto.

The Pila de San Juan

In 1840, the municipal government in Suchitoto built the first public *pila de agua* on the banks of the San Juan River, on the outskirts of town. It came to be known as la Pila de San Juan (Saint John's water tank). According to oral tradition and its own archaeological remains, it was about two meters high, more than five meters long, and approximately four and a half meters wide. It was made of masonry and stone slab. Its main and initial purpose was to help mitigate the hardships of water scarcity during the dry season, and provide Suchitoto with a modern and more dignified venue to obtain drinking water. Next to this water tank, public spaces for bathing with *guacal* (a small container) were built. This is the most typical way of bathing in rural El Salvador. It is done by taking water with a small container from a source, and pouring that water over one's body to clean it. Several laundry sinks were also built on the same site so that people could wash clothes on location, instead of fetching the water and doing the laundry at home (figures 5.1 and 5.2).

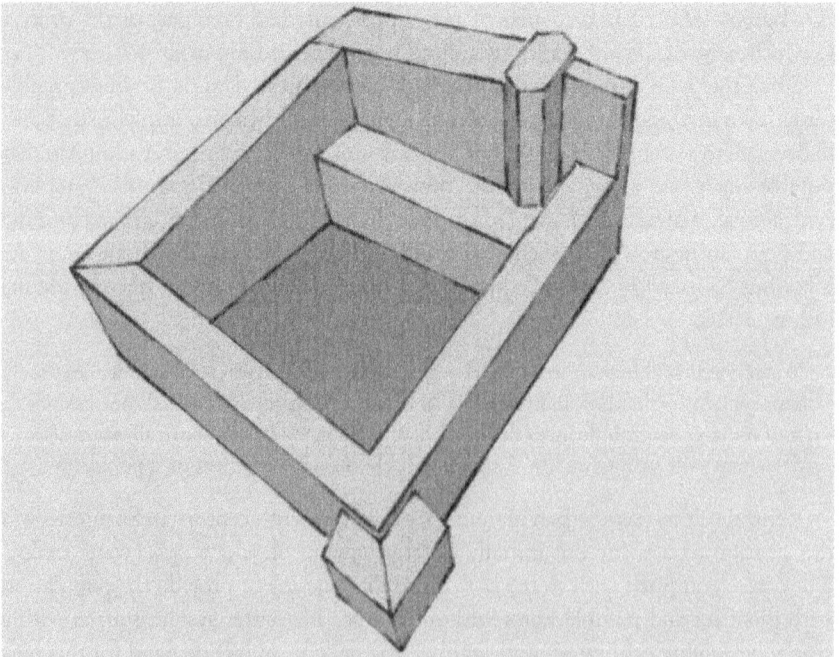

Figure 5.1. Drawing of La Pila de San Juan by architect Hugh Bitz

Figure 5.2. Author (center) accompanied by José René Melara-Vaquero (left) measuring the outside dimensions of the ruins of La Pila de San Juan in 2011

For over a decade, this was the only modern water dispenser in the whole village. In June 1999, an elderly cultural consultant told me that according to her grandparents "the water from the Pila de San Juan had been abundant and very clean."

In nineteenth century Suchitoto, water fetching was both an important domestic and public activity, and was carried out mostly by women. "Wealthy families would hire water fetchers or employ their own servants to get water for bathing, cooking, cleaning, gardening, and drinking," A. Beltran (a key cultural consultant) told me in May 1998. According to written records, industrial usage of water for manufacturing bricks, roof tiles, adobe, and pottery was usually carried out near the water sources, particularly the rivers. Initially, la Pila de San Juan was complementary to other water sources, since most people continued to get their water from wells, nearby waterholes, springs, creeks, and rivers. Eventually, however, the frequency and number of people using la Pila increased. With time and usage, it started to deteriorate and a series of new social problems emerged. By 1849, lack of adequate maintenance instigated protest from community members who used la Pila to meet their daily water needs. The following text comes from an old document I found and restored in 1998; it is located in the precariously kept municipal archive in Suchitoto.

Article 70 from the September 4, 1832, state law established that municipalities are responsible for keeping public sites, such as plazas and marketplaces, and even rivers, clean and tidy. Water is a vital necessity for human beings. Towns with greater population have organized commissions for the administration of such a vital resource. In Suchitoto, however, we have none of that; therefore it is necessary to organize one [such a committee].... Although there is a water tank on the banks of the San Juan River to provide considerably abundant water for the community, the tank is terribly abandoned. Far from providing a solution to the water shortage it has instead made it worse.... People who come for water and want to get it from the faucets often fight over the right to get it first. Therefore, it is necessary to take urgent action in this regard. This problem cannot be solved if you are not well-informed about the precarious conditions of la Pila San Juan. Sincerely, Manuel Flamenco. (Book of Municipal Records, 1849: 43, quoted in De Burgos 1999: 107)

On March 26, 1849, members of the municipal council, Francisco Revelo and José Herrera, brought to the attention of the city council the urgent necessity to repair the Pila de San Juan. Following that meeting, the mayor authorized the immediate reparation of the water tank (De Burgos 1999: 107). During the following three years, not much improved. The same water tank continued to serve a rapidly changing and growing community of water users. A new public preoccupation with the administration of water sources had emerged.

The Emergence of New Water Meanings

With the building of the Pila de San Juan, water had been centralized and bureaucratized for the first time in the history of Suchitoto. It was moved from its previous natural and domestic sphere to a public and bureaucratic space that was both conceptual and physical (De Burgos 2010). Although water as nature had been part of the social life of Suchitoto, Suchitotenses thought that with the new technology water needed to be socially tamed. It was no longer perceived as a natural symbol (Douglas 1996). It had become a social symbol with the transformative power (Wolf 1999) to change the physical and the conceptual space it previously occupied. As a new conceptual and physical landmark, la Pila de San Juan inscribed new meanings into the seemingly wild space. Water was the medium of the natural area, which, through their own cultural agency, Suchitotenses had transformed into a cultural space called la Pila de San Juan (see Sauer 1974). Harnessing water and placing it into domestic containers, as they had traditionally done in the past, was not the same as harnessing it and placing it into a giant communal water tank. In its new cultural/public space, water started to acquire a prominent political significance. Beyond its primary function—to dispense water—the Pila de San Juan had become an important meeting place, a cultural venue for people to exchange domestic, political, economic, religious, medical, and social information (De Burgos 1999). The place of a water source was used for significant social interaction in Suchitoto until the late 1960s.

As la Pila de San Juan increasingly became an important social feature in Suchitoto, it no longer functioned as a place where people went exclusively to get or use water. It became a place where complex intersubjective experiences merging social and ecological actions were enacted in many forms. La Pila de San Juan became a place where social and economic tensions were objectively visible, publicly reified, and personally manifested. Some tensions were expressed in physical confrontations among water users. One such incident took place in 1852, when tensions arising from the inadequate maintenance of the water tank escalated into a bloody fight. The archival document I include below describes this incident.

> Having the honor of belonging to this community, and as president of the municipality it is my duty to inform you of the following: The needs of the community are plenty and urgent; and according to law, municipal authorities have the responsibility to solve these problems. We have been having water shortages due to [the] poor intelligence with which people use the water tank built in 1840 by the municipality. Such water tank is currently broken, and its precarious state is causing disorder and confrontation among people who gather there to get water. There have been even incidents of blood fights. (Book of Municipal Records from 1852, quoted in De Burgos 1999: 107)

Claiming that water shortages are "due to [the] poor intelligence with which people use the water tank" clearly indicates that the shortage was caused by cultural behavior and not by natural events. In the summer of 1865, municipal authorities reported another water shortage caused by road blockage due to floods (De Burgos 1999: 110). In fact, there was plenty of water in la Pila de San Juan and still more available from natural sources. According to cultural consultants, even in the mid-twentieth century, water in Suchitoto was by all accounts abundant. Nearby rivers, springs, and ravines provided everyone with plenty of clean water. The flow of the numerous rivers from the area was greater than it is today. People from eastern parts of Suchitoto had a hard time crossing the strong flows of the Paso Hondo, Palancapa, Sucio, Quezalapa, and San Juan Rivers. Today, remains of old bridges are vestiges of the mighty flow of some of these rivers and the abundance of water. Near the lower land in the Aguacayo canton, an old bridge remains in the almost dry river bed of Las Señoras River (Baiza 2010).

According to the above text written by the then mayor of Suchitoto, the disorder and violent confrontations were caused by the precarious state of la Pila. Notwithstanding this, some may argue that the fights were more likely to be triggered by ordinary competition over a limited resource. The popular phrase, "Whiskey is for drinking and water is for fighting over" (of unknown origin, but often attributed to Mark Twain), supports this assumption. It relegates water to having merely biological importance to humans. It implies that underlying these belligerent actions there was an ecocentric and utilitarian intention. I argue, however, that these confrontations elucidate the emergence of new understandings of, and engagements with, water in the community due to changes in technology. New water conditions engendered new meanings, which consequentially established novel social relations and innovative intentional engagements with water. As a thing, water is profoundly infused with social value (see Appadurai 1986).

Confrontational interactions based on water cannot be seen strictly as eco-logical. Human actions are never based on mere organism—environment relations wholly contained within a conceived natural world. Rather, as claimed by Ingold (1987: 103), human engagement with water is "the dialectical process between both intentional and behavioral actions." The act of drinking water cannot be understood as mere biologically determined behavior, one devoid of control by a knowing subject (Ingold 1987). Human engagement with water, however prag-matic, is always driven by meanings (Geertz 1973) and purposes (intentionality). Nevertheless, water's centrality for human survival, generally speaking, gives it the status of an instinctually vital substance. Naturally, as living organisms, humans biologically depend on water. As culture bearers, however, water's vitality is more than an instinctual experience or a biological requirement for living. Empirically, people also fight and even kill for whiskey as well as for many other nonvital substances. These kind of struggles, it seems to me, are clearly driven by cultural significance, meaningful purposes, and political intentions. The new technologi-cal framing of water brought the cultural conditions for an intentionally political engagement with it.

Ontologically, for the first time in Suchitoto, people were no longer bending down to reach water from a spring, or tediously pulling a heavy bucket out of a well. Water users were now distinctively obtaining it from a metal device, which, as simple as it may seem by today's standards, was enough to change the relational ontology between water and people. Before water was publicly and bureaucrati-cally harnessed in la Pila de San Juan, people related to it in a more communal fashion. Different kinds of social meanings permeated the ethos of water usage. Water was conceived as a communal and natural God-given substance. It had not been publicly politicized or bureaucratized. People would fetch it from naturally occurring sources in the wild with little human intervention, or extracted it from domestic wells dug and built near their homes. In the past there was no sense of institutionalized authority linked to water access and management. La Pila de San Juan, however, was linked to a socially recognized authority for its function-ing and maintenance. Water was now linked to new social meanings of modern and civilized access to water. The distinction municipal authorities made between the animal and human usage of water as instinctual as opposed to moral allowed people to be equally interested in its new symbology as they were in its essential materiality. Suchitotenses, it seems to me, saw themselves as moral beings in a hu-man—environmental relationship to water. (See Strang 2009 for a similar discus-sion on the cultural meanings that people encode in water as part of their material surroundings.)

Hegel (1977) in a phenomenological account of the human–environmental engagement claims that humans project the self into the environment and reincorporate this projection into their daily life. At the same time, the new social meanings of water in Suchitoto were projected and incorporated into already existing structures (Sahlins 1995) and hierarchies of power, moral obligations, ideas of decency, personal preferences, and class distinctions. For instance, in a municipal document from 1852, a resident of Suchitoto in reference to la Pila de San Juan stated the following:

> [F]urthermore, the place lacks bathrooms for respectable individuals; since due to decorum, these people cannot go into the bathrooms for the commoners. Our community does not lack well-dressed individuals, who also have great public spirit. Therefore, it would be very appropriate to provide separate bathrooms for them, as this is an exigency of decency and good manners. (Anonymous, Book of Municipal Records, 1852: 8, quoted in De Burgos 1999: 108)

Although la Pila de San Juan had built-in infrastructure for bathing and for washing clothes, beyond the pragmatic value of these services, some users seem to have been more concerned with the corresponding symbolic significance between those newly created spaces and the locally perceived status of the individuals using them. The spatial configuration of water usage at la Pila de San Juan was now embedded in social relations. Ideas of status, decorum, decency, privilege, and personal preferences were now informed by a new conceptual and physical space created around the biological need for water and its social significance. All this was brought about by technological innovation. Discrepancies between the material conditions of la Pila de San Juan and conflicting ideational expectations of the emerging meanings of water made class tensions more visible in this ecosocial setting. (For a similar discussion, see Harris 2005.) For instance, the only bathing space at la Pila de San Juan was largely exposed. There was only one wall separating the space for getting water and doing the laundry from the area for bathing. People had to bathe collectively, exposed to one another. As stated in the document above, due to decorum, people of higher status could not go into the bathrooms where common people went. Clearly, the impediment was not physical. It was by all accounts symbolic and moral. Water usage in this instance was a way of demarcating class membership and social status.

The conceptual separation and valorization of personhood based on class distinctions was now articulated in the context of the social use, and the public spatialization of water (Gordillo 2004; Hirsch and O'Hanlon 1995). As stipulated in the document above (De Burgos 1999: 108), the moral need to acquire and express social status through the public usage of water was conceived as "an exigency of decency and good manners." The new social space of water had meanings for the "well-dressed individuals, [with] great public spirit" different from the meanings it had for the commoners. The former were thought of as more morally deserving due to a self-proclaimed public recognition of their community altruism, and the latter as simple, unprivileged, ordinary people. Whether new facilities were built to accommodate these individual preferences is not known. No archaeological record, oral reference or written documents have been found to establish that fact.

Technology particularly contributed to transforming water into a politically contested symbol (Cohen 1979). It went from being a taken-for-granted substance to a valued public entity, which could be possessed both as physical and symbolic property. At the core of this ecological appropriation were the intentional actions of Suchitotenses. As argued by Ingold (1987), human confrontation with objects of the natural environment lies not in their manipulation and transformation, but in their appropriation. "For it is nature appropriated, and not necessarily trans-formed that is engaged in social relations" (Ingold 1987: 126). As claimed by Ingold (1987: 136), "tenure engages nature in a system of social relations." There-fore, through the appropriation—not the transformation—of nature, new eco-logical contexts are created; consequentially, new social relations between people and objects as well as between people emerged. Water, as an ecological substance, has a social life because the moral and the intentional are the domain of human purposes (Geertz 1973). Human intentionality toward water in terms of its us-age, tenure, exploitation, dependency, and social and political significance is of necessity grounded on its very materiality, one that dialectically mediates between humans as biological entities and persons as cultural agents. This seems to be the case because, as claimed by Ingold (1987), the boundaries between the social and the ecological correspond to that between the intentional and the instinctually driven components of human actions. Human beings are simultaneously involved in systems of both social and ecological relations, constituting them respectively as conscious persons and as individual organisms. Descola and Pálsson (1996) similarly point out that the relationship between humans and the environment is mutually constitutive. That is to say, people could no more exist as persons outside society than they could as organisms outside the ecosystem (Ingold 1987: 126).

La Pila de San Juan was the reified medium through which Suchitotenses were simultaneously involved in both systems: the social/intentional and the ecologi-cal/instinctual. It is precisely this dialectical interaction that makes it possible for a natural object to be simultaneously both a public symbol and an ecological substance. In Suchitoto, symbolic expressions of water were and continue to be as strong as the practical interest everyone has in it. The centrality of la Pila de San Juan for Suchitoto was both pragmatic and symbolic. Despite class antagonism and other social tensions, Suchitotenses in the nineteenth century did not live in an arid region "where access to water lies at the heart of much conflict" (Heathcote 1998: 1). Water technology created new circumstances that were, nevertheless, morally understood in terms of previous presuppositions. As Sahlins (1995: 67) wrote, "[P]eople act on circumstances according to their own cultural presup-positions, the socially given categories of persons and things." Finally, it has been argued that the formal qualities and characteristics of an object are crucial in that they provide a common basis for the construction of meanings (Geertz 1973; Ingold 2000; Strang 2004, 2009). Hence, water, as an ecological object possessing a given number of qualities and characteristics, was socially constructed in both ecological and moral terms by Suchitotenses in the nineteenth century. At the center of this construction, was la Pila de San Juan, a landmark of an important social transformation of water.

Water in the Twentieth Century in Suchitoto

Archival data shows that by the turn of the twentieth century the number of *pilas* had increased in Suchitoto, moving water out of the city periphery into its urban core. For example, on July 4, 1903, residents from the centrally located Santa Lucía neighborhood requested the installation of a public *pila*. With city council approval, on December 25, 1903, potable water was introduced to this neighborhood with funds obtained from contributions made by residents. With the proliferation of more water technology in the city, a different kind of social consciousness about water usage, tenure, and waste seemed to have emerged in the twentieth century. In 1908, for instance, city council decreed that wastewater drained from the Santa Lucía barrio *pila* was going to be utilized to clean the municipal slaughterhouse. Also in 1908, a pipe broke in the Concepción barrio and the potable water pouring out of it was seen as wasted. With the aim of using this water, a concerned citizen wrote the following petition to city hall.

> I, Luis Parada, adult, with all due respect I expose the following: Since the public pila located in the Concepción neighborhood is broken, constantly pouring water out, and threatening the health of residents, I would like to request permission to reroute that water through some gutters into my property, which is next door to the aforementioned pila. (Book of Municipal Records, 1908, quoted in De Burgos 1999:112)

Luis Parada's letter shows an important aspect of the changing attitude toward water in Suchitoto at the turn of the twentieth century. An emphasis on the more pragmatic aspects of water seemed to have emerged. Issues of health and the need to avoid wasting water clearly indicate a new social understanding of water as a limited resource. An increasing population and the introduction of pipes to the urban core had modified, once again, the social and ecological relationship between humans and water. The system of *pilas* was now part of the urban landscape. Harnessed water was no longer located in the semiwild space, as was the case with la Pila de San Juan. Through subterranean pipes, water was invisibly and almost magically coming to the city. By 1917, there were six *pilas* in Suchitoto. They were located in the neighborhoods of Santa Lucía, la Cruz, San José, Concepción, el Calvario, and Acupán. By the mid-1920s, the municipality had introduced an incipient system of water pipes to supply water to some public buildings, private homes, and several *chorros públicos* in different parts of the city. *Chorros públicos* were different from *pilas* in that they were not water tanks any more. Water came directly from a central source and was distributed through the pipes. The introduction of the pipe system became a new marker of economic class, and power, and a personal symbol of modernity. The affluent were able to afford private *chorros* in their home while the rest of the population had to fetch the water from the *chorros públicos*, which, nevertheless, provided free access to potable water. The most memorable *pila* in the early twentieth century was known as la Caja de Agua. It was located in el Calvario Plaza. By late 1960, la Caja de Agua had been closed and a children's park built in its place. Blanca Bográn, Juana Avalos, Lidian Leiva, and other elderly cultural consultants told me in June of 1998,

> Water from la Caja de Agua was so abundant. It had some very nice gutters made of slab rocks. These gutters would take the water to another water tank, around a hundred meters away. That was a very long tank. It was approximately twenty meters long and around a meter tall. It had some structures for doing the laundry by hand. Women would come with huge load of clothes to be washed. That place was always full of people [who] were fetching water, while others, especially and women and children, were using laundry sinks.

As the population in Suchitoto increased, the demand for water augmented. In order to solve ongoing pragmatic water problems, successive municipal governments in Suchitoto searched for new strategies. They created by-laws to regulate its access, usage, and tenure. This last aspect is of central importance to the events that have unfolded over the past half century.

The National Office for Water and Sewerage

Up until the 1960s, potable water service in Suchitoto and everywhere in El Salvador was operated by municipal governments. In 1961, however, the Civil/Military Directory—a transitional government—created the Administración Nacional de Acueductos y Alcantarillado (ANDA; the national office for water and sewerage). By law all water, sewerage systems, and easements belonging to the state, municipalities, and autonomous institutions, as well as property registered on behalf of departmental development boards, were transferred to the jurisdiction of ANDA (ANDA 1961). In most cases, municipal water and sewerage systems were transferred without remuneration from ANDA, under the assumption that these systems were going to continue to serve the community (Moreno 2005). ANDA was given the legal power to access water bodies to conduct surveys, studies, and measurements in order to purchase, use, and treat surface water or groundwater

(Moreno 2005)). With the creation of ANDA, a long history of municipal water management in Suchitoto had ended. In the 1960s, ANDA introduced a new network of pipes in Suchitoto and used what was already available from the previous municipal water management.

Water in Suchitoto in the Twenty-First Century

Ironically, despite its abundance, access to clean, permanent, and inexpensive water is increasingly becoming a serious health and political problem for everyone in Suchitoto due to demographic, political, and ecological changes. The most emblematic case of political and ecological alteration in Suchitoto is the once mighty Lempa River. In 1976, Suchitoto's best agricultural lands were flooded with the creation of the electrical dam El Cerrón Grande. Hundreds of homes as well as potable water sources were also lost. As a result, an artificial lake, Lake Suchitlán, was created. This lake is now the greatest reservoir of freshwater in El Salvador. It is an important source of revenue through tourism and local fishing. However, all the wastewater and contamination from the metropolitan area of San Salvador and other nearby cities is also deposited in this lake. At least seven rivers that disembogue in it are, literally, sewerage systems. Some of these contaminated rivers include the Acelhuate coming from Salvador, the Suquiapa from Santa Ana (the second largest city in the country), las Cañas from Ilopango and Soyapango, the Tomayate from Apopa, the Nejapa from Aguilares, the Sucio from Zapotitàn and San Juan Opico, the Quezalapa from Tenancingo, and the Michapa from Cojutepeque and Tejutepeque. Although the processes and infrastructure for providing water to households has in general improved, the quality of wells, waterholes, springs, and rivers has drastically decreased. Industrial and domestic use of agrochemicals and cattle has also contaminated water sources. All of this is a serious issue because most people in Suchitoto rely of these sources for potable water. Suchitoto is not alone in this regard. Most of El Salvador is facing similar water problems.

According to Moreno (2005), a 2003 independent survey revealed that 58.5 percent of Salvadorean households obtain their potable water through pipes inside and outside their home but still on their property, while the remaining 41.5 percent of households have no residential service. According to the same source, more than 659,834 households have to rely on the neighbor's pipe, public *pilas*, streams, wells, rivers, springs, and water-vendor trucks. ANDA's water service coverage is marked by geographical differentiation. Out of the total number of households with residential water service, 78.3 percent are concentrated in urban areas and the remaining 21.7 percent in rural areas. Furthermore, 41.5 percent of the households with no residential water service have to travel great distances to have access to this vital liquid. Moreno (2005) claims that this gap in potable water coverage between urban and rural areas, as well as between people with residential water services and those without it, is an indication of the enormous challenge faced by ANDA to increase the availability of water resources and to develop the necessary infrastructure to achieve universal coverage in the country. In the past decade, however, even people with residential water service have had to fetch their water from elsewhere at times because the service is irregular. Sometimes, households with residential water service are without water for up to a week or so.

Except for the capital and other large cities, in Suchitoto and elsewhere in El Salvador the most common sources of water other than residential water service are wells (11.9 percent); public *pilas* or *chorros* (10.3 percent); spring, river, or stream (6.8 percent); and neighbor's pipe (6.7 percent) (Moreno 2005). These water sources are most commonly used by rural dwellers: 91.2 percent of rural households obtain their water from these sources. Urban residents who do not have service to their homes mainly use *pilas* and *chorros públicos.* According to Moreno (2005), of all the Salvadorean households who get their water from wells, 81.3 percent are located in rural areas and only 18.7 percent in urban areas. In contrast, out of the total of households that obtain their water from public *pilas* or *chorros,* 64.6 percent are found in the urban areas, and only 35.4 percent are households in the rural regions. Getting water from the neighbors' pipes seems to be a more balanced situation, where 54.6 percent of the households who do so are in urban settings and 45.4 percent are in rural areas (Moreno 2005).

Dependency on pipe water in Suchitoto has increased because 90 percent of the country's rivers are currently polluted with industrial chemicals and waste dumped from factories and intensive agriculture (personal contact in 2009 with Ricardo Navarro, of the Centro de Tecnología Apropiada).

The Privatization of Water

"In those years [early twentieth century], our ancestors could not even imagine that water could ever be sold or bought" (Mariana M., key cultural consultant from Suchitoto. In 2001, ANDA created an inventory of private water developments and implemented water management programs to regulate private production of subterranean sources. Almost four decades after ANDA was created, the introduction of the Ley de Adquisiciones y Contrataciones de la Administración Pública (LACAP; procurement and contracting law of public administration) started a new trend toward decentralizing water and sewerage services (Moreno 2005). This new disposition allowed some municipalities to manage their water and sewerage systems independently of the central administration. Rates for potable water and sewerage, as well as other items and services sold or delivered by ANDA and its concessionaries, must be, in theory, reasonable and submitted to the ministry of economy for approval (ANDA 1980). ANDA may also participate in public–private partnerships pursuing the same goal (ANDA 1980)

For more than a decade, in El Salvador, as well as other countries in the region, a number of economic and institutional reforms have taken place. Since the early 1990s, ANDA has been subjected to a so-called modernization program financed by the Inter-American Development Bank (IDB) and the World Bank. This program seeks to establish an internal market of water in which water rights would be reduced to a business transaction (Moreno 2005). Driven by financial interests of the IDB, and implemented by national governments, these reforms aim at reducing the powers and functions of the state by promoting the liberalization and openness of the economy, and the privatization or mercantilization of state assets and the public sector services (Arias 2008; Ibarra 2001; Moreno 2005). The actions to privatize public services in this process of neoliberal reform include the selling of state-owned enterprises and assets, which end up in the hands of transnational corporations. Today in El Salvador these corporations have a sturdy oligopoly on markets for telephones, electricity, pension funds, and so on, without a supervisory body able to regulate their functioning and protect the consumers (Arias 2008). The dynamics of privatization in El Salvador have led to the auction of hotels, refineries, cement factories, coffee roasting plants, banks, electricity distribution, thermal power plants, and telecommunication companies, among others enterprises. Lately, however, transnational companies such as Agua Cristal and Coca Cola Water, among others, have also attempted to get into the water business by trying to privatize water resources and water services (Arias 2008; Moreno 2005: 4). Hall, Lobina and de la Motte (2003: 1) argue that the definition of water as "a commodity by multilateral organizations in the early 1990s has allowed a few transnational corporations, supported by the World Bank and the IMF [International Monetary Fund], to become centrally involved in management of public water services in poorer countries." These companies seek to move an essential public good, one that is indispensable for the existence of all living beings, into the realm of the free market. Given great political and economic disparities in El Salvador, this would translate into denying to the majority of the population their inalienable right to water, and hence to life. In 2009, Suchitoto resident Alberto Durán told me that Suchitotenses dreadfully fear how "water can be commercially appropriated and made into a profitable commodity, which only some will be able to afford." In Suchitoto, access to potable water has always been a marker of social differentiation. Water has always been less accessible to people in rural areas and to the urban poor. Suchitoto is not alone in this regard. Worldwide, water scarcity has never been as endemic as it is today (Barlow 2008).

The Plan de Modernización del Sector Público (modernization program of the public sector) as defined by President Cristiani during his 1989–94 government, was taken up in the second Alianza Republicana Nacionalista (ARENA; nationalist republican alliance party) administration between 1994 and 1999. However, although neoliberal rhetoric overtly sought the modernization of the state, government's enterprises are made deliberately inefficient so the poor quality of their public services can be used as justification to privatize ANDA, and other governmental institutions (Moreno 2005). Indeed, President Antonio Saca slashed ANDA's budget by 15 percent in 2005, to its lowest level in a decade, in a country where more than 40 percent of the rural population have no access to drinking water (Ibarra 2001; Moreno 2005). In 2007, members of the ANDA Union accused the government of President Antonio Saca of promoting a plan to discredit ANDA's managing competence to justify its privatization. In fact, on Monday July 2, 2007, President Saca was scheduled to announce, in Suchitoto, his government's plan to privatize water in El Salvador. The plan was officially known as the Política de descentralización (decentralization policy). Many people in Suchitoto do not believe that privatization will solve the water crisis but that, on the contrary, it would make it worse. Water activists from Suchitoto believe that beyond extracting and distributing, natural resources must be protected.

New social relations of power brought new political symbols of water to Suchitoto where the value of water is now based on its scarcity. Suchitotenses today fear how the most basic requirement for life is gradually being appropriated and changed into an obscenely profitable commodity. Therefore, the day President Saca was supposed to announce his privatization plan, hundreds of people from Suchitoto peacefully demonstrated against Saca's intentions. Public protest led by the Sindicato de Empleados y Trabajadores de ANDA (SETA; water workers union) and the La Asociación para El Desarrollo de El Salvador (CRIPDES; rural communities association for the development of El Salvador), and a number of other groups, prevented President Saca from arriving in Suchitoto. Coordinated street blockades turned back Saca's presidential caravan. He was forced to return by helicopter to the presidential house in San Salvador, where he then held a press conference to formally announce the new policy. Meanwhile in Suchitoto, riot police opened fire on demonstrators with rubber bullets, tear gas, and pepper spray, injuring around seventy-five people. Throughout the day helicopters circled Suchitoto and San Salvador. The riot police did not withdraw from the scene until well into the afternoon (La Prensa Gráfica 2007). Fourteen of the peaceful demonstrators were brutally beaten and arrested by the Salvadorean national police and charged under an antiterrorism law. If convicted, they could have faced up to sixty years in prison (La Prensa Gráfica 2007). The charges fell under the jurisdiction of El Salvador's 2006 Ley Especial en contra de Actos de Terrorismo (special law against acts of terrorism), which was championed by the U.S. embassy in San Salvador.

Human rights experts in El Salvador and abroad uniformly concluded that the Suchitoto protest was lawful and denounced the terrorism charges (Committee in Solidarity with the People of El Salvador [CISPES] 2008). Social justice and environmental organizations from all over the world expressed their indignation with the arrests. Horizons (2007), a Canadian nonprofit social justice organization also published a document containing the names of the people arrested in 2007.[1] In July of 2007, in response to grassroots pressure organized by the CISPES and allied solidarity organizations, more than forty members of the Salvadorean congress signed a letter sent to President Saca questioning the application of the antiterrorism law in the case of the nonviolent Suchitoto protestors. A handful of progressive Salvadorean congressional representatives also sent personal letters to the president reiterating their concern for the state of human rights in El Salvador and urging President Saca to respect basic civil liberties, including freedom of political expression. After more than six months of investigation into the events of July 2, 2007, the Salvadorean government was unable to substantiate its original terrorism accusations. On February 14, 2008, El Salvador's attorney general requested that charges of acts of terrorism be dropped against fourteen peaceful demonstrators from Suchitoto.

Although all charges were formally dismissed on April 16, 2008, two weeks later, on May 2, nineteen-year-old Héctor A. Ventura was brutally stabbed to death at night in a house where he and a fourteen-year-old friend were sleeping in the community of Valle Verde in the jurisdiction of Suchitoto. The fourteen-year-old survived the attack. Ventura was stabbed in the head and the heart. He was one of the fourteen demonstrators who had been arrested in Suchitoto in 2007. Ventura's murder came two days after he had met with the mayor of Suchitoto and had agreed to speak about the Suchitoto case for water as a public good during the "Day against Impunity" event, a political activity planned for the first anniversary of the arrest of the four Suchitoto demonstrators (U.S.–El Salvador Sister Cities 2008). Ventura's water-related murder is now the object of the same impunity he was about to denounce. The tumultuous social life of water in Suchitoto is inscribed in the multiple social and symbolic meanings that have accompanied its concrete vitality throughout its own moral, pragmatic, and political history.

Conclusion

In this chapter, I have examined and interpreted the ways in which material conditions, social relations, and ideational configurations in Suchitoto infuse water with a social life. By exploring the historical trajectory of water in this community, I have elucidated how water, as a natural substance, was culturally transformed into a politically, economically, and lethally contested object. As the most materially important and emblematic public good, the cultural valuation and pragmatic signification of water in Suchitoto has ranged from being a metaphor for prosperity, social prestige, and life, to an object of greed, wealth, and social conflicts. Regrettably, such public contestation has already resulted in the irreparable death of young Suchitotenses.

I have also shown how the introduction of novel water technology in Suchitoto in the nineteenth century fundamentally transformed people's understanding of and engagement with water in that community. La Pila de San Juan, as technological innovation, changed forever the water experience for Suchitotenses and created the conditions for the emergence of new social and ecological meanings of water. Because in the past clean water was abundant, and water paucity was only a seasonal inconvenience due to logistical, not ecological factors, Suchitotenses thought of water as an unlimited resource. Contamination and waste were, initially, almost insignificant. However, as ecological conditions of water gradually changed in Suchitoto, so did its social meanings and pragmatic engagement.

On a more pragmatic perspective, the water crisis in Suchitoto and in El Salvador in general is exacerbated by the economically powerful obstinacy with which the World Bank and the IDB continue to push the Salvadorean government to adopt the catastrophic recipe of water privatization. Fearing that the most basic requirement for life is being threatened by unscrupulous business initiatives, Suchitotenses now emphasize the concrete, pragmatic, ecological, and biological aspects of water. Because people as intentional agents live in places that are simultaneously ecological and social spaces, Suchitotenses have been able to historically and conceptually engage water as both a concrete object and a symbolic experience. As a substance that is literally essential to all living organisms, water is, nevertheless, also culturally experienced and embodied (Strang 2004: 4). This is the case because even in the most fundamentally biological and instinctual relation to humans, water is never experienced in a purely mechanistic or instinctual way. Comparatively, Csikszenmihályi and Halton claim that even hunger and sexual drives always appear in consciousness transformed and interpreted through the work of signs one has learned from one's culture (1981: 5). Thus, beyond its biological exigency, drinking or fighting over water is always accompanied and mediated by socially learned meanings as shared cultural intentions resulting from particular historical contexts (Boaz 1982). In this sense, ecological and moral interrelations in Suchitoto have dialectically shaped the social life of both water and people, fusing them into a single ecomoral experience where meanings and engagements, actions and intentions, objects and persons, are redefined by material technology, social relations of power, and antagonistic understandings of water as either an inalienable right or as a profitable commodity that can be coveted and unscrupulously possessed.

Finally, Strang (2004: 129) claims, "given the reality of human need for water, and its economic importance, it is unsurprising that ownership of water is intensely contested." Although this is true in some instances, not all water-related confrontations are driven by purely econometric or utilitarian purposes. Keesing (1990) argues that dismissing the symbolic motives and values through which humans orient their lives as hiding some covert ecological rationality is to misconstrue drastically the nature of the human we are trying to understand. As an ecological and a social object, water from la Pila de San Juan and water from the urban pipes in contemporary Suchitoto was engaged with by Suchitotenses as individual organisms and as conscious persons, whose actions—however solemn, sober, right, wrong, benevolent, simple, complex, hasty, erratic, rational, irrational, nasty, violent, selfish, social, generous, greedy, and angry—nevertheless had the stamp of culture.

Notes

1. "We, the undersigned Canadian organizations with great preoccupation and indignation denounce before the national and international community, the arbitrary arrest and continuing detention of Marta Lorena Araujo Martínez, President of CRIPDES, Manuel Antonio Rodríguez Escalante, who was driving the CRIPDES vehicle, Rosa María Centeno Valle, Vice-President of CRIPDES, Héctor Antonio Ventura Vásquez, María Aydee Chicas Sorto, CRIPDES journalist and photographer, Sandra Isabel Guatemala, José Ever Fuentes, Patricio Valladares Aquino, Clemente Guevara Batres, Santos Noel Mancía Ramírez, Marta Yanira Méndez, Beatriz Eugenia Nuila y Vicente Vásquez, captured on Monday July 2, 2007 in Suchitoto, during a peaceful protest to oppose water privatization. They had not even gotten out of the vehicle, and were arrested for nothing more than heading out toward the demonstration. Those detained were transported by helicopter to Cojutepeque, and en route, were subjected to psychological torture with threats of being thrown out from high altitude. (CoDevelopment Canada; KAIROS) Canadian Alerta Minera Canadá/MiningWatch Canada; ASALCA—La Asociación Salvadoreña Canadiense-Toronto; The Atlantic Regional Solidarity Network (Nova Scotia); Red Ciudadana Salvadoreña en el Exterior; Grupo de Apoyo a los Pueblos de las Américas (GAPA); ASCORCAN Asociación Salvadoreña Canadiense de Ottawa y Región de la Capital Nacional; Inter Pares; Salvaide; Red Ciudadana Guatemalteca; Mennonite New Life Centre of Toronto)." This was published on KAIROS' website (Horizons 2007).

References

Administración Nacional de Acueductos y Alcantarillado (ANDA). 1961. Ley de la Administración Nacional de Acueductos y Alcantarillados. Tipo de Documento: Ley Decreto No.: 341. Diario Oficial No.: 191Tomo No.: 193.

Appadurai, Arjun.1986. Introduction: Commodities and the Politics of Value. In *The Social Life of Things: Commodities in Cultural Perspective*, edited by A. Appadurai. Cambridge: Cambridge University Press.

Arias, Salvador. 2008. *Derrumbe del Neoliberalismo: Lineamientos de un Modelo Alternative. Colección Pensamiento Crítico*. San Salvador, El Salvador: Editorial Universitaria.

Barberena, Santiago I. 1977. *La Historia de El Salvador*, vol. I. San Salvador, El Salvador: Dirección General de Publicaciones.

Barlow, Maude. 2008. *Blue Covenant: The Global Water Crisis and the Coming Battle for the Right to Water*. New York: New Press.

Boaz, Franz. 1982 [1940]. *Race, Language, and Culture*. Chicago: University of Chicago Press.

Browning, David. 1971. *El Salvador, la Tierra y el Hombre*. San Salvador, El Salvador: Dirección Nacional de Publicaciones.

Castro, Barón. 1978. *La Población de El Salvador*. San Salvador, El Salvador: UCA Editores.

Cohen, Abner. 1979. Political Symbolism. *Annual Review of Anthropology* 8: 87–113.

Committee in Solidarity with the People of El Salvador (CISPES). 2008. Committee in Solidarity with the People of El Salvador. Washington, DC.

Csikszenmihályi, Mihaly, and Eugene Halton. 1981. *The Meaning of Things: Domestic Symbols and the Self*. Cambridge: Cambridge University Press.

De Burgos, Hugo. 1999. *Suchitoto: Ciudad y Memoria*. San Salvador, El Salvador: CONCULTURA, Dirección General de Publicaciones.

———. 2010. Placing Illness in its Cultural Territory in Veracruz, Nicaragua. In *Locating Health: Historical and Anthropological Investigations of Place and Health*, edited by Erika Dyck and Christopher Fletcher. London: Pickering and Chatto Publishers.

Descola, Philippe, and Gísli Pálsson. 1996. *Nature and Society: Anthropological Perspectives.* London and New York: Routledge.

Douglas, Mary. 1996. *Natural Symbols.* London and New York: Routledge.

Gazeta de El Salvador. 1858. 7 (23). Corresponding to July 24, 1858, Decree N 7. Government of El Salvador, San Salvador.

Geertz, Clifford. 1973. *The Interpretation of Cultures.* New York : Basic Books.

Gordillo, Gastón. 2004. *Landscapes of Devils: Tensions of Place and Memory in the Argentinean Chaco.* Durham, NC: Duke University Press.

Hall, David, Emanuele Lobina, and Robin de la Motte. 2003. Public solutions for private problems? - responding to the shortfall in water infrastructure investment. Public Services International Research Unit (PSIRU). http://www.psiru.org/publications?sector=All&country=All&subject=All&author=10807

Harris L. 2005. Negotiating Inequalities: Democracy, Gender, and the Politics of Difference in Water User Groups of Southeastern Turkey. In *Turkish Environmentalism: Between Democracy and Development,* edited by F. Adaman and M. Arsel, 185–200. Aldershot: Ashgate Publishing.

Hegel, G.W.F. 1977. *The Phenomenology of Spirit.* (Translated by A.V. Miller). Oxford: Clarendon.

Hirsch, Eric, and Michael O'Hanlon. 1995. *Space and Place.* Oxford: Clarendon Press.

Ibarra, Ángel. 2001. Hacia la Gestión Sustentable del Agua en El Salvador. *Unidad Ecológic Salvadoreña, Federación Luterana Mundial,* Foro Regional de Gestión de Riesgos. (1st ed.). San Salvador, El Salvador.

Ingold, Tim. 1987. *The Appropriation of Nature: Essays on Human Ecology and Social Relations.* Iowa City, IA: University of Iowa Press.

———. 2000. *The Perception of the Environment: Essays on Livelihood, Dwelling and Skill.* New York: Psychology Press.

Horizons. 2007. We, the Undersigned Canadian Organizations. Cached at http://www.zoominfo.com/#!search/profile/person?personId=1218158430&targetid=profile. Original post on www.horizons.ca, Sept. 15, 2007.

Keesing, Roger. 1990. Anthropology Symposium (unpublished notes and personal contact). McGill University, Montreal.

La Prensa Gráfica. 2007. Gráfica, Editions from July 2 and July 3 2007.

Lardé y Larín, Jorge. 1952. *Geología salvadoreña.* San Salvador, El Salvador: Ministerio de Cultura.

———. 1957 *El Salvador. Historia de sus Pueblos, Villas y Ciudades,* vol. 4. San Salvador, El Salvador: Dirección de Publicaciones e Impresos.

Moreno, Raúl. 2005. El Marco Jurídico para la Privatización del Agua en El Salvador. In *Brot Fur die Welt,* edited by Diakonisches Werk der Evangelischen Kirche in Stuttgart.: Menschen Recht.

Rubio Sánchez, Manuel. 1976. *Historia del Añil o Xiquilite en Centro América.* (1st. ed.). San Salvador, El Salvador: Ministerio de Educación, Dirección de Publicaciones.

Sahlins, Marshal. 1995. *Historical Metaphors and Mythical Realities: Structure in the Early History of the Sandwich Islands.* ASAO Special Publication. Ann Arbor: University of Michigan Press.

Sauer, Carl O. 1974. The Fourth Dimension of Geography. In *Selected Essays, 1963–1975.* Berkeley, CA: Turtle Island Foundation.

Strang, Veronica. 2004. *The Meaning of Water.* Oxford: Berg Publishers.

———.2009. *Gardening the World: Agency, Identity and the Ownership of Water.* Oxford: Berghahn Books.

U.S.–El Salvador Sister Cities. 2008. Recently Acquitted Political Prisoner Assassinated. Press release June 10, 2008. http://elsalvadorsolidarity.org/joomla/index.php?searchword=ventura&option=com_search&Itemid=

Wolf, Eric 1999. *Envisioning Power: Ideologies of Dominance and Crisis.* Berkeley: University of California Press.

Chapter 6

NOT SO BORING

Assembling and Reassembling Groundwater Tales
and Technologies from Malerkotla, Punjab

Rita Brara

∞

In varied idioms—"milking the cow to the last drop" or transforming "water into poison"—farmers, policy makers, environmentalists and social scientists are ruing the environmental and economic consequences of groundwater-dependent agriculture in India. The unrelenting movement from lift pumps that drew on shallow and rechargeable waters to submersible pumps that tap deep aquifers to reckon with ever-descending water tables is readily discerned in a linear view (Moench 2003; Rodell, Velicogna, and Famiglietti 2009). The *New York Times* vivifies the time-line of Punjab aquifers as having been created when dinosaurs still roamed the earth (Somini Sengupta, September 30, 2006, "India Digs Deeper But Wells Are Drying Up"). While the aquifers may indeed be characterized as palaeo or fossil waters, evoking similarity with nonrenewable fossil fuels, the histories and technologies of groundwater deployment for irrigation in India stir up more-recent date lines.

Some basics are not disputed. No country deploys more groundwater than India (World Bank 2010). Most of the country's groundwater (83 percent) is used for irrigation and drawn up by tubewells or assemblages described in local parlance as pumping sets. While wells have long been exploited for agriculture in northern India, the large-scale deployment of groundwater occurred in the 1970s during what is dubbed the Green Revolution. The cultivation of high-yielding varieties of crops requiring chemical fertilizers as well as pesticides, and the new pumping technologies, gave a fillip to both groundwater use and food production.

In the agricultural state of Punjab, the number of tubewells rose seven fold in about forty years—from approximately 200,000 pumps in 1970 to nearly 1,400,000 in 2010, enabled by new and evolving market technologies as well as subsidized electricity (Government of India 2007a). So it was not surprising that 103 of Punjab's 135 blocks or administrative subdivisions were declared to have overexploited groundwater in 2004 (Government of India 2007b). Yet, what is meant by overexploitation or even exploitation, perhaps, of groundwater from a farmer's perspective, cannot be fathomed without an acquaintance with the historical, technological, and social shaping of groundwater use.

Farmers, along with other laypersons, work with crude indicators of groundwater depth, decline, and availability. But even within the realm of scientific debate and environmental policy, the definition of overexploitation of groundwater is related to economic and environmental limits (Custodio 2002). In this context Stavric (2004: 1) observes,

> The limits [of] how much water can be extracted from a finite groundwater aquifer are economic and environmental. When water is pumped out faster than the natural processes recharge it, the water level in the aquifer drops, and the distance the water must be raised to the surface increases. Eventually, either the energy costs rise to the point that exceeds the value of the water, or the water quality falls below acceptable levels. At this point pumping must cease.

Although many aspects of what can be called overexploitation are not new in hydrogeology, this concept is still poorly defined and subjected to varying interpretations by different kinds of specialists, managers, policy makers, and the public.

As new hydrogeological methods access data from gravitational changes on earth through satellites to estimate the nature and extent of underground aquifers, the debate on groundwater use becomes more sophisticated (Rodell et al. 2009). Distinctions between exploitation of shallow waters and deep aquifers are measured more accurately over large swathes of land, replete with seasonal changes in real time. At the policy level, such data begin to influence talk of the relative values of shallow waters that are more easily contaminated by pesticides and fertilizers, with the comparatively pristine quality of deepwater aquifers, for instance.

Or again, as in the state of Punjab in India, while some policy makers declare that electricity tariffs, fixed at a flat rate determined by the motor's horsepower, aggravate the problem of groundwater extraction from aquifers, politicians with their ears to the ground think otherwise. Agriculture, the farmers tell them, is no longer viable without the component of power subsidy by the state even as they explore alternatives to what is, from their point of view, an increasingly uncertain and decreasingly remunerative occupation. Is the story of groundwater doomed, or is agrarian life still possible in the Punjab?

In this chapter, I attempt to convey a picture of farmers' attempts at enhancement of water quantities through varied groundwater equipping technologies in a delimited tract over more than a century. I focus on the Malerkotla subdivision of Punjab's Sangrur District. Formerly a part of the Muslim princely state of Malerkotla, this subdivision has 287 villages spread over nearly seventy thousand hectares. The region consists primarily of Indo-Gangetic alluvial plains. The area has an average annual rainfall of 670 millimeters, which falls mostly in the months of July, August, and September, and is distributed, on average, over thirty-one rainy days. But nearly one in every three years has less than 80 percent of this rainfall amount.

The town of Malerkotla was the field site of my doctoral enquiry into the kinship and agrarian practices of Punjabi Muslims. After the partition of the subcontinent in 1947, Malerkotla remains one of two areas with a concentration of Punjabi Muslims in the Indian Punjab. I made my first acquaintance with Malerkotla in 1978 and carried out fieldwork that spilled into the following years. I last visited Malerkotla in 2010. I continue to explore dimensions of ecological, religious, and social life here through documents and conversations in Punjabi, Hindustani, and English. Italicized words in the account that follows are from the local language or rather the admixture of languages that are spoken here.

The figures of the 2001 census indicate that the rural population of the Malerkotla subdivision of Sangrur District was about two hundred fifty thousand and the number of residents in Malerkotla town was approximately one hundred thirty-five thousand. Agriculture is still the main livelihood in this region and vegetable cultivation continues to be especially important in the urban periphery. A caste that specializes in raising vegetables for the market is known locally by the name of Muslim Kambohs. The industry of these gardeners is widely celebrated and has earned Malerkotla the epithet of the vegetable capital of Punjab.

Welling Up Memories: Malerkotla Watersheds—1890 to 1947

Just a decade short of the twentieth century, in what was then the princely state of Malerkotla, the groundwater depth was recorded at levels varying from thirteen to thirty-nine feet (Lall 1892a). In some areas, the groundwater surface could almost be touched by hand. Now the net annual withdrawal of groundwater is reported to be 181 percent more than the net annual recharge, leaving scarce room for further extraction (Government of India 2007b). This phenomenon is termed "overdevelopment" in the official literature. But what does overdevelopment or overexploitation imply for the farming contexts of those who pursue groundwater-based agriculture as a livelihood? For farmers, in contrast with groundwater experts, water requirements are measured in terms of the need for specific crop irrigations and are typically expressed in statements such as, "A motor of five horsepower is enough for me to irrigate my fourteen acres," or "Now I need a submersible because the groundwater table has dropped considerably." Their attempts are not aimed at the maximum abstraction of water but are intended to meet crop needs for water with affordable equipment.

They select the crops they grow, however, on the basis of what is currently remunerative and in sync with the household's groundwater-equipping arrangements. Thus there has been a significant movement away from growing wheat, bajra (Pennisetum americanum), and maize to just wheat and rice by larger farmers and those with access to irrigation facilities. Yet the capital-intensiveness of the new groundwater pumps and deep tubewells sets a limit on the extent of groundwater extraction for the majority of small farmers. From a vantage point in the rurban periphery of Malerkotla town, dotted with the tiny farms of vegetable gardeners and the larger farms of cereal growers a few kilometers a way, the local discourse grapples with the problem of receding water levels. The heterogeneity of groundwater equipping combinations in the present is striking, though the viability of these combinations in the not-so-distant future appears questionable to the farmers themselves. Farmers therefore attempt to alter the cropping pattern to one that accords with their groundwater arrangements even though a few sink deeper wells using submersible pumps and others look at combining agriculture with supplementary modes of livelihood.

Although the former 167-square-mile princely state of Malerkotla was largely bereft of rivers, its soils were viewed as hospitable to rainfed agriculture. Dug wells had introduced the possibility of a fight against drought. In 1890, the settlement officer for the state, appointed by the colonial British power, noted that Malerkotla's *jagirdars* (feudal lords) had often sunk wells on arable lands in order to attract cultivators and enhance their own revenues from agriculture. Especially when the rent was taken in kind from the cultivator, the feudal lords viewed the sinking of a well as an excellent investment (Lall 1892a).

As the land-to-man ratio decreased, the feudal lords began contributing less and less to the well's cost. In 1890, the lords' share of building a well had fallen to one-sixth of its average cost. The smaller feudal lords refrained from such capital investments altogether, which began to be treated as part of the cultivator's interest. The changing demographics led to agrarian unrest and occasioned the settlement officer's enquiry into the rights of cultivators and feudal lords vis-à-vis arable land in Malerkotla State. Cultivators had to be assured that that they indeed had the right to sink wells in their lands and that such wells would be recognized as their property (Lall 1892b).

About 16 percent of Malerkotla's arable lands were irrigated by dug wells and the water was pulled up by means of a rope and buckets with bullock draft power in 1890 (Lall 1892a). The total number of wells in use that year in this princely state was 1,151. About three-fourths of the wells had a bucket each and the remainder had two each. A well with one bucket was deemed sufficient to irrigate about eleven acres. Buckets were thus a proxy for volumetric measurement. But in assessing the enhanced crop yields from irrigated lands, the settlement officer found the volumetric method too cumbersome and preferred to fix a lump sum on each well to compute the enhanced returns on irrigated area. Additional complexities arose as the augmented irrigation capacity documented in the land records did not always translate into irrigated land in the fields. Reckoning with the computation of groundwater usage was already proving to be an administrative hurdle in the 1890s.

By 1914, the number of masonry wells in the Malerkotla state were still rising slowly, increasing by 17 percent to 1,372 in about twenty years (Muhammaddin 1914a). The large majority of farmers relied on rainfed agriculture for a livelihood. Yet almost triumphantly, the second settlement officer of Malerkotla declared that 21 percent of the cultivated area of the state was now protected by masonry wells. Such was his rosy assessment of the march of groundwater development in the state nearly a century ago.

On the one hand, cultivators at Malerkotla had realized that irrigation was a dire necessity with the continuous rise in population. But on the other, the feudal lords walked away with a large proportion of the harvest on irrigated lands. The situation remained restive until the independence of the country when the cultivator's right to his land, to his wells, and, vitally, to his harvest, was clearly established and freed from feudal levies (Brara 2006).

The water history of the Malerkotla state was evidently contoured by the fact that as a semiautonomous, native princely state, it lay outside the directly controlled colonial territories of the Punjab during the British period. Although an irrigation canal constructed by the British—the Sirhind Canal—ran through the Malerkotla territory (and a branch was named the Kotla branch after the ruler of Malerkotla), the chief had decided against defraying a part of its cost to the colonial state (Muhammaddin 1914b). This action entailed that residents of the Malerkotla region could not reap the benefits of the technological investments in canal irrigation until after the independence of India in 1947.

Groundwater after Independence: 1947–2010

In the following sections, I divide the study of groundwater in the post-independence period by distinguishing state endeavors in the arena of irrigation by means of deep tubewells from private efforts at groundwater extraction. By adhering to this broad contrast, I hope to delineate the lines of movement in technologies and management strategies over time for state-initiated and private attempts at irrigation in the region of Malerkotla.

The higher number of state tubewells in the erstwhile state of Malerkotla in the period following independence in 1947 grew out of Malerkotla's comparatively late and partial induction into the canal irrigation network of the newly formed Punjab State. Private investment in irrigation, however, proceeded apace from simple and less-capital-intensive technologies to relatively more-complex and expensive pumping set equipment in the Green Revolution years from 1970 onwards.

The distinction between state and private efforts at irrigation is not hard and fast, or should one say water-tight, since the working of state tubewells increasingly has a private component and private groundwater-equipping arrangements bank on the provision of subsidized electricity by the state. Even now, what farmers desire is a combination of private and state-supported arrangements for irrigation that allow them to reckon with contingencies and current economies.

Groundwater Provisioning by State-Owned Tubewells: 1947–2010

Agriculture in Malerkotla continued to be dependent on bullock-drawn water from wells and monsoonal rain well after the independence of the country in 1947. Subsequently, the princely state of Malerkotla merged with the other Cis-Sutlej princely states of Punjab to form a state named the Patiala and other East Punjab States Union (PEPSU). In 1952, under a program titled "Indo-US Programme for Technical and Administrative Cooperation," villages in the princely state of Malerkotla, especially those that fell under the Nawab of Malerkotla's direct jurisdiction, were selected to benefit from the construction of deep tubewells.

In 1955, deep tubewells constructed by the state with American know-how and what was then state-of-the-art vertical centrifugal pump technology, constituted the first engagement of Malerkotla's farmers with electric pumps for drawing groundwater. The coming of tubewells was hailed in Malerkotla's villages because it also heralded the dawn of electricity in these villages.

Deep tubewells were drilled under this program of Indo–United States cooperation at seventy-one locations in the Malerkotla state. A pump record document was drawn up for each pump, replete with the permanent pump installation report, pumping test record, and diagrams for assembling and dismantling the pumps. This record is still updated by the irrigation department at the subdivision level. I gathered, for instance, that Tubewell 71 at Nathaheri in this block was sunk at a depth of 326 feet and was worked with a motor of fifteen horsepower. The discharge from the pump was observed to be forty thousand gallons per hour in 1956. In 1966, the discharge was reduced to about eighteen thousand gallons per hour because the water level had begun sinking. The pump continued to work till 1978 with repairs, but could not be fixed after that year.

The state of PEPSU was merged with the territory of the state of East Punjab in India in the year 1956. The princely state of Malerkotla, too, was transformed into an administrative unit of East Punjab at this juncture. It was now integrated as a subdivision of the Sangrur District in this state. Both tubewells in the former princely state of Malerkotla and those that were later set up by the Punjab government's irrigation department came to constitute part of the state's water infrastructure for farmers.

Of the total of seventy-one tubewells set up in Malerkotla in 1955, often of the brand name Layne, fifty-seven were still reported to be working in 1984. By 2005, this number of functioning tubewells was down to twenty-seven. In Songali village, farmers irrigated their fields with water from a Layne tubewell in 2010. I was told that it sufficed for watering the fields of thirty-odd families living there, irrigating about two acres of land comfortably in ten hours or so.

As the water table declined, some pumps were converted from a two-stage to a three-stage pump, and lowered farther into the ground by twelve feet or so. But at Katron and Adampal, for example, these tubewells had to be abandoned because the groundwater level had decreased to a hundred feet and seventy-eight feet, respectively.

Even where the Layne pumps no longer worked, they were remembered for their unflagging performance in the past. The pumps set up prior to 1956 were invariably spoken of as a gift from the American people. The Layne pump had proved itself over time in this area. Often, all it needed was oil and lubrication and the only replacements necessary were of the shaft and of bushes that had worn out.

In 1975, about twenty years after the Layne pump's introduction in Malerkotla, replaceable parts of the pump had begun to be produced in the nearby towns of Ludhiana and Patiala. At meetings with farmers, I learned that the new manufacturers were unable to ensure the quality of the parts and the accuracy of their performance. Here I gathered that the gunmetal used for the Layne pumps, which did not wear out easily, was not so readily substitutable.

Social Organization of the State Tubewell

The deep tubewells, worked by vertical, centrifugal pumps, that were installed and run by the state's irrigation department, came to be described as government sarkari (tubewells), in contradistinction to the relatively shallow pumping sets that were privately set up in the 1970s and later. Approximately thirty to forty farming households, with an average holding size of about four hectares, could bank on the water from a state-run tubewell operated by a system of supervised water use. A state functionary worked out the roster for the equitable distribution of water (warbandi), kept an eye on the functioning and maintenance of the pump, and computed the payment for water use that had to be made over to the state. Water timings in this tubewell regime were rotated between days and nights in blocks of four hours. The dawn-to-dusk agrarian cycle had become a thing of the past.

State tubewells were a reliable source of water for irrigation in areas that did not have canals. The state functionary who worked the tubewell was officially termed a tubewell operator, but was described in local parlance as a tubewell patwari. The term "patwari," in its original usage, is the word for a village-level state functionary who collects land revenue from farmers, maintains the local-level land records, and is the official face of the state in the locality. That the term was carried over to the new functionaries of the state who operated tubewells reveals the sense in which state functionaries were perceived by farmers as similar regardless of whether they were in charge of water or of land.

While ostensibly carrying out their official functions, tubewell operators, like the erstwhile patwaris, sometimes appeared to local farmers as arbitrary controllers of the right to state services—here the groundwater at the pumping point (Campbell 1995). The tubewell operator, it was held, often facilitated the interests of the larger landlords and did not look to the special requirements of small and marginal farmers.

Once private pumping sets came into the market, water from state-run tubewells gradually turned into a supplementary rather than an exclusive source of water, especially for richer farmers. Energies and capital were no longer invested in tubewells. The state increasingly argued that water provisioning was costly because of the number of personnel employed in the state's irrigation set-up. The state tubewells were not discontinued, but were not repaired with alacrity either.

In a move aimed to raise the effectiveness of state-level functionaries, each state tubewell operator was later made in-charge of three or four pumps. Since it was not possible for the tubewell operator to be at four locations simultaneously, however, his reliance on local farmers to perform the daily regimen during his absence visibly increased. Informally, the farmers operated the pump themselves by prior agreement, while the official operator showed up off and on or whenever he deemed it was urgent.

Some tubewell operators reiterated that their attempt was to satisfy the farmers' water demand and to keep the maximum number of wells functioning. They claimed that if they did not put in their best, their department and the farmers incurred a loss. Other, older tubewell operators averred that the restructuring, which now entailed traversing long distances, was indeed trying but that a pensionable government job was worth keeping despite the inconvenience.

Small landholders without capital considered the system of tubewell water-sharing preferable to the rampant privatization of water that left the small farmer in a worse situation than before—a declined water table and rich farmers who were no longer interested in pursuing repairs of the state tubewell. In two villages where the state tubewells had stopped working, small landholders found that they had no recourse other than to buy water from their richer neighbors. By contrast, tubewell water in the past had been a boon for them since it was subsidized and much cheaper than the price demanded for water in the village by the new waterlords.

New Uses for Abandoned Tubewells

Under a recent scheme, the Punjab state allowed farmers to pool together half the financial resources required to revive old tubewells and work them on a roster system, supported with a matching grant from the state government. This scheme reintroduced the cooperative functioning of groundwater-supply institutions. It is a measure supported by current rethinking on the subject (World Bank 2010) and builds on the tradition of shared well irrigation in Punjab (Tiwary 2010).

In small farms in the vicinity of Malerkotla town, tubewells operated on this basis worked satisfactorily, although contingent water-equipping arrangements from private bore wells were kept going as well. Threats to the functioning of an equitable roster, I was told, arose from the clout of big landlords who tried to muscle their way into a larger share of the water or to attune pumping hours to their interests. Or, as in a few villages, the smallholders did not have the money to buy a submersible pump even after pooling their resources. They had no choice but to buy water or rent out their lands. Groundwater pumped by submersibles at Malerkotla was sold to them at about $0.50 per hour. At this point, groundwater for agriculture transformed from being a reasonably priced public good to a market commodity.

No farmer who had assured access to the state's tubewell water gave up his tubewell connection or license voluntarily. Presently, it was argued, the charges for water drawn from state tubewells may be higher than the cost of water drawn from private tubewells, since the state subsidized the electricity cost of the private tubewells by charging a flat rate based on the horsepower of the motor. But farmers considered it plausible that the state could increase, all of a sudden, the electricity tariff, making it economical for them to exploit state tubewell water again, despite the shortcomings.

Private Wells and Pumping Sets: 1947–2000

Though private wells, as noted above, were part of the irrigation landscape at Malerkotla even in 1890, the majority of farmers in the Malerkotla region depended primarily on rainfed agriculture at that time. Only the better-off could draw on well water pulled up by bullocks for irrigation even in the period following 1947. Until 1955, only a few farmers and fields had assured access to state tubewell water or canal water. Bullocks came to be replaced by tractors only in the 1970s at Malerkotla, as in the rest of Punjab. The water pump was introduced into the masonry well as a novel technology in the same decade, leading to the now-familiar large-scale utilization of groundwater. This trend was boosted by the higher water requirements of the high-yielding seed varieties that formed part of the Green Revolution (Kumar, Sivamohan, and Narayanamoorthy 2010). Irrigation by pumping groundwater, high-yielding varieties of seeds backed by chemical fertilizers and pesticides, as well as a minimum support price for wheat (and later rice) assured by the state had transformed the agroeconomic climate for Malerkotla's, and indeed Punjab's, farmers. Thus, along with wheat the cultivation of paddy proved to be the most remunerative, even though the latter had not been traditionally cultivated in the area. While standard economic geography textbooks maintained that rice was a crop grown in wet, humid regions, such thinking was overturned with the advent of tubewell irrigation.

Currently, a diversity of private arrangements for lifting groundwater for irrigation is apparent in the Malerkotla region. More often than not, the multiple assemblages of groundwater provisioning seek to both optimize the deployment of subsidized state-supplied electricity and ensure that the watering of crops is possible during periods when the state's intermittent supply collapses. Old groundwater-extracting arrangements continue to have supplementary uses as a back-up if one assemblage fails. Even in the few villages of Malerkotla that are canal-irrigated, farmers who are tail-enders in the canal water distribution system find it handy to have additional groundwater-equipping resources.

Below I outline the vocabularies and drilling practices dealing with the sociotechnical provisioning of groundwater encountered over the longue durée at Malerkotla before turning to newly introduced technological terms relating to more-recent irrigation practices and pumping set assemblages.

Khui *or Open Wells*

A *khui* or *khoo* is a traditional open, masonry well in Punjab. These wells were hand-drilled using an instrument called a *boki* and overseen by a professional called an *ustaad* (master in charge of the drilling). Dug wells were inherited along with the arable lands; all the heirs who had a share in the ancestral land also had the right to access water from these wells. Unlike lands, however, these wells were not easily partitioned. And so long after the arable lands had been divided and subdivided over and over again, wells were often held and operated jointly.

Even today, few open wells are completely dysfunctional if water can still be extracted. Bullocks that were formerly used to draw up water have been replaced by tractors as a source of traction in the first flush of uses for this machine. By about the 1970s, however, farmers began to install private pumps and diesel engines for lifting water in these open wells. Their use now lay in augmenting the winter water supply. Often, new drillings adjoined old wells because it was believed that it was the site of a proven underground water source of an assured water quality. Old wells also were deployed to recharge the groundwater table.

Bore *Khui*

A *bore khui* is a well dug by a drilling or boring machine and fitted with a pump. With the coming of private water pumps, as a first step old wells or *khuis* were deepened and converted into what were called *bore khuis*. Farmers were able to reduce their expenses on drilling here, developing on the investment of ancestors who had already dug fairly deep into the ground. Often, members of a family who had rights to the water of a shared well installed separate motors in the well itself.

I saw five motors fixed in one such well. Here the traditional well was adapted to the new boring technology such that drilling costs were saved to the extent of the well's built-up depth for five branches of the family. In another instance, I encountered five electric connections servicing the lands of five brothers, leading out of one pumping set. Cooperative use of groundwater and minimizing capital expenditure were evident concerns among farming families. The remaking of wells underscored how agrarian livelihoods continued to build and rebuild incrementally on the capital investments made by ancestors.

The off-season for agriculture was considered to be the major drilling season. The person responsible for the boring operation is still known as an *ustaad* (master), though he now uses a powered drill. There was a boring machine in practically every village in the Malerkotla region and I was told that its hiring and operation was a lucrative enterprise. At this time, a drill cost about $10,000.

Boring Rituals

Groundwater in local reckoning was thought of as "a thing that belongs to Allah" (Allah *ki cheez*). In this view, satellite imagery or thermal hydrogeological measurements could only determine the location or volume of water, but not place it there. And so it was believed that Allah's assistance had to be sought before setting up a well. Khwaja Ali (also called Khwaja Pir) was the intercessor for seeking such supernatural help from Allah.

First, the farmers offered rice grains and a little mustard oil, along with a prayer to Khwaja Ali, to make the attempt at procuring water successful. Then, using the traditional water-drilling instrument (*boki*), mud was continuously drawn up to a depth of ten feet. What was called the cutter pipe was inserted only after this initial, ritual drilling. Water was then poured continuously as this pipe was pushed down. If it encountered gravel, the drilling was slowed but the pipe pierced easily through a layer of sand. A geological strata of clay (locally referred to as the *pandu ka kara*) was considered relatively difficult to bore.

Once the tubewell was drilled and the water made available, farmers cooked one and a quarter kilograms of a sweet preparation (*halwa*), offered it first to Khwaja Pir, and then distributed it to family and neighbors. A bore can now dig to a depth of 325 feet if the installation involves a submersible pump, but the rituals have remained more or less the same.

The Pumping Set: A Sociotechnical Assemblage

Groundwater use is growing a shared albeit new language that conveys a vivid description of swiftly developing terms for groundwater-equipping technologies, personnel, and practices that builds on former Punjabi usages and words borrowed from the English language. The local discourse enables an understanding of the social trajectories of water as it is drawn from ever-deeper depths by newer machines assembled in what are referred to as a pumping set. Snatches of conversation like, "The water level is 50 foot so he needs a bigger dia [diameter] motor," or "His *bore khui* has a monoblock" made me appreciate the extent to which technological terms permeate the groundwater talk of farmers.

Apart from the drilling equipment and the pump, the energy source was an essential part of the pumping set assemblage now, since animal power had become more or less a thing of the past. What made the pumping set a rapidly evolving sociotechnological assemblage (Latour 2005) was the market's capacity to adapt its components to Malerkotla's agrarian contexts. I try to develop this aspect of pumping set assemblages by briefly unpacking the social contexts within which the add-ons make sense. In the following sections, I specify four local contexts that contour the social espousal of new groundwater technologies: (1) the uncertainty of electricity supply, (2) declining water tables, (3) the high capital costs of submersible pumps, and (4) the considerable reliance of farmers on local technicians to maintain increasingly complex technologies. The social signs of a gradual retreat from groundwater use is still nascent but discernible in the search for livelihoods outside of water-guzzling agriculture.

Electricity Supply: The Energy-Water Nexus

The electricity supply at Malerkotla was limited and choppy and its timings often were unpredictable. Sometimes it was available for eight hours, at other times only for five hours. Nevertheless, it was the subsidized availability of state-supplied electricity, though irregular and characterized by fluctuating voltages, that made the pumping of groundwater in the Malerkotla region economically feasible, though with standby arrangements. And so a generator that ran on diesel was considered a necessary investment and kept operational. It added to the cost of the groundwater equipment and use but was a vital part of water provisioning.

The supply of electricity was embedded in a highly charged political regime. While a shortfall in production and breakdowns in transmission were predictable under local conditions in the common reckoning, the pricing of electricity for agriculture was enough to make or break governments at the state level. It could cause mayhem to the subsistence and profit margins of farmers.

At present, the tariff for electricity is charged on a flat rate by the state government, depending on the horsepower of the motor. From the point of view of the country's central planning body, this practice did not make for prudent use of water or energy (Government of India 2007a). Critics maintained that farmers often claimed that they had the state-approved ten horsepower motors but actually deployed motors with higher capacities (ranging from fifteen to twenty horsepower). The larger farmers often admitted this privately but claimed that they had no choice because of the lowered water table. However, they thought alternative proposals for charging on the basis of the volumetric measurement of water were also impracticable because of the vast number and dispersal of such wells (World Bank 2010).

Whether the pricing of electricity, measured by the consumption of units, would be reintroduced, was a recurring subject of anxiety and conjecture among farmers in Malerkotla. Keeping in mind the ups and downs of electricity pricing and in an attempt to conserve their capital investment, farmers had not removed their meters once electricity came to be charged on a flat rate based on the power of the motor. Nor were the coveted licenses to utilize the state-subsidized electricity for agriculture ever surrendered or revoked (Government of India 2007a).

The Recent Add-on: The Automatic Starter

Apart from the standby generator, automatic starters were a twenty-first century add-on to pump assemblages. Since the electricity made available for agriculture showed up suddenly, sometimes at odd hours of the night, the automatic water starter was a convenience device. Some older farmers had formerly hired additional help to turn on their water connections after dark. Yet, this technology has had its own share of hazards in this milieu. Two accidents occurred when workers went down to check the well's motor at a time that the supply of electricity had been turned off. When it came on unexpectedly, the automatic starter was set off and tragically electrocuted two workers.

State-Supplied Electricity vs. Diesel

At present, farmers in Malerkotla (as in the rest of Punjab) prefer to power their water pumps with state-supplied electricity because of the lower expenditure entailed in the state's flat-rate billing system. Yet the diesel engine, though currently sidelined, had been a significant component of the pumping set assemblage. A diesel engine, also called a *bumba* locally, often ranged from ten to sixteen horsepower in the pumping set assemblages at Malerkotla.

The early pumping sets in this region, during the late 1960s and early 1970s, were outfitted with diesel engines because not all the villages had electricity when pumping sets entered the market. Even in villages that were electrified, farmers required an approved connection or license for electricity from the state to work water pumps, which were not easy to procure without political clout or greasing the palms of local-level state functionaries. Furthermore, since farmers had to reckon with the unpredictability of the power supply and frequent outages, the diesel-fuelled engine was still considered helpful.

The Decline in the Water Table and the Newer Technologies

Monoblocks

A twelve-horsepower diesel engine, I learned, works best if the level of water is sixty feet below the surface but not if the water table drops farther. The maintenance expenses are said to go up if it is used for greater depths. What materialized as an effective technology in the new scenario is the monoblock. A monoblock is a pump that combines the impeller and the motor in a compact whole that requires lower maintenance than the diesel engine. A five-horsepower monoblock was considered adequate for irrigating about four acres of farmland in the Malerkotla region about a decade ago. Now monoblocks of a higher horsepower are considered desirable since the water table is going down rapidly; though these are expensive they cost less than the submersible water pump.

Submersibles

At a groundwater depth of sixty feet, and if they can afford it, farmers prefer to install submersibles. Otherwise, I was told, the process of digging for water becomes a recurring operation since the water table has fallen by a foot per year in some tracts and by as much as five feet annually in other stretches. The water table is variable within the region, ranging from fifty to a hundred feet at present. It was formerly higher near the canal areas, even where these canals did not irrigate the adjoining lands. But whatever the existing level, it is decreasing fast.

Investing in submersible pumps that come with a guarantee of drawing water from a depth of 150 feet is how rich farmers are equipping themselves against the plummeting levels of groundwater. Technologies that allow extensions, as add-ons, that go still deeper into the earth, drawing water from depths of up to 325 feet, are also part of the new designfeet.

A submersible pump of seven and a half–horsepower is thought to be adequate for irrigating ten acres but the newer motors offer more than what is needed for lifting water presently. The promise of reducing the recurring costs is the selling point of these pumps. The life of a submersible pump is believed to range from seven to fifteen years; it is currently priced at about $2,000–$2,500, on average.

Over and above the power of the motor, submersible pump manufacturers advertise that they are outfitted to tolerate considerable voltage fluctuation, showing the extent to which the market is oriented to predictable shortcomings in the state's supply of electricity as well. In local experience, voltage fluctuations still constitute the primary cause of motor burnouts, despite the claims of advertisers and the effort of designers. According to a local mechanic, the simplicity of repairs is now being proffered as a new sales pitch.

Strikingly, the assemblages offered up by the market are not especially oriented to saving electricity for the country and are not incentivized in that manner. Nor is water conservation a declared focus of the groundwater assemblages encountered here.

Capital as a Limit to Groundwater Mining

The pumping of groundwater at Malerkotla rests on energy-equipping arrangements that combines private and state resources to sustain the agricultural enterprise. The supply of subsidized electricity, though unpredictable, remains the bulwark of private pumping operations and especially so in the wake of a lowered water table. The movement from open wells to *bore khuis*, diesel pumping sets, monoblock motors, and now submersible pumps, at one level, appears to be linear and chronological. But at any point of time, a cross-sectional analysis shows that these technologies are in concurrent use at Malerkotla. Former logics have not, irreversibly, lost their reasons for existence.

The high capital cost of setting up submersible pump assemblages entails that small and marginal landholders still have to use a variety of incremental groundwater-equipping technologies. For the majority, the limit to groundwater exploitation by submersible pumps is set by the capital-intensiveness of this technology. The ecological argument for saving the earth's finite resources of groundwater for future generations is not the farmer's first priority here and yet the lack of capital set a cap on abstracting groundwater for a large number.

The provisioning of groundwater is still seen to gain from building on the labor and capital invested by ancestors as old, open dug wells were transformed into an intermediate stage where the wells were deepened, and therefore reduced drilling costs for the current landowners. These wells, now outfitted with monoblock or, where affordable, submersible pumps, are treated as an economic investment for the present and future generations.

But what of the ecological future of Malerkotla's coming generations? The monoculture of wheat and paddy on large irrigated holdings has noticeably caused the degradation of the soil. Salinity levels, too, have recorded an increase as water is pumped from lower depths in some tracts. In boring deeper and plumbing for submersibles, the richer farmers seem to reject the writing on the aquifer. Yet, in seeking diversification within agriculture and livelihoods outside of agriculture for some of their children, both rich and poor farmers are beginning to affirm the limits of groundwater-dependent agrarian livelihoods in the future. Relationships between apparently opposed perceptions have been crafted in the space of households and conceived as fluid arrangements of both desire and capital (Appadurai 1986) that make for new assemblages of the sociotechnical.

The negotiation of transformations out of water-constrained agriculture is evident in decisions at the household level. For instance, there have been attempts by members of farming families to diversify into seed shops or agricultural machinery renting or repair services without giving up on agriculture. Like pumping set assemblages, such combinatorial modes within families are regarded as synergistic, flexible, and therefore viable in the present context (Brara 2002).

Furthermore, vegetable growing as it is practiced in the rurban margins of Malerkotla affords glimpses into how tiny farms that consume less water than cereals remain profitable. Strategies here include the preference for vegetables with short cultivation spans, varieties that require less-frequent watering, the use of drip irrigation, and the complementary development of plant nurseries. Small farmholders, who can scarcely afford costly irrigation technologies, raise seedlings of both the local and hybrid varieties of cauliflower, for instance, that are in popular demand outside Malerkotla as well. The quality of indigenous cauliflower was appreciated for its long shelf life and clean appearance throughout the region.

The hybrid varieties of summer vegetables—such as cucumber—which consume large quantities of water in the summer months, are left for cultivation by the water-rich landowners. But even large farmers at Malerkotla reiterate that agriculture, financed by private and full payment for electricity, did not offer the prospect of being profitable any more. The only reason farmers can continue with rice cultivation is because of state subsidies in the provisioning of electricity and a minimum support price. They readily grant that the cultivation of rice is the main cause for pushing down the water table. A change in the cropping pattern, in their view, however, would need to be bolstered by a remunerative support price by the state for other crops. But some large farmers, cognizant of water troubles, have already taken to growing *kinnow* (citrus) and guava trees that need to be watered only once every eight days or so.

The "Motor-Mistry" Is a Component of the Assemblage

Practically every village in the Malerkotla region now has a "motor-mistry." The word mistry is employed for a skilled worker, and in this context, for a technician who understands and repairs the motors that run water pumps. The term "mistry," like the term "tubewell *patwari*" mentioned earlier, carries over a preexisting term that has been modified to denote a new but related profession.

These technicians (motor mistries), like the *ustaads* or tubewell *patwaris*, provide the interface between technology and its local use. They guide the farmers through the maze of pumping assemblages offered in the regional market. Given the range of genuine and imitation pumping equipment, the motor-mistry doubles as a farmer's local consultant and the de facto guarantor of the equipment, sometimes with an unwritten service contract. These mechanics are preferred to a manufacturing company's repair person because they can be reached round-the-clock. The sociotechnical assemblage of the pumping set is apparently constituted by the interrelations of people and motors.

A farmer told me: "An engineer is a *kitabi* (bookish) fellow but these repairmen, they have their own *kitabs* (books); they are oriented to local conditions and keep costs low." By contrast, the sales agents of the manufacturers, it is felt, disappeared after the sale and their guarantees imposed conditions in fine print that made them unworkable from the farmers' point of view.

Most technicians had undergone training in industrial schools and only a few said they had picked up the skills for the more-frequent problems in this area on the job—by disassembling motors and pumps in a period of about six months. There was a seasonal cast to the motor-mistry's work as well, as he perforce followed the agrarian rhythm. New motors were frequently installed in February–March because farmers hoped to be able to pay for them after the harvest in April.

That the language of technicians, pumping sets, engines, motors, and pumps changed swiftly underscored the need for knowledge and practical acquaintance with new products and technologies. Some of this technological talk was circulated and assimilated by farmers, but by and large they banked on the interpretations of local technicians. What was assembled and reassembled dovetailed with the current socioeconomic realities, till further changes warranted sociotechnological rethinking.

Conclusions

In the course of this chapter, I have recounted the language in use and the evolving social life of groundwater objects and personnel that partake of the character of sociotechnical assemblages (Latour 2005). Wells and energized pumping sets for groundwater appear in this narrative as dynamic assemblages connecting the lives of farmers, technicians, policy makers, and ancestors.

What is underscored in this account is the symbiotic relationship of private and state assemblages in equipping farmers with groundwater, including electricity supplies. Ironically, capital shortages and the inefficiencies of electricity generation and transmission set a limit on groundwater withdrawal, despite the availability of submersible pumps for lifting groundwater. The unintended effect, perhaps, limits an ecological tragedy of larger magnitude—an even-faster receding water table in the region.

By gathering experiences around the construction of pumping set assemblages in the main, and from the margins of old and new wells of the past as well as the sprinkler technologies of the future, groundwater arrangements come alive in their diversity. The different assemblages come into view as variants within an environment rather than as a uniform set of technologies and practices pertaining to groundwater both spatially and over time (cf. Laet and Mol 2000).

The installing of a well or tubewell by farmers too is cast as a social event that engages kin and nonkin assemblies, local and foreign technicians, ancestors, state personnel, supernatural beings and, of course, components of groundwater equipment. These events and occasions are precisely the sites where technologies and tales are assembled and reassembled. It is at these social sites that things, people and personages come together to partake in the life-and-water provisioning of farmers over the longue durée imbuing farmer efforts with reason, ancestor-oriented emotion, and micropolitics oriented not only to a past, but also to an uncertain social regenesis and future. The personnel of the state, too, feature here as actors who can be moved and cajoled to enable the harnessing of groundwater from wells through actions that lie both inside and outside the regimes of the official.

Practices pertaining to groundwater at the local state–farmer interface, it turns out, are both constraining and enabling, within the formations of democratic and even populist sentiment. As farmers pursue multiple, combinatorial, and flexible modes of groundwater provisioning—combining state and private water-equipping resources—what is accumulated is an agrarian treasure trove of experiences and histories, languages and thoughts, around groundwater that can help to cope with new and unforeseen risks (Brara 2002).

Such many-sided perspectives on water issues contrast with one-dimensional approaches. Policy making is too often concerned with one dimension of efficiency—energy saving—or one type of environmental risk, for example, a declining water table. But the diverse practices of small and large farmers juggle groundwater assemblages in the here and now, even while building on the efforts of former generations with an eye to the future.

Their tales and technologies are not simply about boring.

Notes

Once again I would like to thank the residents of Malerkotla for sharing their lives and livelihood stories so generously. This chapter would not have been possible without the unstinting assistance of Nizamuddin from Malerkotla's Tube-well Irrigation Office.

References

Appadurai, Arjun. 1986. Introduction: Commodities and the Politics of Value. In *The Social Life of Things: Commodities in Cultural Perspective*, edited by A. Appadurai, 3–63. Cambridge: Cambridge University Press.

Brara, Rita. 2002. Ecology and Environment. In *Companion Encyclopedia of Sociology and Social Anthropology*, edited by Veena Das. New Delhi: Oxford University Press.

———. 2006. *Shifting Landscapes: The Making and Remaking of Village Commons in India*. New Delhi: Oxford University Press.

Campbell, D.E. 1995. Design and Operation of Smallholder Irrigation in South Asia. World Bank Technical Paper 256. World Bank, Washington, DC.

Custodio, E. 2002. Overexploitation: What Does It Mean? *Hydrogeology Journal* 10 (2): 254–277.

Government of India. 2007a. The Report of the Expert Group on Ground Water Management and Ownership. The Planning Commission, New Delhi.

———. 2007b. Ground Water Information Booklet Sangrur District, Punjab. Chandigarh: Ministry of Water Resources, Central Groundwater Development Board, North Western Region.

Kumar, M. Dinesh, V.K. Sivamohan, and A. Narayanamoorthy. 2010. Irrigation Water Management for Food Security in India: The Forgotten Realities. Institute for Resource Analysis and Policy.

Lall, I.C. 1892a. *Assessment Report of the Malerkotla State, 1888–92.* Delhi: I.M.H. Press.

———. 1892b. *Report on the Settlement of Malerkotla, Kalsia and Pataudi States: 1888–92.* Delhi: I.M.H. Press.

Latour, Bruno. 2005. *Reassembling the Social: An Introduction to Actor-Network-Theory.* Oxford: Oxford University Press.

Moench, M. 2003. Groundwater and Poverty: Exploring the Connections. In *Intensive Use of Groundwater: Challenges and Opportunities,* edited by R. Llamas and E. Custodio, 441–455. Lisse, Netherlands: Swets and Zeitlinger.

Muhammaddin, Chaudhri. 1914a. *Assessment Report of the Malerkotla State, 1910–13.* Lahore: Civil and Military Gazette Press.

———. 1914b. *Final Report of the Second Regular Settlement of the Malerkotla State, 1910–13.* Lahore: Civil and Military Gazette Press.

Rodell, M., I. Velicogna, and J.S. Famiglietti. 2009. Satellite-Based Estimates of Groundwater Depletion in India. *Nature* 460: 999–1002.

Stavric, Vladimir. 2004. Aquifer Overexploitation and Groundwater Mining. Paper presented at a BALWOIS Conference on Water Bodies Protection and Ecohydrology, Ohrid, F.Y. Republic of Macedonia, May 25–29.

Tiwary, Rakesh. 2010. Social Organization of Shared Well Irrigation in Punjab. *Economic and Political Weekly* 45 (26): 208–219.

World Bank. 2010. Deep Wells and Prudence: Towards Pragmatic Action for Addressing Groundwater Overexploitation in India. Report 51676. Author, Washington, DC.

Chapter 7

KENYAN LANDSCAPE, IDENTITY, AND ACCESS

Swathi Veeravalli

∽

> Without water we are nothing ... even an emperor, denied water,
> would swiftly turn into dust. Water is the real monarch and we are
> all its slaves.
> —Rushdie, The Enchantress of Florence

When an opportunity arose for me to go to Kenya to assist in a World Bank–funded research collaboration between Oxford and Stanford Universities, I readily agreed. After numerous discussions with my supervisor at Oxford, it made sense to align my master's thesis with this project, which was focused on the issue of whether the enhanced sustainability of water services and increases in the economic productivity of water use contributed to poverty reduction. With a background in identity research, I soon discovered that the link between water and identity was not often explored. I did not understand why this was so, and thought the Kenyan landscape would be an ideal setting in which to examine this relationship further. What impact does identity have on access? Were there certain identity groups in this setting who had more access or less access, and what were the causes and implications of these differences?

The complex relationship between humans and water has been constructed and deconstructed throughout time. In this chapter, I propose a reenvisioning of the multifaceted ways in which water mediates, structures, and forms relationships within both environmental and social landscapes. I argue that the interplay between environmental and social landscapes produces both real and perceived notions of access to water supplies, and that critical aspects of the social life of water in this setting are rendered visible by the contradictory perceptions of diverse water users.

Water as Core or Periphery?

The connection between humans and water is perhaps one of the most fundamental relationships. However, the frequent use of the term "human–water" to describe it implies that the relationship is both anthropocentric and unidirectional. By reordering the words, the concept of a water–human relationship can be used to infer a more multidimensional relationship in which water can be understood to have its own agency. Thus, by reordering the structure of a human–water relationship to a water–human relationship, however seemingly awkward this phrasing may be initially, we transform this analysis to one that emphasizes the importance of understanding the various processes and practices that render water as central to many relationships. This hybrid framework, referred to here as hydroidentity, provides the foundation for the analysis that follows.

According to Swyngedouw (1996: 5), "[W]ater is a 'hybrid' thing that captures and embodies processes that are simultaneously material, discursive and symbolic." Water connects various sociopolitical-spatial relations that embed "a series of multiple power relations along ethnic, gender and class lines" (Swyngedouw 1996, as cited in Swyngedouw 2006: 5). What and how are these relationships developed? Rushdie (2008) suggests that water is at the core of these power relations; it shapes and forms societal relationships as they network, interact, and extend outward toward the periphery. The complexity between water and society (Mosse 2003) can be both powerful and unevenly distributed (Swyngedouw 2004, 2006), and thus it can be difficult to understand the fluidity of forces that mutually shape and are shaped by water. In order to explore various societal relationships *surrounding* water, in this chapter I explore the impacts of water by means of the notion of access, with access briefly defined as "the right or opportunity to use or benefit from [water]" (Oxford Dictionaries 2011).

Hydroidentity: A Theoretical Overview

Since this analysis proposes reenvisioning the way in which water can impact and influence its surrounding landscape, existing paradigms remain insufficient to frame the research question; for this reason, an alternative is necessary. This analysis introduces the concept of hydroidentity, which is premised on the dynamic relationship between identity and access to water and the coevolution of both as they occur within various landscapes.

Identity can be defined as the membership characteristics of a particular social group. Membership within a particular group can be arguably ascriptive (Kellas 1991) and members of a social group can largely possess similar interests. Identity cannot be defined as a homogenous entity, however, because it dynamically comprises a range of variables. Thus, identity has a variable definition and simultaneously varies both temporally and spatially (Anderson 1991; Gilroy 1993; Gunaratnam 2003; Kellas 1991). Based on its fluidity and variability, I employ poststructural insight in order to better understand the discursive nature of identity. Anderson (1991: 6) suggests that the nation can be considered an "imagined political community." Identity can be seen in a similar vein. This analysis contends that there are imagined networks of ties that are intricately reproduced depending on the spatiotemporal context. Thus, to understand identity is to understand a snapshot of a lengthier process within a member's social landscape.

By positioning identity in this manner, I attempt to avoid ascribing a static conceptualization that would otherwise hide nuances. In Gilroy's (1993) thesis on the hybrid nature of identity, he emphasizes the apparent tension that emerges from striking a balance between commitment to a group with larger overarching social commitments and obligations between particular groups. Gilroy refers to this process as the double consciousness of identity. His points are useful for this analysis as they demonstrate the variability of identity as contingent on the landscape. Whilst analyzing partisan identification in heterogeneous African societies, Norris and Mattes (2003) demonstrate that identity matters, and is determined by the weight of social cleavages. They define social cleavages as cues of social class, religion, and center-periphery relationships. Norris and Mattes' stress on the importance of social cleavages is useful for this analysis because it offers a starting point for the research itself: Can identity within a peri-urban community shape or determine an individual's access to water supply and services?

The concept of hydroidentity as I use it here also requires an understanding of landscape as the outcome of a dialectic process involving "nature" and "society" (Mitchell 2002). As Latour (1993) and others have suggested (see Smith 2008), nature and society comingle and resultantly, a hybrid network between human and nature is produced. Hydroidentity is thus a type of hybrid network that occurs within a particular spatiotemporal context situated between both environmental and social landscapes.

Hydroidentity potentially could be applied at any scale, but for the purposes of this analysis, I will apply hydroidentity theory to the case study of a peri-urban community in the Rift Valley Province of Kenya. I limit my examination in this case to two variables, identity and access, and to the key set of subvariables shown in figure 7.1. Identity is exceedingly hard to conceptualize within the Kenyan context because it has been inextricably reworked, reimagined, and reproduced within the postcolonial landscape. But given the short timeframe of this study, I was able to identify a set of four key subvariables (land tenure, occupation, language, and socioeconomic class) in an attempt to capture as much complexity as possible. Based on the poststructuralist framework adopted by this analysis, I recognize that discourses and narratives are enmeshed in lived experiences as well as in institutional and social power relations that have a variety of emotional, material, and embodied consequences for individuals and groups; for this reason, they are

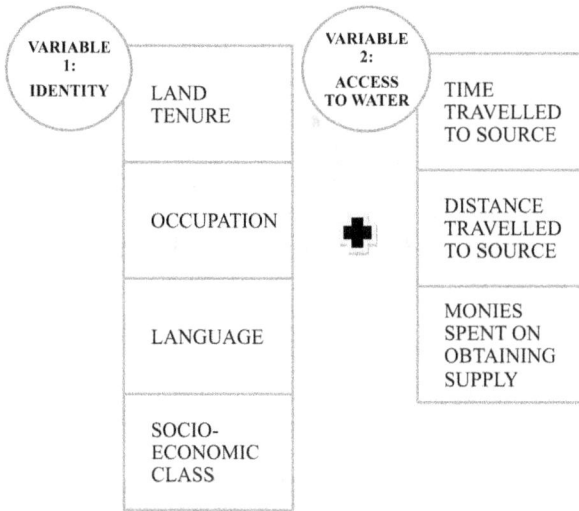

Figure 7.1. Kenyan hydro-identity variables and subvariables

coconstituted and intermingle and inhabit the other's experience (Gunaratnam 2003). This realization permits a balance between both quantitative and qualitative research based on situating the voices of Kenyan participants.

Kenyan Water Scarcity

In 2002, the Kenyan government passed a new water act. This act shifted the responsibility for water service provision from the government to local communities under the auspices of functional decentralization.[1] The government based its actions on its difficulty in providing sufficient access to water services for all its citizens. It now turned to small-scale water service providers to fulfill its role (Government of Kenya 2002). Reliance on these network operators to provide water to individuals meant that water was to be provided at a profit and assumed that communities would be able to pay for water services.

Water Act reforms at the national level have ramifications for all local communities but of most concern in this chapter are the ramifications for marginalized peri-urban and rural communities. With water accessibility rates currently ranging anywhere between 40 percent to 48 percent (Economisti Associati 2007; Government of Kenya 2007; Jorgenson 2005), scanty service provision as well as poor management of water resources characterize water service delivery throughout Kenya (Water Sector Reform Secretariat 2009). Additionally, water and sanitation services are unevenly distributed, with the Rift Valley and Lake Victoria regions having the highest percentages of underserved populations (USAID 2008). Given the vast variation in access rates and the government of Kenya's continuing fundamental objective of achieving equitable access to water resources in a decentralized manner, very little information currently exists on the impacts of such water policies at the community level within the Rift Valley.[2] Implicit within this analysis is the importance of understanding which individuals receive more or less access, and determining why. Thus, to understand identity is a step in understanding the relationship between individuals and their access to water. A gap in the literature currently exists in framing and understanding the links between identity and landscape, specifically in the East African context. Do all community members have equal and equitable access to water supply and services within their communities? If not, what are the factors encouraging or impeding access? A better examination of the physical landscape and environment is necessary in order to truly understand the social landscape.

As a highly productive agricultural zone, the Rift Valley is considered to be Kenya's breadbasket. In actuality, the environment within which the research was conducted had experienced and was in the midst of an agricultural drought (BBC 2009; Kenya Red Cross 2009). Agricultural drought, caused by changes in precipitation and soil moisture, affects the quality and quantity of crop production (Jones 1997). The following excerpts are taken from Kenyan and international news sources with regard to the recent drought within the Rift Valley.

> Across the province [Rift Valley] considered the country's grain basket—agricultural officials are reporting significant crop failure. The larger Nakuru area ... [is] the hardest-hit ... overall, we expect at least ninety-five percent maize crop failure across the larger Nakuru areas. (IRIN 2009)

> The drought is so severe that it has compromised food security in areas such as Rift Valley Province and Central Kenya, which are normally highly productive. During the long rains season—March to May—the country received very little rain with some areas of northern Kenya and the greater Kajiado receiving none at all. (Wachira 2009)

> Kenya will have to rely on massive importation of wheat to meet a looming shortage in the country. As harvests continue to stagnate, more and more farmers are contemplating a shift to other crops come next planting season.... Formerly key wheat farming areas of Narok, Uasin Gishu and Nakuru are in a sorry state today. (Kapchanga 2009)

> More than one million Kenyans risk facing hunger because of a prolonged drought, the UN has warned. The lack of rains has caused crops to fail and cattle-herders are also struggling to keep their animals alive. The worst affected areas are in the country's semi-arid south-east regions as well as some parts of central Kenya. (BBC 2009)

The area was experiencing severe regional crop loss and agricultural devastation attributable to a lack of precipitation and low soil moisture. Regional seasonal anomalies in 2008 created a culture of unpredictability and thus crops could not be planted and ultimately harvested. The impacts of water scarcity within the Kenyan case study are thus crucial to understanding the area's water–human relationships.

The Kenyan Landscape

In order to find the most suitable area to conduct my research and examine the impacts of water scarcity, identity, and access, I traveled to the Rift Valley Province in July 2009 to various communities. I visited the areas of Nakuru, Koibetek, and finally Lake Baringo in search of the ideal site. In all locations, I met with the staff of local district water offices and their environmental engineers, to discuss my sample requirements. I was in search of a community that had at least three different language groups, recent immigration, and various forms of improved water sources. I sought and found an interpreter who could speak, at a minimum, four languages, in order to ensure full understanding of the interview questions by all respondents.[3] In order to get the cooperation of the local populace and ensure that my translator and I had the permission of local authorities, we continued to meet with the local water committees as well as the community head official to explain our research intentions.

Bahati

We ultimately settled on the location of Bahati—a word meaning "luck" in kiSwahili—because of its social diversity as well as its population of approximately one thousand people—a sufficiently large population to accommodate our sampling strategy. Bahati was originally a colonial settlement but was sold back or reallocated to the local population in the 1960s following Kenyan independence. Since no major violence occurred within the community during the riots of 2008, the location also seemed relatively safe to conduct research on diversity without putting either the respondents or the researchers in harm's way. Mass rioting had occurred throughout the country in 2008 after the swearing in of President Kibaki. With accusations and allegations of voting misconduct and manipulation, conflict among ethnic groups escalated to violence nationwide. One respondent (Respondent 51) indicated that a potential reason that no actual violence occurred was that, within the region, migrant workers of a different identity group left and the groups that remained were relatively homogenous. Since then, the area has experienced consistent in-migration, mostly by migrant workers looking for temporary jobs.

The site was approximately ten kilometers from the town of Nakuru (figure 7.2). We traveled by local transportation—small white minivans known as *mata-*

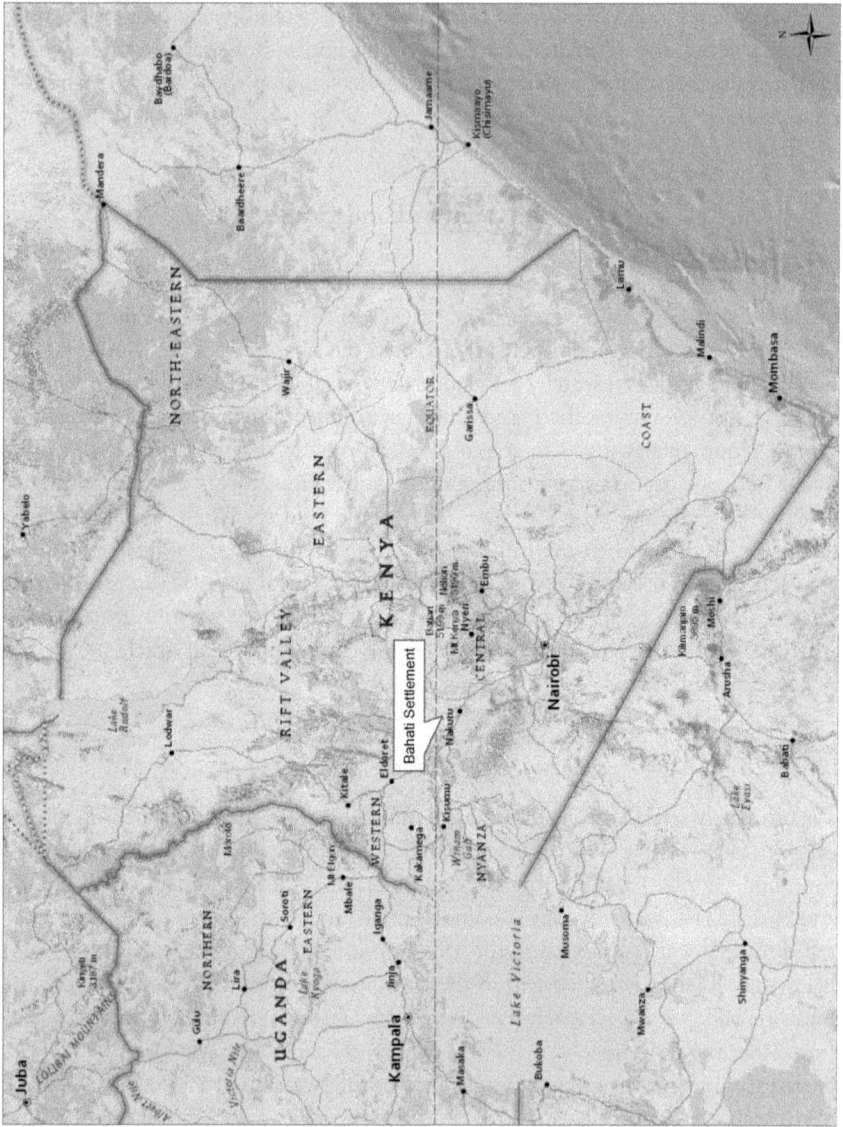

Figure 7.2. Bahati settlement area, Nakuru, Kenya

tus—to and from the site daily between the days of July 7, 2009, through July 14, 2009. Semistructured interviewing was chosen as the best type of methodology because it incorporated a guide and allowed the interviewee to fully elaborate on his or her responses without being restricted to a rigid set of questions (Bernard 2002). We conducted forty-seven semistructured household interviews and four institutional interviews within Bahati. For the purposes of this chapter, only the household interviews are analyzed. Approximately 70 percent of the respondents were women; 60 percent of all respondents were between the ages of twenty to forty; typical household composition was variable but was usually some iteration of the nuclear family. Because the interview was designed with a qualitative framework, respondents were able to elaborate in detail why they chose to answer a question in a particular manner. Bernard (2002) indicates that two methods of analyzing qualitative data include narrative analysis and discourse analysis. We used both in the analysis of data collected. We used narrative analysis to discover regularities and irregularities in how people answered, and discourse analysis to evaluate the close study of speech terms. We coded data using a data matrix that allowed for both qualitative and quantitative analysis.

The interview required individuals to describe their own degree of access to water and their subjective impressions about variations of access to water supply and services within the community. Identity was self-ascribed, with respondents identifying themselves as belonging to one or more of various groups (table 7.1). The access subvariable categories, however, were preselected in order to ensure consistency for analysis (see table 7.2).

Table 7.1. Identity subvariables

Land	Socioeconomic Class	Language	Job
Renter	Rich	Kikuyu	Permanent
Homeowner	Poor	jaLuo/Luhya	Migrant/Temporary
		Kisii	
		Turkana	
		Other	

Table 7.2. Access to water subvariables

Time traveled to source	Units in minutes, hours
Distance traveled to source	Units in meters, kilometers
Monies spent on supply	Units in Kenyan shillings

Results

Quantitative Analysis

We used statistical analysis to quantitatively assess the coded data. Testing and analysis focused on determining whether the four categories of identity (land security, language, job security, and socioeconomic class) were significant controls on determining access to water supply (time traveled to source, distance traveled to source, and money spent on supply). We treated the four variants of identity as independent variables, and treated access to water supply as a dependant variable. Using SPSS, we employed both parametric and nonparametric tests in order to understand the relationships between identity and access at variable and subvariable levels. Specifically, we ran nonparametric Chi-square test $\chi2$ and 1-way ANOVA tests to examine the research question but could not deduce statistical significance. That is, it was not possible to verify statistically whether identity had a positive or negative correlation to access.

There are many possible reasons for this lack of statistical significance, beginning with the limited sample size of forty-seven households and the limited number of access subvariables (time, distance, money spent) by comparison to the much more complex set of identity subvariables. We did not make an attempt, for instance, to gather quantitative information about the actual amount of water being used by different households, or about household use patterns (e.g., levels of domestic versus agricultural consumption), or the consistency of available supplies by household. It is possible then, that the access data do not fully capture the reality on the ground. In my opinion, however, the lack of correlation can more plausibly be interpreted as evidence that very little inequality of access, in fact, did exist during the study period. As discussed more fully below, the lack of precipitation meant that water tanks were empty and therefore provided their owners with no access advantages. Without diminishing the results of the data, it is clear that more research and analysis are needed in order to truly quantify the relationship between identity and access to water supply.

Qualitative Analysis

In addition to asking the respondents about their own access to water, we also asked them for their opinions about other identity groups' access within Bahati. In total, we asked respondents fifty-one questions about water access. Table 7.3 details several examples of questions that we used to explore the concept of other identity groups' access compared to the individual respondent's. We constructed percentage tables to determine what the aggregate common answer was; we further explored this point across each of the four identity groups. We conducted discourse and narrative analysis across each identity group, looking for variations and discrepancies within each group. As previously stated, we divided groups on the basis of four main identity categories—language, socioeconomic class, job security, and land security—and further subdivided within group-by-group dichotomies. By analyzing both the discourse used within the narratives as well as the narratives themselves, several significant patterns emerged; these are discussed in the following section.

Table 7.3. Sample questions from household questionnaire

(26) Can other groups in the community access water easily? Please identify those groups.

(29) Do you think other groups in the community have equal, more, or less access to water than you?

(31) What languages do those groups speak?

(32) Do you feel all groups around you have the same type of access to water services as you do?

(33) What are the identities of those groups?

(41) Who do you think has the most access to water supply/services in your village?

(50) Who do you think has the least access to water supply in your village?

Perceptions of Unequal Access

In terms of what they suggest about inequality of access to water, the contrast between quantitative and qualitative data is startling. As mentioned previously, during the time when this research was conducted, the region was experiencing immense water shortage, with the usual long rains being almost inconsequential that year. Kenya usually has two sets of rainy seasons, short and long. In essence, because of the immense water scarcity, it was impossible to deduce which group had more or less access because during that specific period no identity group had an apparent access advantage. Clear and nuanced variations of perception of access between identity groups emerged, however, as distinct patterns within the data. It is imperative not to privilege either the qualitative or quantitative analysis, but merely to understand the complexity revealed by their contradictions.

Answers to some of the questions dealing with perceptions of accessibility provided interesting data that appeared stratified along identity lines. When asked which identity groups had the most access to water, the general discourse consistently signaled the Kikuyu language speakers as "appearing" to have most access. This perception was consistently reported across all socioeconomic, land, and job security groups, and across language identity groups except for Kikuyu speakers themselves. Kikuyu speakers across all categories did not identify themselves as having the most access, and often signaled out specific individuals who had access to storage tanks as those with the most access. Another widespread perception was that the most privileged individuals within the community were the "rich Kikuyu and Kisiis." They were considered rich because "they could allegedly afford storage tanks" I also interviewed socioeconomically poor Kikuyu and Kisii speakers, however, who did not possess storage tanks.

Through many interviews, it was intimated that both the renters and migrant workers were considered the most marginalized groups with the least access to water within the community. The ethnic groups jaLuo and jaLuhya often occupied both these identities. However, I also conducted interviews with jaLuo and jaLuhya individuals, and some of these respondents were home-owners or had permanent jobs and did not, therefore, fit into the marginalized category to which others had assigned them.[4] And although they were depicted as being the most marginalized, certain migrant workers believed that access was evenly distributed, whereas those individuals with more permanent jobs believed that access to water was unevenly distributed within the community. The theme of misperceived notions of access between groups continues with several renters who believe there was no variation in accessibility, whilst a smaller minority of homeowners felt the opposite.

As demonstrated by these results, there appeared to be several shared yet contradictory perceptions concerning which group(s) had the least or most access. The contradictory results of quantitative and qualitative analyses render visible the dichotomy between differential access, as it occurs at any given time, and perceptions of access. Despite the lack of hard quantitative evidence about access levels, the qualitative data demonstrate a perception of increased or decreased access contingent on identity. Fully relying on the qualitative component or relying solely on statistics ignores the shades of variance that structure access to water in relation to identity within rural Kenya. The relevance of perception is solidified through examination of individuals' role-related behaviors compared with the intergroup relations. From this it is possible to infer that the multiplicities of identity could affect access to water supplies and services. It is beyond the scope of this research to determine this conclusively, however; more research needs to be conducted in this and similar settings in order to further test these relationships.

Rendering Visible Water's Role within the Landscape

Periodic water scarcities within the region can be understood as a mechanism that affects the relationship between access and identity by limiting opportunities for differential access. This fact highlights the importance of understanding the region's environmental landscape—the natural environment composed of the ecosystem, atmosphere, and climate. Without rain, there can be no rainwater harvesting and all Bahati residents share equal difficulty in obtaining individual access to water. Based on the response of those interviewed, this was regardless of whether the individual had access to personal storage tanks or a neighbor's. Access to water is obviously contingent on water existing within the region.

In a sense, then, the environmental landscape, and water more specifically, can be seen as exercising agency in this context in relation to the construction of identity and class ideology. The balance of power around water supplies within Bahati may be consistently mediated through periodic lack of access to water; further analysis is necessary to test this hypothesis.

A Manufactured Sense of Inequality?

Although drought appears to have limited inequalities of access, it can be argued, conversely, that drought conditioned and heightened the *perception* of water inequality among different identity groups. Within the social landscape, several respondents indicated that they thought identity group *x* definitely had more access, consequently implying that they had less. They also were able to identify groups who they perceived had the absolute least access but who still consistently referred to their own relative sense of water scarcity. An imagined power asymmetry existed as access was perceived to be unevenly distributed throughout the community and self-marginalization was evident. The following quotations are excerpts from interviews and highlight the narrative of perception of different identity groups having more or less access:

> Kikuyu who have settled have built storage tank. Luos usually buy from vendors. Luos just moved recently so always having problems. Kisii buy land. Luos can't afford to buy land. (Respondent 4, a female shopkeeper in her mid-thirties from the "other" language category)

> Those who have settled have water tanks. Renters are more recent. (Respondent 5, an elderly women in her mid to late fifties from the Kisii language category)

> Some have less. Residents have better advantages and cheaper [access] because they get [water] at 5 Kenya shillings (USD $0.05) per twenty liters because [they] can buy from each other.... Tenants would pay 10 shillings. (Respondent 17, a male renter in his mid-thirties)

Cogency of Hydroidentity

I now return to the framework of hydroidentity, which I define as the balance of power in a particular spatiotemporal context that structures and affects water–human relationships as they occur within both the environmental and social landscapes. Within the setting of Bahati, no apparent physical conflict over water access appeared to have taken place in the community during the time we conducted research. And despite the perceptions of unequal access, the majority of respondents indicated, in response to direct questioning, that they had not experienced conflicts over water during or immediately preceding the period when the research was conducted.

Why was this so? Although it is beyond the scope of this chapter to explain why direct conflict had not occurred, it is plausible that the respondents generally had positive feelings regarding various identity groups, despite the sense of unequal access. This was demonstrated when respondents were asked to share their feelings about the general diversity of identity within the village. This question was especially significant because Kenya had experienced nationwide violence in 2008 due to ethnic conflict and land tenure issues. When Bahati respondents were asked whether diversity was helpful or harmful within the community, approximately 80 percent of all forty-seven respondents replied that diversity was helpful.[5] Their reasons for the benefits of diversity ranged from that it allowed for "cheap labor," to being able to "borrow water from our neighbor" (Respondent 40), to "generally helping the community." Another potential reason could be that the balances of power among various groups were mitigated through the social cohesiveness around water services. This correlates with a more recent phenomenon about sixty kilometers south of Bahati:

> [T]he villagers have vowed to make the well a uniting factor and have put aside their political and ethnic differences to cultivate peace … we now meet often near this well; this time to talk about peace rather than initiate violence. It is this well that has been the source of conflict and it shall now be a source of peace. (IRIN 2010)

Water shapes and mediates various social relationships and community dynamics by connecting a variety of sociopolitical–spatial factors (Swyngedouw 1996). The weight of the combination of these factors is dynamic and may increase or decrease in a particular spatiotemporal context. In the case of Bahati, it is not clear whether hydroidentities helped residents avoid conflict, or whether conflict was avoided despite a tendency for residents to construct perceptions of inequality where very few inequalities could be demonstrated to exist.

Conclusion

Contradicting the parochial concept of an African landscape laden with tribal conflicts, this research has demonstrated that various complex processes operate to condition both the social and environmental landscapes. The dynamic nature of identity continuously fluctuates and adapts with the environmental and social landscape, responding to variations in water accessibility. Hydroidentity proceeds to work through these water–human relationships by producing and being produced by the environmental and social landscapes. Hydroidentity theory posits that a balance—or imbalance—of power stems from both these landscapes; this balance of power may be then sustained through the spatiotemporal nature of access to water supply or services. The meanings and imaginings of perception and scarcity could potentially either provide palliative effects that stabilize relationships or aggravations that destabilize communal relationships resulting in localized conflict over water. This needs further research to be determined conclusively, however.

By using hydroidentity as a lens to view both societal dynamics and the environmental landscape on which they exist, it becomes possible to see the relationship between identity and access more clearly. Within Bahati, there does not appear to be a direct correlation with identity increasing or decreasing access to water supply or services. This relationship, however, also requires further analysis because it proved indeterminable within the short timeframe during which the research was conducted.

What did become evident was the tangible relationship between perceptions of access that correlated with the various identity groups. Inequalities in gaining access to water are shared equally across all identity groups; in this manner, access did not neatly overlay identity as many of the respondents believed. This contradiction was highlighted by the startling difference between what the quantitative analysis revealed compared to the qualitative analysis. The social imaginaries of various respondents tended to overlook these shared inequalities. It becomes plausible, then, to abstract constructed notions of power that do align neatly with the language identity group as being the perceived precursor for an increase or decrease in access.

In this manner, water's agency becomes apparent. The quantities of rain falling on the Bahati landscape impacts social life with its variability and conditions perceptions of access. Water's role, functionality, and effects on the environmental landscape, with less water for irrigation and household use, are and have been well understood through research and analysis. But water's role extends beyond whether there is quantitatively more or less of it. What is less understood is water's capacity to influence social dynamics (Mosse 2003). Water's role within social life is apparent in Bahati. Although Bahati residents do not account for the ways in which less or more rainfall affects inequalities of access, the same individuals construct their own imaginings over how decreased precipitation automatically correlates with certain identity groups having less access. This perception of the inequality of access becomes perceptible only through qualitative probing. In this manner, water has a dual role: first, the impacts of its quantities, and second, the impacts of its qualities. This double consciousness of water strongly encourages further examination of global and regional water–human relationships.

Notes

I dedicate this chapter to the community of Ahero, Bahati, for allowing me into their spaces; their voices serve as my inspiration. I would also like to thank the following individuals: For guidance: David Anderson, Allida Black, Jenna Davis, Rob Hope, and Christine McCulloch. For support: Huber Technology, UK; Norman Muchori; the team at Rural Focus Development; Tumbutui Samuel; St. Cross College, Oxford University; Stanford University; and Steven Waweru, Nakuru District Water

Officer. For bringing piped water to Ahero: Engineers Cheruiyot and Gicheru, Rift Valley Water Service Board. For keeping me in line: Andrei, Alvar, Cass, Fiona, Gemma, Matt, Sara, and Sonia. For Kenyan love (and some bongo): Ben, Clarry, Enock, Essie, Lilly, Lilande, and Njuguna. For the love of water.

1. Functional decentralization occurs when a state devolves many of its major functional responsibilities, including but not limited to the delivery of services such as water (Jaspers 2003).
2. Part of the Water Act's reforms included the goal of reaching at least 50 percent of the underserved in rural areas with safe and affordable water by 2015, one of the Millennium Development Goals (MDG), and thereafter to move to sustainable access for all by 2030 (Government of Kenya 2007).
3. The interpreter I found spoke more than four languages, including English, Kikuyu, kiSwahili, and Meru.
4. I classified respondents based on self-ascribed identity rather than the identity others may have assigned them. Although some respondents reported uniform access patterns for particular identity groups, reliance on self-assigned identities revealed a more complex pattern.
5. The very nature of this overtly political question needs to be further examined and understood. Several biases would need to be overcome in order to determine the depth of the communal perception of diversity.

References

Anderson, Benedict. 1991. *Imagined Communities: Reflections on the Origin and Spread of Nationalism.* London: Verso.

Bernard, H. Russell. 2002. *Research Methods in Anthropology Qualitative and Quantitative Methods.* Walnut Creek, CA: AltaMira Press.

BBC (British Broadcasting Corporation). 2009. Kenya Drought Worsens Hunger Risk. http://news.bbc.co.uk/2/hi/africa/8211753.stm

Economisti Associati. 2007. Small-Scale Private Service Providers of Water Supply and Electricity: Survey and Mapping Initiative. A Comparative Review. Author, Bologna, Italy.

Gilroy, Paul. 1993. *The Black Atlantic: Modernity and Double Consciousness.* Cambridge, Boston: Harvard University Press.

Government of Kenya. 2002. The Water Act 2002 Public Consultation: The National Water Resources Management Strategy. Ministry of Water and Irrigation, Nairobi.

———. 2007. The Water Act, 2002: The National Water Services Strategy, 2007–2015. Ministry of Water and Irrigation, Nairobi.

Gunaratnam, Yasmin. 2003. *Research Race and Ethnicity Methods Knowledge and Power.* London: SAGE.

IRIN. 2009. Kenya: Massive Crop Failure in "Grain Basket." http://www.irinnews.org/Report.aspx?ReportId=85788

———. 2010. Communities Forge Their Own Peace in the Rift Valley. http://www.irinnews.org/report.aspx?Reportid=90882

Jaspers, Frank G.W. 2003. Institutional Arrangements for Integrated River Basin Management. *Water Policy* 5: 77–90.

Jones, J.A.A. 1997. *Global Hydrology Processes, Resources and Environmental Management.* Singapore: Addison Wesley Longman Limited.

Jorgensen, Peter. 2005. Poverty Targeting in Kenya Water and Sanitation Programme. Danida Water Sector Workshop, Accra, Ghana.

Kapchanga, Mark. 2009. Kenya: Unleavened Misery for Wheat Farmers. http://allafrica.com/stories/200908101363.html

Kellas, James G. 1991. *The Politics of Ethnicity and Nationalism.* Hong Kong: Macmillan.

Kenya Red Cross. 2009. Kenya Drought Appeal No. 1 Alleviating Human Suffering. http://www.kenyaredcross.org

Latour, Bruno. 1993. *We Have Never Been Modern.* Boston: Harvard University Press.

Mitchell, Don. 2002. Cultural Landscapes: The Dialectical Landscape—Recent Landscape Research in Human Geography. *Progress in Human Geography* 26: 381–389.

Mosse. 2003. *The Rule of Water: Statecraft,Ecology and Collective Action in South India.* New Dehli: Oxford University Press.

Norris, Pippa, and Robert Mattes. 2003. Does Ethnicity Determine Support for the Governing Party? The Structural and Attitudinal Patterns of Partisan Identification in 12 African Nations. In Annual Meeting of American Political Science Association, Philadelphia, August.

Oxford Dictionaries. 2011. Access. http://oxforddictionaries.com/definition/access?region=us&rskey=JRU7Lt&result=3

Rushdie, Salman. 2008. *The Enchantress of Florence.* London: Jonathon Cape.

Smith, Neil. 2008. *Uneven Development Nature, Capital, and the Production of Space.* Athens: University of Georgia Press.

Swyngedouw, Erik. 1996. The City as a Hybrid: On Nature, Society and Cyborg Urbanization. *Capitalism Nature Socialism* 7 (2): 65–80.

———. 2003. Scaled Geographies: Nature, Place and Politics of Scale. In *Scale and Geographic Inquiry: Nature, Society, and Method,* edited by Eric Sheppard and Robert McMasters, 129–153. Oxford: Blackwell.

———. 2004. *Social Power and the Urbanization of Water: Flows of Power.* Oxford: Oxford University Press.

———. 2006. Power, Water and Money: Exploring the Nexus. In *Human Development Report Office Occasional Paper,* edited by Human Development Report. Oxford: United Nations Development Programme.

USAID (U.S. Agency for International Development). 2008. KENYA Water and Sanitation Profile. *Advancing the Blue Revolution Initiative.* USAID.

Wachira, Muchemi. 2009. Kenya: Food-Country's Next Crisis After Water and Electricity. http://www.nation.co.ke/News/-/1056/637224/-/ulmj6i/-/index.html

Water Sector Reform Secretariat. 2009. Water Sector Reforms. Nairobi: Ministry of Water and Irrigation and Tana Water Services Board.

Part III

⋘

Urbanization

Issaka Kanton Osumanu begins this section by reminding us that 50 percent of the world's population now lives in urban centers. Given that only 19 percent of the world's population was living in urban areas in 1920, and only 29 percent in 1950 (United Nations Department of Economic and Social Affairs 1969: 67), this represents a recent and dramatic transformation of human living patterns. Osumanu acknowledges that urbanization throughout the world is associated with economic growth and improved living standards for many, but that it also has led to the proliferation of urban slums and new forms of entrenched poverty associated with poor access to clean drinking water and sanitation services. The challenges faced by the mid-sized cities of northern Ghana are less dramatic in scale and severity than those afflicting many of the world's megacities, but are representative of the types of problems being faced in less-developed countries throughout the world. Osumanu relies heavily on a complex, statistical analysis of quantitative survey data to describe and offer explanations for the particular distribution of inequality that occurs in this setting and the association of poverty with access to water. This pattern of distribution cannot be predicted or understood on the basis of the limited, macrolevel data gathered by national government surveys, which provide averages for entire urban areas without distinguishing among different socioeconomic groups. Osumanu's analysis demonstrates that basic water and health services remain beyond the reach of large numbers of urban poor, and that current government policies and programs need to be significantly altered in order to address the problem. Given the rapid pace of urbanization in Ghana and in other less-developed countries, and the current state of the world economy, these problems can be expected to grow worse in the immediate future.

Poor people in urban centers face multiple challenges and are often living under conditions that represent an abuse of basic human rights as defined by the United Nations—rights to employment, education, housing, and health services, as well as to water. Many of these people's health issues, such as diarrheal diseases, are caused mainly by unsafe drinking water and lack of adequate sanitation facilities. According to the World Health Organization, diarrheal diseases cause the deaths of one and a half million children annually; the count is much higher if one includes the deaths caused by malnutrition resulting from chronic diarrhea (Prüss-Üstün et al. 2008: 7). Other waterborne diseases such as intestinal nematode infections, lymphatic filariasis, trachoma, malaria, and schistosomiasis cause more than two billion infections and millions of death annually (2008: 8). Many of these illnesses are referred to as diseases of development because their main causes are environmental disturbance and degradation associated with deforestation, dams and irrigation projects, and rapid urbanization (Townsend 2009).

In chapter 9, Sarah Smith points out that rapid urbanization is also a major cause of dengue which has recently become "the most widespread mosquito-borne virus in the world, with approximately two billion people living in areas of transmission" (this volume, p. 180). A 2007 epidemic in Cambodia resulted in 407 deaths and the hospitalization of 39,851. Unlike the diseases caused by contaminated water supplies, the dengue mosquito vector, Smith writes, "prefers laying its eggs along the walls of human-made containers filled with clean water" (this volume, p. 182). Smith's main concern in her chapter is not with the relationship of poverty and health to inequality of water access, but rather with the inability of state health authorities in Cambodia to provide urban populations with viable remedies for dengue infection. Health officials, following a standard biomedical approach in their assessment of risk and choice of remedy, recommend the use of temephos, an organophosphate larvicide, to disinfect water-storage containers. Medical testing has not demonstrated any negative health effects from use of temephos, but the larvicide alters the taste and clarity of water in ways that make it unacceptable to local residents. The unwillingness of state health officials in this setting to work with local residents to devise culturally appropriate remedies against the spread of dengue in storage tanks compounds other state and provincial government failures to protect the quality of local drinking water supplies. Rapid development and industrialization have caused the contamination of the local river, and for political reasons the state has been unwilling to provide the community with piped water infrastructure.

In chapter 10, Liam Leonard reminds us that urban populations in the developed world are also vulnerable to water supply failures. In 2007, residents of Galway, Ireland, were faced with a cryptosporidium outbreak caused by a combination of flooding, unregulated urban development, and inadequate wastewater treatment facilities. Hundreds of people were hospitalized; the political turmoil and blame game that ensued made it clear that the city's water supplies had been under risk for some time, but that no action had been taken. As part of the European Union (EU), Ireland has helped to develop some of the best water policy in the world but its ability to implement the terms of the EU Water Framework Directive (WFD) has been constrained by the economic collapse that followed the Celtic Tiger period. Government spending on water infrastructure had never kept pace with the dramatic rise in personal and corporate incomes, and by 2007, when the period of accelerated growth had come to an end, cities and towns throughout the country found themselves facing severe infrastructure renewal problems at a time when they were least able to afford them. Many other examples of infrastructure failures in the developed world are also closely associated with economic recession and cuts in government spending, as the cases of Walkerton, in Canada (Paar 2010), and Highland Park, in the United States (Miller 2007), demonstrate. Given the current world economic situation and the debt crises in Europe and the United States, Leonard's chapter is a sobering reminder of the vulnerability of the urban north to water supply failures and their consequences for our health and the environment.

References

Miller, Liz (Director). 2007. The Water Front. Documentary film. Red Lizard Media, Tampa Bay, FL (Producer).

Parr, Joy. 2010. Local Water Diversely Known: The E. coli Contamination in Walkerton 2000 and After. In *Sensing Changes: Technologies, Environments, and the Everyday, 1953–2003*, 163–188. Vancouver, BC: UBC Press.

Prüss-Üstün, Annette, Robert Bos, Fiona Gore, and Jamie Bartram. 2008. *Safer Water, Better Health: Costs, Benefits and Sustainability of Interventions to Protect and Promote Health.* Geneva: World Health Organization.

Townsend, Patricia. 2009. *Environmental Anthropology: From Pigs to Policies.* Long Grove, IL: Waveland Press.

United Nations Department of Economic and Social Affairs. 1969. *Growth of the World's Urban and Rural Population, 1920-2000. Population Studies No. 44.* United Nations publication, Sales No. E.69. XIII.3.

Chapter 8

Health Challenges of Urban Poverty and Water Supply in Northern Ghana

Issaka Kanton Osumanu

∞

Introduction

At the turn of the twenty-first century, about 50 percent of the world's population was estimated to be living in urban areas and by the year 2025, the number is expected to reach 55 percent (United Nations Population Division 2005). Much of this urbanization is occurring in developing countries where annual growth rates have recently averaged 2.3 percent, which is far in excess of the developed world's rate of 0.4 percent. By the close of the century, more people will be packed into the urban areas of the developing world than are alive on the planet today. This, undoubtedly, has enormous health consequences for people living in urban areas of Africa.

In the advanced industrialized countries, urbanization not only has been closely associated with economic and cultural development, but also has conferred major benefits to health and a quality living environment by lowering the unit costs in the provision of housing and environmental services (Songsore 2003a). For many developing countries, rapid urbanization has overwhelmed the capacity of governments to provide essential services for the sustenance of the ever-increasing urban dwellers. This development is worsened by the nature of urban growth in these countries. Most urbanization in the developing world is occurring in areas extending beyond municipal boundaries and is driven mainly by individual residential development, usually of squatter settlements, as well as by increasing informalization of commercial activities (Osumanu 2009).

Africa, in particular, exemplifies in stark reality many of the worst difficulties of urban poverty, health, and ecology (Clarke 1993) because it has the highest population and urban growth rates in the world, combined with the lowest economic growth rates. Sub-Saharan Africa is home to many of the world's poorest countries. There are significant differences in poverty rates between Africa and the other regions of the world, with Africa having the highest rates of incidence, depth, and severity of poverty in the world (Ali and Elbadawi 1991). More than half of Sub-Saharan Africa's urban population lives on an average income of less than $20 per person per month. The facets of deprivation in Africa are many, complex, and dynamic in nature, and characterize both rural and urban areas. Related to poverty, access to piped water, a proxy for urban health achievement, paints a similar picture of a very high degree of deprivation in urban Africa. The most recent available data for thirty-two countries in Africa suggest that some 39 percent of the urban population of Sub-Saharan Africa is connected to a piped network, compared with 50 percent in the early 1990s (table 8.1). Public stand posts, also supplied by utilities, are the second-most widely used source, serving 24 percent of the population.

Table 8.1. Percentage of Africa's urban population accessing various water sources

Period	Piped Water	Stand Posts	Wells, Boreholes	Surface Water	Vendors
1990–1995	50	29	20	6	3
1996–2000	43	25	21	5	2
2001–2005	39	24	24	7	4

Source: Banerjee et al. (2008: ix).

The available evidence points to a high degree of urban poverty and exclusion of a large portion of residents from economic opportunities and good living conditions in many African cities. The trend suggests that, generally, poverty is increasing in urban areas compared to rural areas. According to the World Bank (2004), the levels of poverty in rural areas of Kenya were 18 percent higher than in urban areas in 1994, for example, but were only 4 percent higher in 1997. Within that period, urban poverty in Kenya increased by a staggering 20 percent while rural poverty increased by only 5 percent, accounting for the closure of the rural–urban poverty gap. The same World Bank report indicated a decline of about 9 percent in the rural–urban differential in poverty in Zambia between 1996 and 1998 as a result of a 10 percent rise in urban poverty, while rural poverty remained at a very high level of about 83 percent. There are also higher rates of mortality and morbidity and poor nutritional status amongst urban dwellers. High population densities, together with unequal access to potable water, adequate sanitation, and refuse collection, have resulted in a large proportion of urban households being at risk of fecal contamination and other environmental hazards (Osumanu 2007). As a general rule, however, urban areas in Kenya, Zambia, and throughout Africa continue to be better off in terms of poverty and access to services than are rural settlements.

In Ghana, a number of studies have emphasized the broad disparities between the north and the south in terms of levels of economic development and the general quality of life (Bening 1990; Songsore 1992, 2003b). All analyses have emphasized the relative lack of development of northern Ghana, comprising Northern, Upper East, and Upper West Regions, in relation to southern Ghana. At the next level of spatial resolution, there is a contradiction between urban and rural, the urban areas having a clear advantage over the rural districts. Yet at the next-lower scale, there are class contradictions in the urban areas between different wealth groups. What remains is to examine the structural patterns of general well-being within northern Ghana as a spatial entity. My main objective in this chapter is to provide an empirical documentation of the structural pattern of inter- and intraregional inequalities in northern Ghana in terms of poverty and access to water supply, highlighting critical health challenges related to today's unprecedented urban expansion. My second objective is to advance beyond the spatial manifestations of inequality to an explanation of the pattern that emerges.

Data and Methods

I rely on two types of data in this study. The first, which provides aggregate information of general well-being, is derived from the Ghana Living Standards Survey and the Ghana Demographic and Health Survey. These surveys have emerged as Ghana's most important tools in the welfare monitoring system and have provided a wealth of information for understanding living conditions in the country. To date, Ghana has carried out five rounds of both living standards surveys (1987, 1988, 1991/92, 1998/99, and 2008) and demographic and health surveys (1988, 1993, 1998, 2003, and 2008).

While aggregate indicators of well-being can serve to paint the overall welfare picture when countries and regions are the units of analysis, an accurate picture of living conditions at the household level requires more microlevel information. Such information can be obtained on the basis of detailed household surveys or community mapping to produce high-quality data. To this end, I gathered the second dataset through the administration of household questionnaires. I relied on classic household surveys in which survey instruments are administered to households, rather than on participatory surveys in which households participate in survey design and in actual gathering of information. The household surveys formed the basis for the money metric approach to poverty analysis, a method that has recently come to dominate poverty assessments in Africa and elsewhere. In this context, the relevant measure of the standard of living for households in the study areas is taken as per capita consumption expenditure (including the consumption of own production), known as the cost-of-basic-needs method (Ali, Mwabu, and Gesami 2002). The method involves identifying household consumption expenditure, covering food and nonfood (including housing).

I measured household size as the number of equivalent adults, using a calorie-based scale from the tenth edition of the National Research Council's *Recommended Dietary Allowances* (National Research Council 1989). This scale has commonly been applied in nutritional studies in Ghana (Ghana Statistical Service 2007). Measuring household size in equivalent adults recognizes, for example, that the consumption requirements of babies or young children are less than those of adults. This scale is based on age and gender-specific calorie requirements (National Research Council 1989). Each individual is represented as having the standard of living of the household to which he or she belongs. It is not possible to allow for intrahousehold variations in living standards using the consumption measure. The standard of living for each individual is measured here as the total consumption expenditure, per equivalent adult, of the household to which he or she belongs, expressed in constant prices.

I conducted the survey in the three regional capitals of northern Ghana—Tamale in the Northern Region, Bolgatanga in the Upper East Region, and Wa in the Upper West Region (see figure 8.1) in October 2009. My survey covered eight hundred household heads or their representatives.

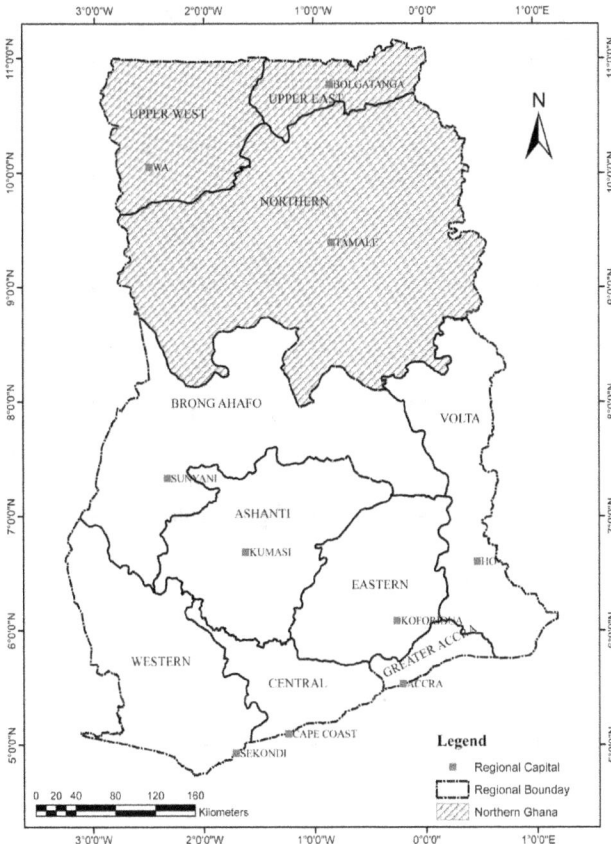

Figure 8.1. Study area, Ghana

I used a multistage sampling procedure for the selection of the subjects. In the first stage, I used a simple random technique to select four residential areas in each town. Second, I proportionally allocated the eight hundred respondents (the total sample size) among the twelve residential areas in the three towns using 2008 projected population figures (Ghana Statistical Service 2002). These methods resulted in the sampled households shown in table 8.2. In the third stage, I created blocks in the residential areas based on the number of houses and interviewers selected households to interview by systematically walking through the blocks and interviewing one household head or their representatives in every twenty or twenty-fifth house.

Table 8.2. Stratification of the study area by residential areas

Tamale		Bolgatanga		Wa	
Residential area	*Sample size*	*Residential area*	*Sample size*	*Residential area*	*Sample size*
Lamashegu	102	Gambigo	54	West Airport	28
Kalpohin	86	Tanzue	35	Kumbiahi	32
Gumani	108	Zaare	25	Dondoli	54
Choggu	168	Dapoore–Tindogo	78	Kpaguri	30
Totals	464		192		144

Source: Based on author's household survey (October 2009).

The households were categorized into three levels according to standard of living, based on the poverty lines used by the Ghana Statistical Service in poverty assessment in Ghana (Ghana Statistical Service 2007) as follows:

1. Low standard households: a lower poverty line of $200 per adult per year. Individuals whose total expenditure falls below this line are considered to be in extreme poverty, since even if they allocate their entire budget to food, they would not be able to meet their minimum nutrition requirements.
2. Medium standard households: a middle poverty range of between $200 and $300 per adult per year. Individuals consuming within this range can be considered able to purchase enough food to meet their nutritional requirements, but are not able to meet their basic nonfood needs.
3. High standard households: an upper poverty line of $300 per adult per year. Individuals consuming at levels above this can be considered able to purchase enough food to meet their nutritional requirements, and to be able to meet their basic nonfood needs.

The Nature of Urban Development in Northern Ghana

Ghana's urban population has been rapidly increasing since the last half of the twentieth century. Between 1950 and 2000, it grew almost threefold, from 15.4

percent to 43.5 percent (Ghana Statistical Service 2002) as a result of rural–urban migration and high birth rates among urban residents. The urban population is currently estimated to be at 50 percent, according to the 2010 WHO/UNICEF report: "Progress on Sanitation and Drinking Water." Urbanization in Ghana dates back to colonial development policies and postcolonial import-substitution strategies. The colonial administration oriented the country's economy toward international trade by specializing in cash crop production and mineral extraction, importing capital and labor to exploit these resources. This led to the development and growth of mining towns, regional administrative centers, and numerous commercial towns, mainly in the southern parts of the colony. In addition, port towns that linked local economies with international markets, finance, and other services also grew and other interior settlements were developed to channel export commodities to the port towns. As commercialization became more important and opportunities arose, many people migrated to port towns and interior settlements. With commercial, financial, and other services well established in the capital city and rural nonfarm entrepreneurship largely destroyed, local governments encouraged the growth of these regional centers, which after political independence became major cities. Thus, while urbanization was confined to the south, the north remained largely rural with relatively undeveloped infrastructure and services.

The initial process of urbanization in northern Ghana began in Tamale in the early 1900s. Historically, Tamale became a preeminent center in the Northern Territory of the Gold Coast in 1907 when it was made the administrative headquarters of that territory. Roads were built to connect it with all parts of northern Ghana, and a port was built at Yapei (on the White Volta River) in 1908 to serve as a river port for the town. Government offices and schools began to appear in larger numbers in the town, which consequently grew rapidly as population was attracted by opportunities for employment and education. The important initiating processes of development in northern Ghana were thus confined to Tamale, the nerve center of the colonial administration. These included the construction of rest houses and government quarters for European and indigenous officials (Bening 1974). After independence, a chain of agglomerative processes were soon set in motion, further consolidating Tamale's preeminent position in the economy of northern Ghana. Many head offices of state institutions and corporations were established because it was the seat of government at the regional level. Added to this was the development of public housing with virtually complete infrastructure of urban services for government officials.

Tamale remained the focus of development in northern Ghana until 1960, when it was divided into Northern and Upper Regions. Bolgatanga was made the capital of the Upper Region, and experienced a pattern of development and subsequent urbanization similar to Tamale. This pattern was repeated again in Wa in 1983 when the Upper Region was further divided, with Wa the capital of the Upper West Region and Bolgatanga the capital of the Upper East Region.

Urban growth in northern Ghana comprises mainly haphazard development with insufficient infrastructure. Like towns in other regions of Ghana, the spatial organization of northern towns follows peculiar geographical characteristics, with an extreme concentration of activities in the central areas. These central areas are dominated by the old compound houses that formed the preurban settlement. Rents are low and amenities are moderately developed. These areas have generally become sophisticated commercial areas, and while informal businesses make the streets lively, there are also signs of "inner city decay." They are inhabited mainly by indigenous households and migrants from rural hinterlands, who often find sleeping places in front of shops at night, and are very densely populated with close to one-third of households living in slums.

Amenities are relatively well developed and rents are higher in the areas immediately adjoining the central indigenous core. These areas may have inner zones similar to the central indigenous core and are dominated by compound houses with a few two-story and three-story houses, especially along the major streets. Business is lively here, but is still mostly small scale. Principally, middle- and lower-middle class households inhabit these residential areas. They are densely populated, comprising heterogeneous ethnic groups, but dominated by indigenous households, and include first-generation immigrants as well as residents who have been there longer.

The third category of spatial arrangement is what can conveniently be called the high-class residential areas. These are areas of relatively better residences that have the highest amenity values and rents. They have generally good building standards with very few households living in precarious housing. These mainly residential areas are increasingly characterized by mixed use, incorporating substantial commercial activities, and are populated principally by residents of the upper and upper middle social strata. Physical development here is better than in all the other residential areas of the urban agglomeration.

The rural fringes have the lowest rents and amenity values and consist of the village enclaves that have been absorbed into the urban agglomeration as a result of rapid expansion through land development. The majority of houses in these areas are built with earth and roofed with thatch. These areas are generally without environmental services and infrastructure, but life here is slightly better than other rural communities in northern Ghana. The inhabitants are mostly a mixture of indigenous households and recent migrants. Fringe areas tend to be more rural in nature, with mainly widely spaced residential houses compared to the core. For populations in these areas, small-scale trade followed by peasant farming and a few (generally agrobased) industries are their main sources of income. Fringe areas also attract people from rural areas, and tend to be diverse, dynamic, and constantly evolving environments. The presence of a few schools, and health and administrative centers, in addition to other socioeconomic factors, may attract further in-migration.

Trends in Urban Poverty, Access to Water Supply and Health in Ghana

Urban Poverty

Over the past two decades Ghana is reported to have witnessed a considerable measure of positive real economic growth. GDP growth rates averaged 4.7 percent over the period 1982–1992 and 4.2 percent between 1992 and 2002 (World Bank 2003). This correlated with a significant reduction in consumption poverty between 1991–92 and 2005–06. Consumption poverty is specifically concerned with those whose standard of living falls below a defined minimum standard at which minimum consumption requirements can be met. This differs from the concept of absolute poverty which is defined as the degree of poverty below which minimal requirements for survival are not being met.

Using a poverty line of $260 per adult per year, the proportion of the population of Ghana defined as poor fell from 51.7 percent in 1991–92 to 39.5 percent in 1998–99 and further to 28.5 percent in 2005–06. This impressive decline in poverty incidence has led to a lowering of the absolute numbers of poor from around 7,931,000 individuals in 1991–92 to 7,203,000 in 1998–99 to 6,178,000 individuals in 2005–06 (Ghana Statistical Service 2007). Whilst agreeing that some measure of poverty reduction has been achieved over the period, there is evidently a significant section of the population that is still classified as poor and living well below the poverty line. It is equally true that there has been a significant poverty reduction for the country as a whole but some regions were completely left out. In particular, the already poorest part of Ghana (the northern savannah area) did not benefit from the country's economic growth.

The picture of economic growth and poverty reduction in Ghana raises questions that support the debate that has existed in the development literature for many years as to the extent to which economic growth reduces poverty. It has been argued that when economic growth benefits everyone in equal proportion, the incomes of the poor also grow at the same rate as the mean income. However, if the economic growth is unequally distributed, the effects on the poor are less (or more) depending on whether the incomes of the poor grow by less (or more) than the average (Deaton 2003).

The decline from 1991–92 to 1998–99 was not evenly distributed geographically. The poverty reductions during the period were concentrated in the city of Accra, located in Ghana's urban coastal zone, and in the urban forest zone located in the middle of the country. In the remaining localities, both urban and rural, poverty fell only very moderately, apart from urban savannah, comprising the three regions of northern Ghana, where the proportion of the population defined as poor increased during the period. However, between 1998–99 and 2005–06, poverty fell significantly in all localities except Accra, which experienced an increase in poverty (see figure 8.2). In the case of Accra, mixed results have been reported. In 1991–92 about 23 percent of the population of Accra fell below the poverty line. This reduced significantly to about 4 percent in 1998–99 but increased, again significantly, to about 11 percent in 2005–06 (Ghana Statistical Service 2007). The pattern observed in Accra has been attributed to the result of a large increase in net numbers of migrants from the poorer regions, especially from northern Ghana, to Accra; net migration was found to be about +310,000 for the Greater Accra Region but −332,000 for the Upper West Region and −219,000 for the Upper East Region, which are considered the poorest regions. North–south migration in Ghana is mainly attributed to unequal development in social infrastructure and employment opportunities that tends to favor the south. However, about 70 percent of this kind of migration is seasonal, during which male members of households move to the south during the dry season when there are no farming activities and return home at the onset of the raining season.

While urban households from the forest ecological zone experienced the largest decline in poverty during the 1990s, the coastal areas have benefited significantly from the country's economic growth since the late 1990s. In addition, after its urban poverty rate increased during the 1990s, northern Ghana experienced a decline in urban poverty incidence from 43 percent to about 28 percent by 2006. Even if urban poverty in northern Ghana has been declining in the past ten years its higher percentage share of Ghana's urban poor is due to the fact that poverty has been declining even faster in the southern part of the country.

There is now consensus among analysts of economic growth and development that a poverty measure needs to consider not only the incidence of poverty (the proportion of people living below a given poverty line), but also the distribution of poverty incidence among those considered as poor (Aryeetey and Codjoe 2005; Sen 1976). The concept of absolute poverty sets the poverty line as the minimum amount of resources required for survival at any given time (Zheng 1998). In Ghana, there is a similar measure known as extreme poverty, which is adjusted from period to period to reflect changes in prices. In Ghana, extreme poverty is defined as those whose standard of living is insufficient to meet their basic nutritional requirements even if they devoted their entire consumption budget to food.

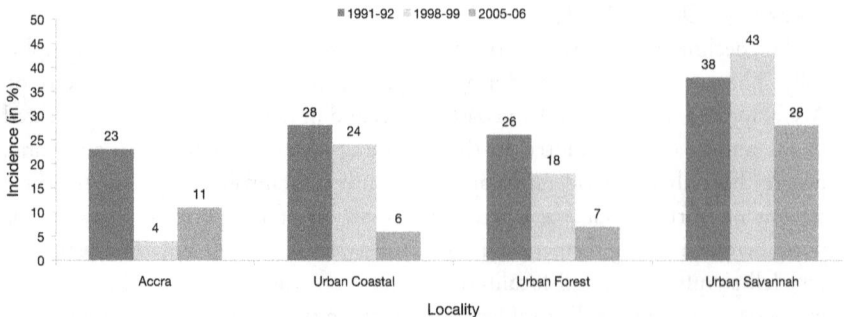

Figure 8.2. Incidence of urban poverty by locality—1991–92 to 2005–6

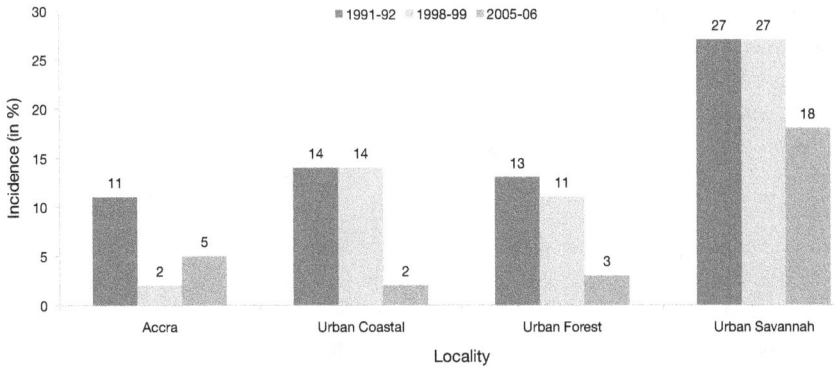

Figure 8.3. Extreme urban poverty incidence by locality—1991–92 to 2005–6

Figure 8.3 shows the trend in the incidence of extreme urban poverty for the four geographic localities in Ghana. At the national level, the incidence of extreme poverty fell from a little over 36 percent in 1991–92 to just fewer than 27 percent in 1998–99 and declined further to a little above 18 percent of the population in 2005–06. With the exception of Accra, there has been a substantial decline in the incidence of extreme urban poverty in all the localities in 2005–06 compared to 1998–99. In the case of northern Ghana (urban savannah), the incidence of extreme urban poverty, which remained at 27 percent between 1991–92 and 1998–99, fell significantly to 18 percent in 2005–06, though still the highest incidence in the country. It may be noted that the contribution of the urban coastal locality to extreme urban poverty in the country in 2005–06 was only 2 percent, having fallen from about 14 percent in 1998–99.

Access to Water

Over a quarter of Ghana's urban population does not have access to clean and potable water (potable water is defined as treated surface and underground water or boreholes). Open wells and untreated surface water are classified as not potable because they are subject to contamination. The higher level of contamination of rivers, streams, and open wells by comparison to boreholes can be attributed to the absence of natural soil protection and filtration functions and the possibly short distances between the input of contamination and water extraction, because households who rely on these sources of water supply live close to them. Most catchment areas of these waters are completely devoid of forest cover and are heavily affected by agriculture and settlement activities, providing ideal conditions for erosion, particularly during heavy rainfalls. Another source of contamination arises from the relatively high density of livestock, including cattle, sheep, and goats. Other factors influencing the quality of water are human activities such as bathing and washing clothes close to the water source as well as the use of multiple containers to draw water.

Figure 8.4 presents the proportion of urban Ghanaians having access to potable water between 1991–92 and 2005–06 according to standard of living statistics. The proportionate changes in access between the survey years were relatively small for the two top quintile groups for the period 1998–99 to 2005–06 after the initial modest increases between 1991–92 and 1998–99. The two lowest quintiles, though, had large increases between the years 1998–99 and 2005–06 as compared to the period 1991–92, when there was virtually no significant change (Ghana Statistical Service 2007). The pattern of access to potable water in Ghana indicates a significant reduction in the urban–rural gap in access to safe water as compared to the situation that prevailed about twenty years ago. Historically, rural areas have lacked access to potable water but recent trends suggest an increasing number of urban populations lacking access to potable water by comparison to their rural counterparts. Changing patterns of use in rural areas reflect increased use of water from boreholes or protected wells and less use of rainwater and water from lakes, rivers, and other surface water sources. These trends are consistent with government interventions that are focused mainly on improving access for rural areas while encouraging private sector partnerships in water provision for urban areas (Osumanu 2008).

One of the main problems in Ghana's urban water supply provision is the inability of the public sector to deliver and maintain basic infrastructure services. Currently, only 60 percent of Ghana's urban population is adequately served by the Ghana Water Company Limited (GWCL), the main water supplier. Poor access to improved water in Ghana is attributed to a number of causes, including weak public sector policies, lack of political will, weak local government capacity, and inadequate financing. Attempts by the government to reform the urban water

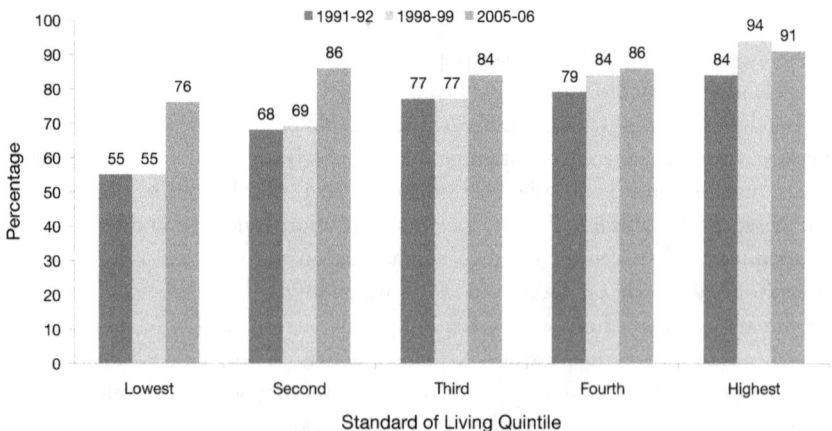

Figure 8.4. Percentage of urban households having access to potable water by standard of living quintile

sector have focused on public–private partnerships in the form of management contract arrangements. These contracts require that tariffs be structured so that cost recovery, and therefore financial sustainability, is ensured. Even though reform of the urban water system is still ongoing, it has not yet had many of the desired results, and it is anticipated to have a negative impact on the poor by restricting their access to clean water supplies as a result of high tariffs. Again, private utilities are perceived by the public as being biased toward meeting the demands of upper- and middle-income groups, and for not being responsive to ensure equality of access for all segments of the population.

Although the unserved and the underserved include all categories of households, rich and poor, in formal and informal settlements, those who suffer most from inadequate water supply are the poor and marginalized households found in informal settlements, because the rich are able to obtain water from other sources such as tanker services. Within urban areas with piped systems, only 15 percent of the poor have access to piped water directly supplied by GWCL, with benefits of scale and cross subsidies. The prevailing sagacity has given rise to parallel economies—the formal water economy based on a public supply network and the informal water economy serviced mainly by small-scale, private sector water vendors (Osumanu and Abdul-Rahim 2008). The provision of water by vendors has been shown to be expensive (Osumanu 2008), irrespective of whether they obtain water from GWCL or tanker supplies; it is generally the case that households served by vendors pay higher charges for water than those directly connected to the piped system. The block tariff system operated by GWCL makes water vendors pay high tariffs as a result of high rates of consumption, with costs passed on to consumers.

State development policies enjoin the government of Ghana to promote just and reasonable access by all citizens to public facilities and services, which naturally include water supply services. The policies also allow for different provisions for different communities with regard to their special circumstances. This further allows service providers to have appropriate mechanisms to optimize social inclusion. Arguments for the utility company's inability to serve informal communities often relates to the unplanned nature of these areas, however, which makes extension of water facilities difficult physically. Another argument relates to the fact that utilities are not authorized to connect residents in unplanned areas. Other reasons are that some of the informal areas are "difficult areas" where actions to recover bills could be a problem and where residents have low ability to pay for connection and water charges. Yet, what is often overlooked is that low-income households can afford and are willing to pay for water services, and often spend a much higher proportion of their income on water (Whittington et al. 1992), a clear indication of their readiness to participate financially in water provision.

To examine regional inequalities in access to potable water, table 8.3 presents the percentage distribution of urban households according to main source of water supply for general use in 2003 and 2008.

Table 8.3. Households by main source of water supply for drinking by locality, 2003 and 2008 (%)

Source of water supply	Accra		Kumasi		Tamale		Bolgatanga		Wa		All	
	2003	2008	2003	2008	2003	2008	2003	2008	2003	2008	2003	2008
Pipe-borne	**86.6**	**84.3**	**77.9**	**81.6**	**69.8**	**72.5**	**20.1**	**25.7**	**17.9**	**20.7**	**72.5**	**73.1**
Indoor plumbing	10.2	10.9	9.0	10.8	5.6	5.8	3.7	3.5	1.6	2.0	8.0	8.5
Inside standpipe	31.0	31.3	28.8	30.6	10.7	22.6	4.8	4.2	3.2	3.1	19.6	21.9
Pipe in neighbor household	28.5	28.7	26.0	26.8	28.9	20.5	4.2	4.2	4.7	4.9	21.4	19.7
Private outside standpipe/tap	10.0	9.0	7.4	8.0	14.1	15.7	2.1	3.6	5.5	4.2	7.3	7.4
Public standpipe	6.9	4.5	6.7	5.4	10.5	7.9	5.3	10.0	2.9	6.5	16.2	15.6
Well	**1.1**	**1.2**	**12.0**	**10.2**	**2.3**	**2.0**	**62.9**	**56.6**	**60.5**	**55.4**	**17.1**	**16.0**
Borehole	0.0	0.1	4.9	3.4	0.7	0.7	47.8	40.8	53.3	48.2	7.5	6.1
Protected well	1.1	1.1	7.0	6.8	0.7	0.7	11.9	13.3	5.3	6.2	7.5	8.0
Unprotected well	0.0	0.0	0.1	0.0	0.9	0.6	3.2	2.5	1.9	1.0	2.1	2.0
Natural sources	**0.1**	**0.1**	**0.5**	**0.5**	**17.3**	**14.3**	**8.0**	**6.5**	**11.6**	**10.6**	**5.4**	**2.5**
River/stream	0.1	0.1	0.3	0.3	5.1	5.0	3.0	2.0	5.3	4.8	2.4	1.8
Rain water/spring	0.0	0.0	0.2	0.2	0.0	0.0	0.0	0.0	0.1	0.1	1.0	0.4
Dugout/pond/lake/dam	0.0	0.0	0.0	0.0	12.2	9.3	5.0	4.5	6.2	5.7	2.0	0.3
Other	**12.2**	**14.3**	**9.6**	**7.7**	**10.6**	**11.2**	**9.0**	**11.2**	**10.0**	**13.3**	**6.0**	**8.4**
Water truck/tanker service	1.3	1.3	1.5	1.0	5.6	6.0	3.6	3.0	2.0	3.5	1.2	0.9
Water vendor	4.0	4.5	3.5	2.5	2.0	2.0	2.5	2.6	2.5	3.5	2.6	3.4
Sachet/bottle water	6.9	8.6	4.6	4.2	3.0	3.2	2.9	5.6	5.5	6.3	2.2	4.0
All	**100.0**	**100.0**	**100.0**	**100.0**	**100.0**	**100.0**	**100.0**	**100.0**	**100.0**	**100.0**	**100.0**	**100.0**

Source: Ghana Statistical Service (2003, 2008a).

The proportion of urban households having access to pipe-borne water increased slightly from 72.5 percent to 73.1 percent over the period, but in most cases, that source is from outside the house. Those using well water reduced from 17.1 percent to 16.0 percent, while households depending on access to natural and other sources decreased marginally from 11.4 percent to 10.9 percent (Ghana Statistical Service 2003, 2008a). The household profile reveals greater access to potable water in the well-endowed south. While more than 80 percent of residents in Accra and Kumasi have access to pipe-borne water in 2008, wells and natural sources constitute the main source of water supply for households in Bolgatanga and Wa. Tamale is the only location in northern Ghana where a majority of residents have access to pipe-borne water (72.5 percent in 2008).

Health Status

The health status of urban Ghanaians depicts a trend that generally correlates with poverty and access to potable water supply. Fewer people tend to be sick in Accra than in other urban localities. The Ghana Living Standards Survey (2008a) reported that while in Accra about 14 percent of persons were sick in the two weeks preceding the survey, in 2006, the corresponding proportion for the other urban localities was about 22 percent. Additionally, Accra had the lowest proportion of persons, at 53.9 percent, who stopped their usual activities due to illness or injury, by comparison to other urban localities at 56.5 percent (the average rate for urban centers outside Accra), with urban savannah recording the highest rates at 67.4 percent (Ghana Statistical Service 2008b). Figures available on infant and child mortality, similarly, are consistent with this observation (see table 8.4). Results from five Demographic Health Surveys show a marked decline in childhood mortality in urban areas over the ten-year period from 1998 to 2008. Infant mortality

Table 8.4. Early childhood mortality rates by region

Region	Neonatal mortality	Postnatal mortality	Infant mortality	Child mortality	Under–five mortality
Western	40	11	51	14	65
Central	47	26	73	38	108
Greater Accra	21	15	36	14	50
Volta	26	11	37	13	50
Eastern	29	25	53	30	81
Ashanti	35	19	54	28	80
Brong Ahafo	27	10	37	41	76
Northern	35	35	70	72	137
Upper East	17	30	46	33	78
Upper West	45	52	97	50	142

Source: Ghana Statistical Service (2008b).

rates reduced from 96 deaths per 1,000 live births, and the under-five mortality rate fell from 105 deaths per 1,000 live births to 75 deaths per 1,000 live births. Differences in mortality by region are marked. The infant mortality rate varies from 36 deaths per 1,000 live births in the Greater Accra Region to 97 deaths per 1,000 live births in the Upper West Region. Differentials in under-five mortality show a similar pattern. For example, under-five mortality ranges from a low of 50 deaths per 1,000 live births in the Greater Accra and Volta Regions to a high of 142 and 137 deaths per 1,000 live births in the Upper West and Northern Regions, respectively.

As expected, nutritional levels of children are higher in the Greater Accra Region than all other regions in the country, with the three northern regions being amongst the lowest (see table 8.5). For example, stunting is highest in the Eastern and Upper East Regions (37.9 percent and 36.0 percent, respectively), and lowest in the Greater Accra Region (14.2 percent). Wasting is more common in the Upper West (13.9 percent), Northern (12.9 percent), and Central (12.0 percent) Regions than elsewhere. The prevalence of overweight children ranges from 1.3 percent in the Upper East Region to 12.0 percent in the Eastern Region. Finally, the proportion of underweight children ranges from 6.5 percent in the Greater Accra Region to 27.0 percent in the Upper East Region.

The disparity in the patterns of poverty as well as that of basic infrastructure across the country clearly portrays the existence of a wide gulf in the standards of living of southern and northern Ghana. Consequently, it becomes evident that the economic conditions in these areas are vastly different. Having discussed the patterns of inequality, it is pertinent to attempt a general discussion of the underlying processes at work in the generation of these inequalities.

The observed pattern depicts a clear contradiction between the urbanized regions of north and the south, the north having all the features of an underdevel-

Table 8.5. Nutritional status of children by region

Region	Stunting (% below – 2 SD)	Wasting (% below – 2 SD)	Overweight (% above + 2 SD)	Underweight (% below – 2 SD)
Western	27.0	5.6	5.9	10.3
Central	33.7	12.0	9.7	17.2
Greater Accra	14.2	5.9	4.9	6.5
Volta	26.8	5.2	7.9	13.6
Eastern	37.9	6.4	12.0	8.7
Ashanti	26.5	9.2	3.7	12.1
Brong Ahafo	25.2	5.4	2.8	13.5
Northern	32.4	12.9	2.0	21.8
Upper East	36.0	10.8	1.3	27.0
Upper West	24.6	13.9	3.0	13.1

Source: Ghana Statistical Service (2008b).

oped region in relation to the south. Factorial ecologies of urban spaces in Ghana, undertaken by Songsore and McGranahan (1993) and Benneh and colleagues (1993), reveal the same contradictions as this case study. Colonial dependency is at the root of the intra-urban inequalities so evident in the Ghanaian spatial economy. Under the impact of colonialism, the internal production structure was divided into a dynamic modern export sector and an underdeveloped subsistence sector that also serves as a labor reserve for the modern sector. In other words, the object of colonial policy was to keep the north undeveloped so that cheap labor for the industry and agriculture of the south could be guaranteed (Bening 1972).

Spatial and class inequalities created to a large extent through colonialism have been exacerbated by the policies pursued by the intermediary strata, who captured state power at independence and continue to use it to foster their class interest. The import substitution strategy of industrialization that was pursued after independence meant that the urban centers, which played a key role in the functional organization of the colonial economy, attracted all the industrial capital, giving rise to a further polarization of development. This resulted in the phenomenal growth of the neocolonial state apparatus and sharpened the north–south dualism established during the colonial period (Ofori-Atta 1978).

Urban Poverty, Water Supply, and Health in Northern Ghana

In this section, I present microlevel information, within northern Ghana, gathered from the household survey. I begin with an analysis of household poverty profiles. In this way, I am able to make well-informed assessments of how the various components of poverty and access to water supply influence the living conditions of residents. The distribution of households by living standards in relation to localities indicates that dramatic differences exist in the poverty levels of households from one locality to another (see figure 8.5). Wa had average scores below Bolgatanga, which in turn scored lower than Tamale. This observation reflects the relative availability of employment and other income-earning opportunities in the localities covered by the study. Tamale is the third-largest urban settlement in Ghana (in terms of population size), after Accra and Kumasi, with a much more vibrant and buoyant economy dominated by commercial and service activities. On the contrary, the majority of the residents of Bolgatanga and Wa are food crop farmers who produce mainly to feed themselves. This finding also confirms the Ghana poverty assessment report, which indicates that the Upper West Region is the poorest region in the country (Ghana Statistical Service 2007). The poverty profile of households further shows that standard of living increases the more educated the head of the household is, an indication of the importance of education as a means to achieve economic success, which in turn is determined by relative access to educational opportunities across the country.

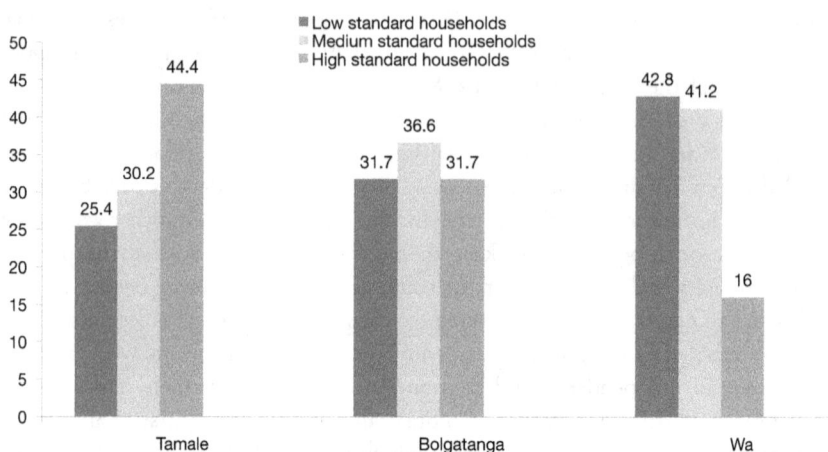

Figure 8.5. Classification of households according to living standards

About 60 percent of the economically active population in Bolgatanga and Wa engages in agriculture. The main products are millet, guinea-corn, maize, ground-nut, beans, sorghum, and dry season tomatoes and onions. Livestock and poultry production are also important. Industrial activities in these areas are mainly small-scale craft industries. This is due to the simple technology involved, the availability of local inputs, and linkages between them and other economic activities. These crafts, varied as they are, include pottery, basketry, and cloth weaving. Leatherworks are also carried out at areas around Bolgatanga and the surrounding villages. One distinct feature of these cottage industries is that they are basically labor intensive and rely mostly on traditional talent and skill.

The absolute poor households are often found in squatter settlements gener-ally located in peri-urban and the hemmed-in areas of urban centers. In northern Ghana, such households live in huts built of mud and roofed with straw or zinc. The main features of the predominantly traditional architecture here are round huts with flat roofs and small windows with poor ventilation. Additionally, poor house-holds tend to be polygamous with an average household size of seven persons.

As already indicated (see table 8.3), most urban households in northern Ghana have to collect water from other sources, unlike their colleagues from the south, where piped water supplied from municipal treatment works is the norm. The percentage of urban households in northern Ghana having access to potable water varies considerably from region to region, however, with households in Tamale having the best access, though mainly from shared rather than private sources.

Although it is difficult to establish the precise interconnections between poverty and access to water use, it is generally agreed that inadequate access to potable water is an integral part of the severely disadvantaged situation of the poor within the global community (Global Water Partnership 2003). In fact, poverty itself is usually defined in socioeconomic terms, and perceived as a condition in which people's livelihood capacity is inadequate to meet their own and their children's basic needs, including the need for potable water. Clearly, access to potable water in urban areas of northern Ghana is a function of the living standards of households (see table 8.6). Overall, 40.6 percent of the households classified as high standard had access to private pipe-borne water, with 19.4 percent having indoor plumbing and another 21.2 percent having private standpipes. In contrast only 15.8 percent of the low standard households had access to pipe-borne water from either private standpipes (2.4 percent), neighbors' pipes (4.8 percent) or public standpipes (8.6 percent), with none having indoor plumbing. Additionally, 18.6 percent of low standard households depend on water from vendors and 7.2 percent still use water from surface sources (rivers, streams and ponds) directly. Households who depend on surface water directly are mostly found in the peri-urban areas. The average distance between settlements and surface water sources is about one and a half kilometers.

The importance of water to livelihoods, as well as survival, health, and quality of life, is implicit in poverty rates among households. The study considered the incidence of diarrhea in children less than five years of age using mothers' two weeks recall. Children from low standard households were reported to have experienced the highest cases of diarrhea (28.4 percent), against 12.2 percent and 2.1 percent of the children from the medium standard and high standard households, respectively. Similarly, households from Wa reported the highest incidence of 24.8 percent against their counterparts from Bolgatanga—21.7 percent, and Tamale—16.4 percent. Diarrhea incidence was associated with the source of water supply for the household (see figure 8.6), with children from households using water from rivers, streams, ponds, and water vendors reported to have the highest incidence of negative health outcomes such as stunting, wasting and low weight (see table 8.7).

Table 8.6. Poverty and access to water supply (%)

Source of water supply	Low standard households	Medium standard households	High standard households
Indoor plumbing	0.0	10.8	19.4
Private standpipe	2.4	14.6	21.2
Pipe in neighbors house	4.8	16.6	18.6
Public standpipe	8.6	22.8	18.6
Borehole	26.6	10.8	4.8
Protected well	19.6	3.6	5.8
Unprotected well	12.2	3.6	0.0
River/stream/pond	7.2	0.0	0.0
Tanker service	0.0	5.6	8.6
Water vendor	18.6	8.2	0.0
Sachet water	0.0	3.4	3.0
Total	100.0	100.0	100.0

Source: Based on author's household survey (October 2009).

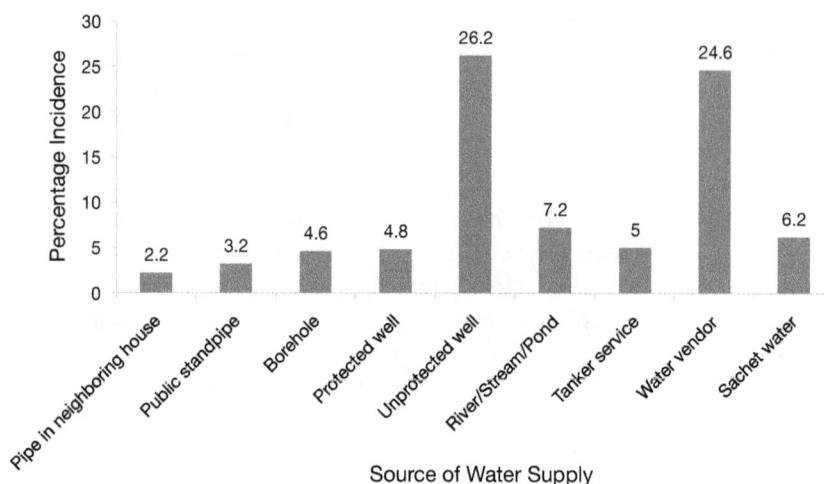

Figure 8.6. Relationship between access to water supply and diarrhea in children

Table 8.7. Nutritional status of children by standard of living and access to water supply

	Stunting (% below − 2 SD)	Wasting (% below − 2 SD)	Overweight (% above + 2 SD)	Underweight (% below − 2 SD)
Standard of living				
Low standard households	37.0	14.1	1.0	28.2
Medium standard households	28.4	12.0	5.9	17.2
High standard households	14.8	5.2	11.6	6.7
*Source of water supply**				
Indoor plumbing	10.2	5.2	13.7	4.2
Private standpipe	14.2	6.8	11.4	7.3
Pipe in neighboring house	17.7	9.6	8.9	9.2
Public standpipe	22.2	11.6	6.9	13.6
Borehole	26.8	14.8	4.2	18.6
Protected well	26.8	13.9	3.0	21.8
Unprotected well	37.9	16.4	1.8	27.8
River/stream/pond	42.6	20.0	0.0	32.4
Tanker service	15.2	8.4	2.4	11.5
Water vendor	32.4	18.9	2.0	11.8
Sachet water	36.0	10.8	1.3	13.1

Source: Based on author's household survey (October 2009).

*For simplicity sake, this table does not provide information about the three-way correlation between nutritional status, standard of living, and source of water supply. The water supply figures given here thus represent an average across households from all three standard of living groups.

As expected, the analysis of child malnutrition, based on the household survey, produced a result that agrees with that of diarrhea in terms of poverty rates among households and access to water supply. However, the distribution of child malnutrition by locality differed slightly. For example, stunting was highest in the Bolgatanga (41 percent) and Wa (36 percent), and lowest in Tamale (24 percent). Wasting was more common in Wa (24 percent) and Tamale (23 percent) than in Bolgatanga (14 percent). The prevalence of overweight children ranged from Bolgatanga (2 percent), Tamale (3.5 percent), to Wa (4.8 percent). Finally, the proportion of underweight children ranged from Tamale (8.5 percent), Wa (12.5 percent), to Bolgatanga (22 percent).

Like the north–south contradiction observed earlier, one notices at the subregional levels an equally sharp contrast between the three urbanized areas of the north in terms of access to water supply and health. In northern Ghana, Tamale performed relatively better than the other urban centers. This is because Tamale had emerged as the regional headquarters of the northern territories in the early 1990s and has since become the most significant administrative, commercial, and business center in the area. The better performance of Tamale, therefore, reflects the concentration of social services in the town.

Conclusion

The positive correlation between urbanization and improvements in the quality of lives of the people living in urban centers is intriguing although the direction of causality is far from clear in Ghana. Water poverty is an important component of poverty generally and lack of access to water is integral to the disadvantaged situation of the vast majority of the urban poor in northern Ghana. However, many people have yet to appreciate water's role in all aspects of productive life and the profound implications of inadequate access to potable water for a country as a whole, let alone its implications for the most vulnerable households. Water's role in poverty reduction is significantly underestimated among decision makers and development practitioners of all kinds.

The analysis presented in this chapter showed clearly the disparities in access to opportunities and standards of living between towns located in southern Ghana and those found in the northern parts of the country. It is equally evident that there are further disparities in northern Ghana, with Tamale being relatively better off than other urban centers in the north. All the study indicators are worst among poorer households living in low-income neighborhoods. For the majority of households the chances of overcoming these conditions are limited. As this study demonstrates, the causes of this particular spatial distribution of inequality are rooted in the colonial era but have been reinforced by the neocolonial policies of national governments that, since independence, have been dominated by elite class interests.

It is imperative that a much better analysis be made of the interconnections between access to water and water-related poverty indicators, and the priority needs of those whose lives are supposed to improve as a result of poverty-reduction initiatives. Experience has shown that reduction of subsidies, charging of fees, privatization of services, and decentralization of maintenance and ownership to community groups often discriminate against poor people unless such policies are introduced in an enlightened, efficient, and equitable manner. It seems reasonable to propose, therefore, that programs be put in place to help those who are in this position to maximize opportunities to improve their living conditions through the provision of improved public sector water services. It would appear that improvements in water supply can enhance general well-being and improve health considerably.

References

Ali, A.A.G., and I. Elbadawi. 1991. *Inequality and the Dynamics of Poverty and Growth.* Cambridge, MA: Center for International Development.

Ali, A.A.G., G. Mwabu, and R.K. Gesami. 2002. *Poverty Reduction in Africa: Challenges and Policy Options.* Cambridge, MA: Center for International Development.

Aryeetey, E., and E. Codjoe. 2005. Income Distribution and Poverty. In *Globalization, Employment, and Poverty Reduction: A Case Study of Ghana,* edited by E Aryeetey. Accra: Woeli Publishing Services.

Banerjee, Sudeshna, Heather Skilling, Vivien Foster, Cecilia Briceño-Garmendia, Elvira Morella, and Tarik Chfadi. 2009. Ebbing Water, Surging Deficits: Urban Water Supply in Sub-Saharan Africa. Africa Infrastructure Country Diagnostic, (AICD) Background Paper 12, Phase I (June 2008). Washington, DC: The World Bank.

Bening, R.B. 1972. Development of Education in Northern Ghana. *Ghana Social Science Journal* I (2): 21–42.

———. 1974. Location of Regional and Provincial Capitals in Northern Ghana 1897–1960. *Bulletin of the Ghana Geographical Association* 16: 54–66.

———. 1990. *A History of Education in Northern Ghana, 1909–1976.* Accra: Ghana Universities Press.

Benneh, G., J. Songsore, J.S. Nabila, A.T. Amuzu, K.A. Tutu, Y. Yangyuoru, and G. McGranahan. 1993. *Environmental Problems and the Urban Household in the Greater Accra Metropolitan Area (GAMA)— Ghana.* Stockholm: Stockholm Environment Institute.

Clarke, J.E. 1993. Polarisation and Depolarisation in Africa. In *Urban Ecology and Health in the Third World,* edited by L.M. Schell, M.T. Smith, and A. Bilsborough. Cambridge: Cambridge University Press.

Deaton, A. 2003. Measuring Poverty in a Growing World (or Measuring Growth in a Poor World). NBER Working Papers No. W9822, National Bureau of Economic Research, Cambridge, MA.

Ghana Statistical Service. 2002. *Population and Housing Census 2000: Summary of Special Report on Final Results.* Accra: Author.

———. 2003. *Ghana 2003 Core Welfare Indicators Questionnaire (QWIC II) Survey Report: Statistical Abstract.* Accra: Author.

———. 2007. *Pattern and Trends of Poverty in Ghana, 1991–2006.* Accra: Author.

———. 2008a. *Ghana Living Standards Survey: Report of the Fifth Round.* Accra: Author.

———. 2008b. *Ghana: Demographic and Health Survey.* Accra: Author.

Global Water Partnership Technical Committee. 2003. *Poverty Reduction and Integrated Water Resource Management.* Stockholm: Global Water Partnership.

National Research Council. 1989. *Recommended Dietary Allowances, 10th Edition.* Washington, DC: National Academy Press.

Ofori-Atta, J. 1978. Income Distribution in Ghana: A Study of Rural Development Strategies. *Ghana Social Science Journal* 5 (1): 1–25.

Osumanu, I.K. 2007. Household Environmental and Behavioural Determinants of Childhood Morbidity in the Tamale Metropolitan Area (TMA), Ghana. *Danish Journal of Geography* 107 (1): 59–68.

———. 2008. Private Sector Participation in Urban Water and Sanitation Provision in Ghana: Experiences from the Tamale Metropolitan Area (TMA). *Environmental Management* 42: 102–110.

———. 2009. Urbanization Challenges in Africa: Creating Productive Cities under Globalization. In *Urban Planning in the 21st Century,* edited by D.S. Graber and K.A. Birmingham. New York: Nova Science Publishers.

Osumanu, I.K., and L. Abdul-Rahim. 2008. Enhancing Community-Driven Initiatives in Urban Water Supply Ghana. Proceedings of the 33rd WEDC International Conference, on Access to Sanitation and Safe Water: Global Partnerships and Local Actions. Accra.

Sen, A.K. 1976. Poverty: An Ordinal Approach to Measurement. *Econometrica* 44: 24–38.

Songsore, J. 1992. The ERP/Structural Adjustment: Their Likely Impact on the 'Distant' Rural Poor in Northern Ghana. In *Planning for African Growth and Development:Some Current Issues,* edited by E. Aryeetey. Accra: Ghana Universities Press.

———. 2003a. The Urban Housing Crisis in Ghana: Capital, the State versus the People. *Ghana Social Science Journal (New Series)* 2: 1–31.

———. 2003b. *Towards a better Understanding of Urban Change: Urbanization, National Development and Inequality in Ghana.* Accra: Ghana Universities Press.

Songsore, J and G. McGranahan. (1993.) Environment, Wealth and Health: Towards an Analysis of Intra-Urban Differentials within the Greater Accra Metropolitan Area, Ghana. *Environment and Urbanization* 5(2): 10–34.

United Nations Population Division. 2005. *World Urbanization Prospects: The 2005 Revision.* New York: Author.

Whittington, D., D. T. Lauria, K. Choe, J. Hughes, and V. Sharna. 1993. Household demand for improved sanitation services in Camas, Ghana: A contingent valuation study. *Water Resources Research* 29(6): 1539–1560.

World Health Organization (WHO)/UNICEF. 2010. Progress on Sanitation and Drinking Water. 2010 Update. WHO/UNICEF Joint Monitoring Programme for Water Supply and Sanitation. Geneva and New York: WHO/UNICEF.

World Bank. 2003. *Ghana at a Glance.* Washington, DC: World Bank.

———. 2004. *World Development Indicators 2004.* Washington, DC: World Bank.

Zheng, B. 1998. Statistical Inferences for Poverty Measures with Relative Poverty Lines. *Journal of Econometrics* 101 (2): 337–356.

Chapter 9

THE RISK OF WATER
Dengue Prevention and Control in Urban Cambodia

Sarah C. Smith

∞

Dengue is the most widespread mosquito-borne virus in the world, with approximately two billion people living in areas of transmission. Dengue is endemic in all provinces in Cambodia, though periodic epidemics do occur when a new serotype is introduced. In this chapter, I will discuss the findings of an ethnographic study of dengue in an urban Cambodian setting that I will call "Boeng," to protect the location and identity of my interlocutors. Specifically, I will explore the human relationship with water in Boeng as it relates to dengue vector breeding and control. I begin with a review of the biomedical interpretation of risk, as compared to the cultural approach taken by Mary Douglas and Aaron Wildavsky (1982). I then discuss local practices around water collection and storage, given the area's exclusion from a reticulated water system. Finally, I examine how the government intervention to restrict mosquito growth in water is incompatible with local understanding of water quality.

This chapter approaches these issues from a critical-interpretive medical anthropology perspective. Margaret Lock and Nancy Scheper-Hughes explain, "the task of a critical-interpretive medical anthropologist is, first, to describe the culturally constructed variety of metaphorical conceptions (conscious and unconscious) about the body and associated narratives and then to show the social, political, and individual uses to which these conceptions are applied" (Lock and Scheper-Hughes 1996: 44). In this context, the narratives of interest are those surrounding water and its link to the transmission of diseases, with particular focus on dengue.

Through examination of the forms of water that constitute risk to the individual body, discord between the individual and the body politic is illuminated.

Dengue and its vector

The mild form of dengue infection, dengue fever, may be asymptomatic. The probability of developing more-severe forms, such as dengue hemorrhagic fever (DHF) and dengue shock syndrome (DSS), is dependent on the virulence of the infection (Burke et al. 1988: 172–180). Approximately 33 percent of DHF patients suffer a progression of disease to DSS. DSS has a case fatality rate of between 12 percent and 44 percent without intervention (Rigau-Pérez et al. 1994: 7–19). With appropriate management, the case fatality rate of DSS can be reduced to less than 1 percent (Monath 1994: 2395–2400). A specific treatment for dengue fever does not exist. Management of symptoms focuses on fluid replacement (Burke et al. 1988: 172–180). and fever reduction with paracetamol (Singhi, Kissoon, and Bansal 2007: 522–535). Approximately one-third of DHF cases require hospitalization (Gubler and Meltzer 1999: 35–70). Infection from one of the four dengue serotypes confers lifelong protection from that serotype and limited protection from the other three serotypes. A short time after infection from one serotype, however, infection from other serotypes is possible (Ligon 2004: 60–65). Immunity is also conferred to young infants (generally birth to age six months (Van Damme et al. 2004: 273–280) through acquired maternal antibodies, though as the infant grows older, these maternal antibodies break down, reducing their protective quality (Pengsaa et al. 2006: 1570–1576). While the probability of developing DHF depends on the virulence of the infecting serotype (Gubler 2002: 100–3), secondary dengue infection, infection with a serotype different from that of the first infection, is a main risk factor for DHF or DSS (World Health Organization [WHO] 1997).

Dengue has been termed an emerging disease. Emerging infections are those that "have newly appeared in the population, or have existed but are rapidly increasing in incidence or geographic range" (Morse 1995: 7). The emergence of dengue has been attributed to environmental changes, rapid urbanization, and population movement.

In Southeast Asia, the dengue virus is primarily transmitted by the female *Aedes aegypti* mosquito (Halstead 1988: 476–481). The *Aedes aegypti* mosquito is small with black and white stripes. The vector most actively feeds in the two to three hours following daybreak and in the few hours before dark, though it is capable of feeding throughout the day when it is inside or on overcast days. An infected female *Aedes aegypti* may easily transmit the dengue virus to several susceptible people: it is common for one mosquito to feed multiple times during a single blood meal, making the vector particularly effective in inciting epidemics (Gubler 1998:

480–496; Gubler and Meltzer 1999: 35–70). The *Aedes aegypti* mosquito is highly domesticated, breeds in clean water containers in and around the house, and thrives in urban, peri-urban, and rural areas equally (Kendall et al. 1991: 257–268). It prefers laying its eggs along the walls of human-made containers filled with clean water (Service 1996), such as those used for storing water.

Dengue in Cambodia

During the 1980s and 1990s, dengue epidemics in Cambodia had a case fatality rate of between 3.6 percent and 15 percent. In the 1995 outbreak, 10,208 cases and 424 deaths were reported nationwide. A 1998 outbreak, which predominantly affected Phnom Penh and nearby Kandal Province, resulted in 16,260 reported cases and 474 reported deaths (Socheat et al. 2004: 8–13; Suaya et al. 2007: 1026–1036). The year 2007 yielded a devastating epidemic in twenty-two of the country's twenty-four provinces. The epidemic caused 39,851 hospitalizations and 407 deaths. The greatest number of cases was reported between April and June of that year, with an average of 6,310 cases per month. The vast majority of cases during this outbreak were among children under age fifteen (97 percent). Siem Reap Province was the most greatly affected (National Dengue Control Program [NDCP] 2007). In 2008 and 2009, the annual number of reported cases returned to previous, more normal, levels of 9,544 and 11,699, respectively, yet Siem Reap continued to have a notably higher incidence than other provinces. In 2008 and 2009, the dengue incidence in Siem Reap Province was 225 per 100,000 and 210 per 100,000, respectively. The three provinces with the next highest incidence rate were Kandal, Phnom Penh, and O. Meanchey, with rates of 175 per 100,000, 175 per 100,000, and 160 per 100,000, respectively (NDCP 2010). Within Siem Reap Province, the District of Siem Reap, which included the research area for this study, the incidence rate in the first ten months of 2008 was 285 per 100,000, as compared to Varin District with the next highest rate of 195 per 100,000. The district with the lowest incidence rate was Banteay Srei at 45 per 100,000 for the same period (NDCP 2008).

According to the 2005 Cambodian Demographic and Health Survey (CDHS), 10.8 percent of postneonatal deaths among infants and children born since January 2002 were attributed, by the mother, to dengue. By way of comparison, 1.6 percent and 2.5 percent of deaths were attributed, by the mother, to malaria and typhoid fever, respectively. Of those deaths ascribed to dengue, 76.1 percent were reportedly diagnosed by a health worker (National Institute of Public Health, National Institute of Statistics, and Orc Macro 2005).

Given the dengue mosquitoes' breeding site preference, water storage containers have been widely recognized as a contributing factor to the spread of dengue by both Cambodian Ministry of Health staff and other researchers (Khun and

Manderson 2007: 139–146; Setha, Chantha, and Socheat 2007: 261–268; Socheat et al. 2004: 8–13). Cambodia's Policy Guidelines for National Dengue and Dengue Hemorrhagic Fever Prevention and Control recognize that vector control is challenged by a geographically limited water system, leaving households with no option other than to store large quantities of water. Under these conditions temephos larviciding of cement tanks and large water jars has emerged as the only dengue prevention and control program in the country (NDCP n.d.). Temephos is an organophosphate deposited in stored water to prevent the development of mosquito larvae. The temephos distribution program was designed to deliver the larvicide annually between May and June, the start of the rainy season, and again between August and October. However, due to limited resources, Cambodia's NDCP selects areas at risk of an epidemic to receive the intervention (Khun and Manderson 2007: 139–146). A system such as this is reliant on accurate and timely surveillance data, but the Cambodian reporting system is impaired by the exclusion of private practice data, delays in reporting, and otherwise incomplete data (Khun 2005: 254). In spite of Siem Reap Province having the highest dengue incidence of any Cambodian province in 2007, 2008, and 2009, none of my informants could recall the last time they had received temephos. During my year of fieldwork from 2008 to 2009, I also observed no temephos distribution. An alternative behavior promoted by the NDCP was the consistent use of tightly fitting covers for water storage containers to limit mosquito access.

Two Approaches to Risk

The concept of risk has been interpreted in varying ways depending on the discipline. For the purposes of this chapter, I will discuss risk from the biomedical and cultural perspectives. Lupton (1999) provides an eloquent description of what she calls the technicoscientific approach to risk, which persists in biomedicine and public health.

For exponents of the technicoscientific perspective, which emerged from and is expressed in such disciplines as science, engineering, psychology, economics, medicine, and epidemiology, risk is largely treated as a taken-for-granted objective phenomenon. The focus of research on risk in these fields is the identification of risks, mapping their causal factors, building predictive models of risk relations and people's responses to various types of risk, and proposing ways of limiting the effects of risk. These inquiries are undertaken adopting a rationalistic approach that assumes that expert scientific measurement and calculation is the most appropriate standpoint from which to proceed (Lupton 1999).

As public health is centered on this technicoscientific approach, public health practitioners typically view risk as "an estimate of probability or likelihood of occurrence based on comparisons" (Trostle 2005: 152). Risk "is the effect of a combination of abstract *factors* which render more or less probable the occurrence of undesirable modes of behaviour" (emphasis in original; Castel 1991: 287). Risk and its contributing factors, in this view, are quantifiably measureable and are expressed in equations such as relative risk and attributable risk.

Epidemiology considers itself "the 'basic science' of prevention," collecting data to "justify a prevention effort" (Gordis 2000: 289). Within the field of epidemiology, there are two approaches to risk: relative and attributable. Relative risk calculates "how many times more likely it is that someone who is exposed to something will develop a certain disease or experience a particular health outcome than (or *relative* to) someone who is not exposed" (emphasis in original; Webb, Bain, and Pirozzo 2005: 129). While this measure can support the association between exposure and health outcome, it does not provide insight into the quantity of a particular health outcome within a given population. Attributable risk allows the epidemiologist to calculate "how much extra disease is actually occurring in the *exposed* group as a result of the exposure" (Webb, Bain, and Pirozzo: 136, original emphasis). Attributable risk also expresses the amount of disease that could be prevented if exposure was also prevented. These definitions of risk are based on biomedical interpretations of health and disease and carry with them biomedically oriented normative connotations.

Individuals and communities that engage in behavior considered risky by biomedical standards are thought to do so in part because of inadequate knowledge. Public health logic contends that when those inadequacies are corrected, the resulting behavior will also be corrected (Douglas and Wildavsky 1982). The objective of health promotion is to influence local epistemology of risk, and thus health and illness, bringing it more in line with biomedical epistemology. An element of health promotion is the "emphasis on self-management of risk and self-care" (Petersen 1997: 197), a liberal push for individuals to assert more agency over their health-related behaviors. Because individuals may be viewed by experts as irrationally choosing actions to mitigate risk, perhaps this established understanding of what constitutes rational decision making needs to be revisited (Douglas and Wildavsky 1982).

Douglas and Wildavsky (1982: 30) offer a strong critique of biomedicine's "modern" and "objective" "selection of dangers." The authors argue, "Anyone who claims to know the right priority among dangers to be avoided and who also pretends that the priority has no basis in moral judgments is making two backward steps toward premodernism" (Douglas and Wildavsky 1982: 30). From a cultural perspective, theories, including those from the sciences, evolve, which inherently means that no single theory can occupy an authoritative position. The privilege that biomedicine has been afforded to discern and prioritize risks is therefore artificial. Additionally, biomedicine's insistence that it is distinct from morality is, from a cultural perspective, an indication of the limitation and lack of self-reflexivity in the biomedical approach.

Thus, in contrast to the objectively defined notion of risk that pervades in biomedicine and public health, the cultural understanding of risk centers around a locally defined knowable threat (Douglas 1992; Ewald 1991). "Risk for the general public is a synonym of menace and danger" (Trostle 2005: 152). Local knowledge of risk is socioculturally determined based on traditional knowledge, history, experiences, and, in the Cambodian context, the collective trauma of thirty years of civil war, all of which are transmitted culturally. Determination of risk is therefore done "less as individuals and more as social beings" (Douglas and Wildavsky 1982: 80).

Risks, then, are categorized based on local understanding. Ethnomedical systems, including biomedicine, utilize different criteria for ranking danger and risk. The action of categorizing creates a risk hierarchy that "may be a productive way to think about how local populations respond to health interventions in contexts where multiple diseases and other health concerns (such as violence, insecurity, stigma, and social risk) coexist" (Nichter 2008: 60). These hierarchies of risk foster notions of risk acceptability in which certain types of activities or behaviors are deemed allowable because the risks they pose are tolerable. Risk, from this perspective, is "a fluctuating, socially seismic field in which definitions of danger, harm, safety, and blame are constantly shifting" (Oaks and Harthorn 2003: 3–11: 4).

According to Mary Douglas, understandings of risk are based on notions of purity and the transgression of pollution occurring across culturally determined boundaries (Douglas 1992: 323). The concept of pollution, either in the literal or metaphorical sense, "implies some harmful interference with natural processes" (Douglas and Wildavsky 1982: 36); hence, pollution presents a risk. This cultural approach to risk reveals different categories of risk based on distinct knowledge types and sociocultural experiences, that of the "experts" and that of the "laypeople." In the context of public health, laypeople are required to translate the population-based risks espoused by public health campaigns and biomedical professionals, combined with local knowledge and experience, into individual risks, which then aid them in determining behavior (Sarangi and Candlin 2003: 115–124). These differing perceptions of risk present conflict between public health officials and communities, particularly in the areas of health education and promotion. As Douglas and Wildavsky write, "Since there is no single correct conception of risk, there is no way to get everyone else to accept 'it'" (Douglas and Wildavsky 1982: 4). It is these "risk disputes [that] express the points of tension and the values conflicts in a society" (Nelkin 2003: viii).

Ethnographic Research in Boeng

I carried out twelve months (2008–09) of ethnographic fieldwork in Boeng, an urban area within metropolitan Siem Reap. I chose Siem Reap as the study area because of the high incidence of dengue in the area. I made the decision to conduct research in Boeng in conjunction with provincial authorities who believed it represented a broad spectrum of socioeconomic groups. The political chief of Boeng and the surrounding environs resided within the study area, and was amenable to my presence. Additionally, a *wat* (Buddhist temple and monastery), secondary school, and foreign research institute were located within Boeng. Within the study area, there were approximately four hundred households, though this number fluctuated as individuals and families moved in and out of Boeng.

Siem Reap has had an integral place in Cambodia's history. Angkor Wat, the heart of the Angkorean Empire that reigned from the year 802 to the mid-fifteenth century, was located within Siem Reap Province and approximately six kilometers north of the site for this study. The architectural wonder that is Angkor Wat has since been designated a World Heritage Site. Approximately one million tourists travel to the area every year to view the magnificent ruins and stay in the tourist section of Siem Reap situated just south of Boeng. As a result, residential areas like Boeng have grown dramatically, with an influx of rural migrants seeking work in the tourism industry. The history of Angkor remains present in collective consciousness of the residents of Boeng, not only for the strength and power of the empire past, but also for the economic benefits that have come as a result of tourism. Under the Khmer Rouge who ruled from 1975 to 1979, the entire urban Siem Reap area, including Boeng, was drained of its residents as the regime sought a national return to the agrarian lifestyle. The area was one of the last strongholds of the Khmer Rouge following the Vietnamese invasion in 1979. Subsequent repopulation of Boeng was done predominantly by "new dwellers. There were not so many old dwellers because they were killed in the Pol Pot regime," according to Yeay Nuon, an elderly nun who was raised in Boeng and left only briefly toward the end of the French Protectorate and during the Khmer Rouge era.

During the time I conducted my fieldwork, residents of Boeng were predominantly ethnically Khmer, with a handful of European residents. Khmer, the national language of Cambodia, was the primary language spoken; older residents educated under the French protectorate, however, often knew and spoke French with me, while younger residents sought to practice their English language skills so they could secure a job in tourism. Given the large tourism industry, the economy in the greater Siem Reap area was based on the U.S. dollar. While the official exchange rate during my fieldwork period fluctuated between 4,029 and 4,121 Cambodian *riel* to $1.00, the rate of 4,000 *riel* to $1.00 was used in daily transactions.

By residing in the area for an extended period, I was able to observe and collect firsthand information on water access and storage across the wet and dry seasons. Much of the data analyzed in this chapter were gathered through participant-observation conducted in public areas, including the Buddhist *wat* and the local secondary school, within residential enclaves, and in individual households. I recorded information about a broad range of activities, but focused especially on those that would provide insight into the human relationship with the built and natural environments and water. These behaviors were especially important because they directly impacted mosquito populations. Local discourse around medical treatments for fever and vector-borne diseases was also investigated; doing so provided insight into ideas and approaches to dengue.

As a complement to participant-observation, I conducted two focus group discussions during the early months of fieldwork to provide preliminary information and identify village residents for interviews and participant-observation. Participants were women between the ages of twenty-five and eighty who lived in the village, all of whom took care of either their own children or their grandchildren. These women all fell into the low or middle socioeconomic class; women in the higher class were invited to participate but did not attend. Participants were recruited using a snowball method whereby I invited a handful of women and requested they bring family, friends, or neighbors who would be willing to participate.

Because I am not a native Khmer speaker, I anticipated challenges in collecting data. On arrival in the field, I began intensive one-on-one language training and spoke English as infrequently as possible. As a result, I quickly became linguistically competent. I also hired a part-time research assistant, a young woman who also worked in a guesthouse and spoke English. She functioned as an interpreter and Khmer language teacher, and provided continual explanation of concepts and terms. Toward the end of my fieldwork, my language skills had developed to the point where I was able to conduct interviews independently; I continued to bring my research assistant with me to ensure the accuracy of my understanding, however.

I conducted direct observation repeatedly in six households selected from the snowball sample, and conducted additional in-depth interviews with prominent villagers, including the village chief, assistant director of the local secondary school, director of the foreign research institute, former local government employees, and the chief monk. I selected other key informants because of a family history with dengue. I attempted to select households across socioeconomic classes, with the exception of the highest socioeconomic class in the study area. Families in the highest socioeconomic class lived in houses surrounded by high gates and sturdy metal fences. The men in these families often worked with the government at a national or provincial level and were rarely home. The women of the households, when I could gain access through the walls and fences, were reluctant to speak with me. My research assistant speculated that their unwillingness was due in part to their husbands' positions in government and in part to their financial status. She indicated that many times foreigners who visited Boeng offered financial or other forms of assistance to villagers. Had the families in the higher socioeconomic groups been seen with me, rumors may have started about their potential financial hardship. Thus, I was excluded from this group due their desire to maintain their reputations.

We conducted interviews in the area immediately outside the houses, an area used for socializing among other activities. Women and children congregated at tables, benches, bamboo beds, or small three-sided wooden huts when breaks from household chores presented themselves or during the hottest part of the day when work was not done. As a result, individual interviews evolved into group conversations in which family members, friends, or neighbors were brought into the discussion by the principal interviewee. While this occurrence altered the dynamic of the interviews, it provided a depth of information and led the conversation down paths that, as a foreign researcher, I would not have been able to achieve. While at times I felt the interviews to have been appropriated by the interviewees, on later reflection I found that I had been exposed to topics and insights I would have not otherwise considered.

The sample for this study was heavily skewed toward female householders. This is in part because of accessibility: women tended to be in the home more often than their male counterparts. However, cultural norms also limit interaction between individuals of opposite genders, especially when the female is unmarried, as I was. Finally, women bore a disproportionate amount of responsibility when it came to caring for sick children and maintaining the house environment, and thus minimizing potential *Aedes aegypti* mosquito breeding sites. I purposively sought out and included male key informants, however, because of the positions of authority or significance they occupied.

Water Collection and Storage

Garn and colleagues (Garn, Isham, and Kähkönen 2002: 36) provide a comprehensive review of public versus private water suppliers in Cambodia; however, in Boeng, piped water was provided by neither sector. There was speculation, especially among the foreign residents of both Boeng and other areas of Siem Reap, that the lack of a water system in Boeng was the result of politics. According to these informants, Boeng had historically supported the Sam Rainsy Party, one of the largest opposition political parties. In 2003 and 2004, the Japanese International Cooperation Agency funded and provided technical assistance for a water system project in Siem Reap. Foreign residents of the area indicated that local and regional politicians were responsible for selecting areas included in the water system, and Boeng was systematically excluded because of its political leanings.

Cambodian residents of Boeng, however, never made mention of the lack of reticulated water or the fact that the area immediately to the south of Boeng, separated only by a single-lane road, did have such a water system. This was perhaps because political violence persisted and there was a very real fear of speaking ill of the government. Additionally, interaction with government officials frequently involved demands for payment, a financial burden few of my interlocutors could afford. Thus, either requesting or demanding the provision of such a service bore the potential of retribution. Moreover, there was an element of pride in building one's own well, especially if a family had sufficient resources to include an electric motor that would pump the water to a *vasang dteuk* (a large metal water-storage container; see below for more detailed description), allowing for access to water inside the house.

Given the lack of reticulated water, residents collected their water from four main sources: wells, the river, the rain, and purchasing purified water. Wells were the most common type of water source in Boeng. Individuals constructed wells at their own expense, though not every house had its own well. Houses within enclaves and residents in blocks of rented rooms often shared access to a single well. In these instances, the well was owned by either the owner of one of the houses in the enclave or the landlord, who may or may not have resided locally. Digging a well could cost between $40 and $60 and up to $200 if one wanted to motorize the pump and connect it to a bathroom within the house. Well water was used for all domestic and business activities, such as bathing, food preparation and cooking, washing clothes and dishes, gardening, construction and other building purposes, and religious activities. It also was used for drinking, both with or without boiling first.

Along the riverbank were a line of houses constructed out of wooden planks, corrugated metal, and thatch. The front portions of these houses were built on top of the bank, right up to the pavement of the road, while the backs jutted out on piles over the Siem Reap River. During the dry season, when the river level was low, there was a meter or so drop from the floor of the house to the water below. But during the rainy season, the water level rose to the height of the houses, sometimes creeping into the residences themselves. Longer-term residents of Boeng spoke of the Siem Reap River with nostalgia. Those who returned or moved to the village in 1979 after the fall of the Khmer Rouge regime recalled that the river ran clear and fish were plentiful. That changed in the following two decades, however, when, according to residents, the area received electricity and there was an influx of rural-to-urban migrants.

The chief of Boeng attributed the dirtiness of the river to the poorer residents who constructed wooden houses along the banks of the river: "Yet, my concern is that there is slum along the river which the people don't understand much about how to live in clean environment. Some people lived there before I became village chief. Some people throw the duck's hair [feathers] into the river." In mentioning those who threw the duck feathers into the river, the chief was referring to a particular family who ran a small business from their house in which they killed, plucked, and sold ducks. The chief continued on to explain that the poorer families living along the river were responsible for the dirtying of the river:

> The people who live along the river … we have to see their living standard. Have you even visited there yet? Their house is built very close together. Some people throw

the hair of the chicken or duck into the river and they also excrete into the river. Therefore, the water is very dirty. We can understand their living by looking at the way that they live which is very dirty. They sleep there, they excrete there, and they throw the hair of the duck there.

At the time of this interview in November 2008, the chief indicated that the provincial government, in an attempt to clean the river and its banks, had planned to relocate these families to another village, at which time, she said, "this river will become the river like before." For the village chief, the poorer residents who lived closest to the river polluted the river. Only when these families were removed to another village would the pollution dissipate and purity be restored. The relocation was meant to happen in January 2009; by the completion of my fieldwork in June 2009, however, these families had not been moved.

Other interventions were carried out to clean the river, principally, constructing a net across the width of the river to prevent rubbish from floating downstream. Boeng and the neighboring areas were located upstream from the tourist area of Siem Reap. A net caught rubbish thrown into the river by both villages, giving the appearance of cleanliness to tourists downstream. There was no regular means of removing the rubbish from the net, and it collected and created an overpowering stench. Children were often seen walking across the rubbish that protruded above the water line, recovering discarded items that might prove useful in the future. In the tourist areas of Siem Reap, there were signs in English imploring people to keep the river clean and not throw their rubbish in the river. There were no such signs in Boeng, however, and the practice of throwing waste in the river persisted.

Children were often seen swimming in the river, jumping naked from the iron bridge that marked the southern end of Boeng. Families living along the river did occasionally wash their clothes in the water, but this was not common because they also had access to wells. Riverwater was never intentionally consumed because the water was viewed as polluted and consumption presented a risk to one's health. Interestingly, fish from the river were caught and consumed and thought to be safe. Some held the belief that wells located on or near the bank drew their water from the river itself. One young man who worked in a tourist restaurant observed that I ate my breakfast every morning at a riverside stall. He commented that I needed to be careful eating there every day as the water used to make the *borbor* (rice porridge) was from the river and could make me sick. The threat that riverwater presented to one's health was very real, given the amount of fecal matter and other waste that entered the river system; if ingested, these contaminants could cause diarrhea, typhoid, or other waterborne diseases.

Many villagers collected rainwater during the six-month-long rainy season and used it for drinking, food preparation and cooking, washing dishes, and cleaning feet and hands as needed. When used for drinking, rainwater was not always boiled; it depended on the physical appearance of the water. (See below for de-

tailed description of water collection and storage practices and local conceptualizations of water quality.) Most residents were adept at reading the sky and
other signs to know whether it would rain. If the clouds "looked like the skin of
a dragon" or if it was windy, regardless of how dark the clouds were, there would
be no rain. According to my informants, the winds also dictated the start and the
end of the rainy season. During the rainy season, the winds came from the west;
and when the winds began coming from the north, everyone understood the rainy
season was over.

Localized flooding occurred in many areas during the rainy season as a result
of poor drainage throughout Boeng. These floodwaters would remain for weeks
or even months until there was a sufficiently long break in rainfall to allow for the
standing water to drain. Traditional houses were intentionally built on piles to keep
the living area dry during times of flooding, among other reasons. Unsealed roads
would often develop large holes as a result of the pounding rain and high traffic. It
was not uncommon to see male villagers filling in these holes with discarded building materials, such as broken bricks and cinder blocks. There was no municipal
service for addressing these roadwork needs, nor was there any expectation that
local government would assist in such repairs. Standing water and puddles were
often indicted as the source of the dengue mosquito by villagers. Given that flooding was a fact of life in Boeng, as with many other areas of Cambodia, villagers, in
their view, could not meaningfully affect the mosquito population: they perceived
the flooding to be out of their control and beyond their capacity to change. This
view of their limited efficacy can be extended to the NDCP or other agency activities around vector control. These interventions, however minimally implemented,
would not impact mosquito populations in any meaningful way because they did
not incorporate one of the more prominent vector breeding grounds, according to
local perceptions.

The campus of the foreign research institute did not flood during the rainy
season because they had the financial means with which to construct a system to
pump any standing water into the river. At a meeting held at the *wat* in December
2008, the chief of Boeng and the surrounding area reported to the small attend-
ing audience that the local government was looking into constructing a drainage
system similar to that of the foreign research institute and the area immediately to
the south of Boeng. The chief justified this project as being health-related, stating that the *dteuk k'mao* (black water) was not good for the health of the villagers.
Though she made no reference to mosquitoes specifically, in a separate interview
she told me that mosquitoes did come from dirty water in puddles and other
flooded areas.

As indicated by the chief, the potential health risks posed by persistent floodwaters extended beyond mosquito breeding. As the water sat on the surface of
unpaved roads, it grew a dark green, almost black algae skin locally known as *slai*.
In certain areas of Boeng, it was impossible to travel a path during the rainy sea-

son without being forced to traverse pockets of *slai*. It was believed that the algae caused sores on exposed skin, leaving round scars on the affected areas.

These floods were an annoyance for villagers, not only because they believed that they brought mosquitoes, but also because they presented challenges to mobility. Most families in Boeng owned a motorbike; during and after heavy rains, the water level could be so high that it caused motorbike engines to flood. Additionally, the mud became extremely slippery, making travel along unpaved streets hazardous. However, these floodwaters also provided a source of food, as water spinach and snails thrived in these environments; children were often seen foraging for such foods.

Finally, water for drinking could be purchased in either smaller, individual-sized bottles or in a larger twenty-liter plastic container. Smaller bottles were less frequently purchased for consumption within the house because they were more costly. They tended to be purchased and offered to monks or nuns during times of festivals, ceremonies, or other blessings. The offering of bottled water to religious figures not only was a means of outwardly demonstrating one's wealth and prosperity to one's peers, but also was intimately tied to the earning or making of merit. In these circumstances, the greater the generosity of the offering, the greater the amount of merit earned. This bottled water was also cleaner, more pure, than other forms of water and was therefore an appropriate offering to make when making merit, or cleansing oneself of ill deeds.[1]

The twenty-liter containers of water came in two types. The first was a white plastic bottle with a smaller mouth at the top for dispensing the water. The second was a transparent blue plastic bottle with a larger mouth at the top and a tap at the base for dispensing the water. In both cases, the purchaser paid an initial amount to buy the container ($4.00 for the white container and $6.00 for the blue container with the tap) and a much smaller amount to refill the container. The cost of refills varied depending on both the vendor and quality of the water as determined by taste and smell, but ranged from $0.25 to $1.00.

In Boeng, well water and rainwater were stored within and around the house throughout the year. There were many different types of water storage containers (table 9.1). The most prolific was the *pee-ung*, a large clay container often located immediately outside the house. During the rainy season a full *tong* (plastic bucket) or *pee-ung* was left near the entrance to the house with a *ptul* (small plastic bowl) floating in the water. This water was used to clean the mud off one's feet prior to entering the house. In Boeng, where only the larger road that ran along the river was paved, feet became covered in a red, sticky mud during the rainy season and needed cleaning before entering a house. Shoes, most frequently rubber sandals, were also washed but were never worn inside residences or religious buildings.

Many houses in Boeng had toilets within the house structure that were connected to septic tanks underground, but did not have running water inside the house from the nearby well. Thus, water was stored in the bathroom for both washing after urination and defecation and for flushing the toilet. Most often, a large

Table 9.1. Categorization of water containers in village Boeng

Name	Size (approx)	Use
ptul	0.5 liters	Small plastic bowls used to retrieve water from a *muntun, tong, kap,* or *pee-ung.* Not used for storing water.
muntun	2–10 liters	Large, round, plastic or metal bowls used for washing dishes or clothes, or bathing infants. May also hold fish purchased at the market prior to cooking. Not used for storing water, but may be strategically placed under house-eaves to collect rainwater.
tong	10–80 liters	Plastic buckets used for collecting and transporting water from well pump to a larger container, though may also be used to store water. May be placed strategically under house-eaves to collect rainwater or located within bathrooms for washing and flushing toilets.
kap	50 liters	Cylindrical cement jar with a dimpled bottom where water may pool if it is inverted.
pee-ung	200–500 liters	Cement pear-shaped water storage containers. May be filled with well water by hand or with a motor. May also be strategically placed under house-eaves to collect rainwater.
loo	—	Cement cylinders used to construct *ang stop dteuk* or septic tanks.
ang stop dteuk	500–1,000 liters	Cement water storage containers with two possible shapes. Cylindrical containers, fashioned from two *loo,* were located outside and had a valve for extracting the water. Water was pumped into the cylindrical containers. The rectangular containers were often built into the bathrooms to store water for personal hygiene and flushing the toilet, and were filled manually.
vasang dteuk	1,000–1,500 liters	Metal water storage drums located on the top of a residential or business structure or on scaffolding. Allowed water to be pumped directly into a bathroom or kitchen.

plastic *tong* with the capacity to store fifty to sixty liters of water could be found in bathrooms. These large buckets were filled either via a motor that pumped water from the well through a hose or by hand. In the first instance, where a motor was involved, men were primarily responsible for connecting the motor and managing the process. In the second instance, where water was manually pumped from the well into smaller twenty-liter buckets and transported by hand to the bathroom, women were primarily responsible, though adolescent boys were also called on to assist. The toilets in the secondary school had large concrete *ang stop dteuk* built into the structures. These were filled using a motor. It was the responsibility of the students to clean and fill the containers, though cleaning was rarely done.

This water stored in and around the house was accessed frequently throughout the day. Table 9.2. summarizes the activities within the house that utilized stored

Table 9.2. Regular household activities using stored water

Activity	*Purpose*	*Frequency*
Food preparation	Washing and cooking rice, washing and preparing meat and vegetables.	Three times per day (approx. 5:00–7:30 a.m., 10:30–11:30 a.m., 5:30–6: 30 p.m.).
Drinking water preparation	Boiling tea or water for consumption.	Throughout the day as needed.
Washing dishes and other cooking equipment	All dishes and other equipment were washed at the same time at ome point after the meal, though snot necessarily immediately.	Three times per day, following meals.
Bathing infants	Children aged approximately eighteen months and older were expected to bathe at the well with the older children and adults. Infants were bathed in bowls closer to or inside the house.	At least two, often three, times per day for a complete bath; as needed throughout the day for a rinse following defecation.
Washing hands and feet	Hands and feet were regularly rinsed because they got dirty during regular activities. This was especially relevant for feet, that often got muddy or dusty during travel on unsealed roads, particularly during the rainy season.	Frequently throughout the day.

water and the frequency with which this water was accessed throughout the day. As a result of this frequent access, effective use of jar covers was limited. Observation revealed that even in those houses where there were covers, they were either ill fitting or incompletely covered the container mouth, or were constructed from material, such as planks of wood, that would permit mosquitoes' access to the stored water. The objective of the cover was not to restrict mosquito breeding, but rather to limit the amount of detritus or other material from falling into the water, which would alter the water's quality.

Water quality

Water quality was locally determined by the senses, primarily vision and smell, though taste was also used in certain cases. Good-quality water was defined as clear. One focus group discussion participant explained that bad-quality water was "water from the well, especially from pump well. It has rust and we can't use it unless we boil it. We boil it, then we can drink it." The comment about rust referred not only to the poor quality of some wells, but also to the presence of red dirt in the water. Another villager said, "Bad water is the water that contains rubbish or waste." Well water often had a reddish tinge from the earth and as a result was viewed as inferior. This was not always the case, however, if the well was dug deep enough. In the event that water from a well was red, a piece of traditional and readily available cloth known as *krama* could be tied around the mouth of the pump to act as a filter. It was also not unheard of for a well that produced "dusty" water to be abandoned or used in limited capacity and another deeper well constructed that would produce clear water. Yet another young man and his wife found that the water from the well at their house was often dirty. Although they could boil it to bathe and drink, because they both worked in a hotel in the tourist area they usually took showers at the hotel. He explained that the water was fine for Cambodians to bathe in, but it would cause the skin of someone like me (a foreigner, who was not accustomed to the water) to become itchy. I never experienced any skin irritation as a result of the well water I used for bathing, however.

As noted earlier, temephos is an organophosphate larvicide deposited in stored water to prevent the development of mosquito larvae. Its presence does alter the appearance of the water, but biomedicine has determined that temephos has no health risks. As a result, NDCP viewed the population's nonadherence as an act of defiance, considering the people to be difficult to work with. This compound was locally called *t'nam* (medicine), though the implication was that the *t'nam* was for the stored water; it was also called by its brand name, Abate. One informant, a young woman with three children, was aware that the temephos needed to be left in stored water for three months, but resisted following this practice: "But we can't keep Abate in the water for the whole three months because it makes the water muddy. I need to clean the jars." She qualified clean water as clear water. When the temephos clouded the water, it was no longer desirable and needed to be discarded. Consequently, the distinct forms of knowledge between the local population and the proponents of the biomedical model resulted in discord. For the residents of Boeng, visual appearance of water was greatly valued over the control of the mosquito population.

Aside from the visual elements of water quality, odor played a key role in assessment of quality. Water that smelled, particularly of *bichayn* (monosodium glutamate [MSG], a chemical compound used frequently in Cambodian and other types of Asian cooking) was unappealing. Taste was also used to determine water quality, though only after the water passed visual and olfactory assessment and was more commonly restricted to purchased water. On one occasion, I refilled my twenty-liter container of purchased water from the most convenient and least-expensive vendor. Later that day, a young woman visited my house and I offered her a glass of water, which she readily accepted. After one sip, however, she put the glass down and asked where I purchased the water. When I told her, she commented that the water from the least-expensive vendors tasted like chemicals and was not good. She admonished me that I should have spent more than $0.25 to buy "good water." While the water was clear and did not smell, the taste betrayed its quality.

Discussion and Conclusion: The Risk of Water

In the context of dengue prevention and control, water presents a risk because it is necessary for vector breeding. In Boeng, political exclusion from the water system created a crisis of water in which households were required to maintain stored water, thus providing a location for the dengue vector to breed. Because stored water was accessed frequently throughout the day, the use of water storage container covers to restrict mosquito breeding was limited. This crisis was compounded by the conflict between local and biomedical knowledge types regarding what constituted water quality, and the chemical methods of mosquito control. Specifically, local knowledge viewed temephos, a larvicide used to control the dengue mosquito, as a contaminant; water treated in this manner was discarded, eliminating the intended benefits. In addition to storing water, the lack of a water system in Boeng led to many residents purchasing water for consumption, a financial expenditure that represented a burden to some.

The anthropological data presented in this chapter illustrate the divergence in perception of risk between the biomedical and ethnomedical perspectives of dengue, a conflict that exists not only in Cambodia and not solely with respect to dengue. The biomedical perspective emphasizes not storing water, or using a tightly fitting cover for storage containers, while also using temephos to prevent dengue vector breeding. The political exclusion from either a public or private water system required residents of Boeng to store water in and around the house. Yet, water stored in and around the house was indicted by ministry of health staff as the primary breeding site for the dengue vector. Local behavior and perception of what constituted good-quality water impaired the effective use of measures to prevent mosquito breeding. These differences in perspective have implications for health promotion. Given that biomedical and local understandings and prioritizations of risk may not correspond, as this chapter has attempted to illustrate, awareness and incorporation of local interpretations of risk into interventions could improve both programmatic outcomes and sustainability. This should not be limited to specific diseases or illnesses, but rather incorporate all locally acknowledged risks. The result would be appreciation on behalf of program implementers of the integrated nature of risks.

Notes

1. Other offerings included prepared food, uncooked rice, or soda, and the burning of incense and candles.

References

Burke, D.S., Nisalak, A., Johnson, D.E., and Scott, R.M. 1988. A Prospective Study of Dengue Infection in Bangkok. *American Journal of Tropical Medicine and Hygiene* 38: 172–180.
Castel, R. 1991. From Dangerousness to Risk. In *The Foucault Effect: Studies in Governmentality*, edited by G. Burchell, C. Gordon, and P. Miller. Chicago: University of Chicago Press.
Douglas, M. 1992. *Risk and Blame: Essays in Cultural Theory.* London: Routledge.
Douglas, M., and A. Wildavsky. 1982. *Risk and Culture: An Essay on the Selection of Technical and Environmental Dangers.* Los Angeles: University of California Press.
Ewald, F. 1991. Insurance and Risk. In *The Foucault Effect: Studies in Governmentality*, edited by G. Burchell, C. Gordon, and P. Miller. Chicago: University of Chicago Press.
Garn, M., J. Isham, and S. Kähkönen. 2002. Should We Bet on Private or Public Water Utilities in Cambodia? Evidence on Incentives and Performance from Seven Provincial Towns. Discussion Paper No. 02-19. Middlebury, VT: Middlebury College of Economics.
Gordis, L. 2000. *Epidemiology* (2nd ed.). Sydney: W.B. Saunders.
Gubler, D.J. 1998. Dengue and Dengue Hemorrhagic Fever. *Clinical Microbiology Reviews* 11: 480–496.
———. 2002. Epidemic Dengue/Dengue Hemorrhagic Fever as a Public Health, Social and Economic Problem in the 21st century. *TRENDS in Microbiology* 10: 100–103.
Gubler, D.J., and M.I. Meltzer. 1999. Impact of Dengue/Dengue Hemorrhagic Fever on the Developing World. *Advances in Virus Research* 53: 35–70.
Halstead, S. 1988. Pathogenesis of Dengue: Challenges to Molecular Biology. *Science* 239: 476–481.
Kendall, C., P. Hudelson, E. Leontsini, P.J. Winch, L.S. Lloyd, and F. Cruz. 1991. Urbanization, Dengue, and the Health Transition: Anthropological Contributions to International Health. *Medical Anthropology Quarterly* 5: 257–268.
Khun, S. 2005. Community Participation in the Prevention and Control of Dengue Fever in Cambodia. Ph.D. dissertation, School of Population Health, University of Melbourne, Melbourne.
Khun, S., and L.H. Manderson. 2007. Abate Distribution and Dengue Control in Rural Cambodia. *Acta Tropica* 101: 139–146.
Ligon, B.L. 2004. Dengue Fever and Dengue Hemorrhagic Fever: A Review of the History, Transmission, Treatment, and Prevention. *Seminars in Pediatric Infectious Diseases* 60–65.
Lock, M., and N. Scheper-Hughes. 1996. A Critical-Interpretive Approach in Medical Anthropology: Rituals and Routines of Discipline and Dissent. In *Medical Anthropology: Contemporary Theory and Method Revised Edition*, edited by C.F. Sargent and T.M. Johnson. London: Praeger.
Lupton, D. 1999. *Risk.* London: Routledge.
Monath, T.P. 1994. Dengue: The Risk to Developed and Developing Countries. *Proceedings of the National Academy of Sciences of the United States of America* 91: 2395–2400.
Morse, S.S. 1995. Factors in the Emergence of Infectious Diseases. *Emerging Infectious Diseases* 1(1): 7–15.
National Dengue Control Program (NDCP). 2007. *National Dengue Control Program Annual Report 2007.* Phnom Penh: Author.

————. 2008. *NDCP Dengue Cases in Siem Reap Province.* Phnom Penh: Author.

————. 2010. *Comparison of Dengue Incidence Rate by Province in Cambodia 2008 vs 2009 during 52 Weeks.* Phnom Penh: Cambodian Ministry of Health.

————. (n.d.). *Policy Guidelines for National Dengue and Dengue Haemorrhagic Fever Prevention and Control.* Phnom Penh: Author.

National Institute of Public Health, National Institute of Statistics, and Orc Macro. 2005. *Cambodia Demographic and Health Survey, 2005.* Phnom Penh: ORC Macro.

Nelkin, D. 2003. Foreword: The Social Meaning of Risk. In *Risk, Culture, and Health Inequality: Shifting Perceptions of Danger and Blame,* edited by B.H. Harthorn, and L. Oaks. London: Praeger.

Nichter, M. 2008. *Global Health: Why Cultural Perceptions, Social Representations, and Biopolitics Matter.* Tucson: University of Arizona Press.

Oaks, L., and B.H. Harthorn. 2003. Introduction: Health and the Social and Cultural Construction of Risk. In *Risk, Culture, and Health Inequality: Shifting Perceptions of Danger and Blame,* edited by B.H. Harthorn and L. Oaks. London: Praeger.

Pengsaa, K., C. Luxemburger, A. Sabchareon, K. Limkittikul, S. Yoksan, L. Chambonneau, U. Chaovarind, C. Sirivichayakul, K. Lapphra, P. Chanthavanich, and J. Lang. 2006. Dengue Virus Infections in the First 2 Years of Life and the Kinetics of Transplacentally Transferred Dengue Neutralizing Antibodies in Thai Children. *Journal of Infectious Diseases* 194: 1570–1576.

Petersen, A. 1997. Risk, Governance and the New Public Health. In *Foucault, Health and Medicine,* edited by A. Petersen and R. Bunton. London: Routledge.

Rigau-Pérez, J.G., D.J. Gubler, A. Vorndam, and G.G. Clark. 1994. Dengue Surveillance: United States, 1986–1992. *MMWR CDC Surveill Summ* 43: 7–19.

Sarangi, S. and C. Candlin. 2003. Categorization and Explanation of Risk: A Discourse Analytic Perspective. *Health, Risk and Society* 5: 115–124.

Service, M. 1996. *Medical Entomology for Students.* London: Chapman and Hill.

Setha, T., N. Chantha, and D. Socheat. 2007. Efficacy of Bacillus Thuringiensis Israelensis, VectoBac WG and DT, Formulations against Dengue Mosquito Vectors in Cement Potable Water Jars in Cambodia. *Southeast Asian Journal of Tropical Medicine and Public Health* 38: 261–268.

Singhi, S., N. Kissoon, and A. Bansal. 2007. Dengue and Dengue Hemorrhagic Fever: Management Issues in an Intensive Care Unit. *Journal of Pediatrics* 83: 522–535.

Socheat, D., N. Chanta, T. Setha, S. Hoyer, C.M. Seng, and M.B. Nathan. 2004. The Development and Testing of Water Storage Jar Covers in Cambodia. *Dengue Bulletin* 28: 8–13.

Suaya, J.A., D.S. Shepard, M.S. Chang, M. Caram, S. Hoyer, D. Socheat, N. Chantha, and M.B. Nathan. 2007. Cost-Effectiveness of Annual Targeted Larviciding Campaigns in Cambodia against the Dengue Vector *Aedes aegypti. Tropical Medicine and International Health* 12: 1026–36.

Trostle, J. 2005. *Epidemiology and Culture.* New York: Cambridge University Press.

Van Damme, W., L. Van Leemput, I. Por, W. Hardeman, and B. Messen. 2004. Out-of-Pocket Health Expenditure and Debt in Poor Households: Evidence from Cambodia. *Tropical Medicine and International Health* 9: 273–280.

Webb, P., C. Bain, and S. Pirozzo. 2005. *Essential Epidemiology: An Introduction for Students and Health Professionals.* New York: Cambridge University Press.

World Health Organization (WHO).1997. Dengue Haemorrhagic Fever: Diagnosis, Treatment, Prevention, and Control. Geneva: Author.

Chapter 10

THE WATER CRISIS IN IRELAND
The Sociopolitical Contexts of Risk in Contemporary Society

Liam Leonard

∞

Introduction

Ireland, with its unspoiled scenery, clean water, and fresh air, has traditionally been known as the Emerald Isle. Recent decades of rapid growth and unregulated development have jeopardized Ireland's natural resources, however. The Celtic Tiger phase of accelerated growth that occurred in the Republic of Ireland reached its peak in 2007. Over the previous decade, there had been an increase in crises related to infrastructural development. While levels of personal wealth increased dramatically in the years leading up to 2007, the accompanying growth in consumption and waste also created a series of crisis and subsequent protests about sewage treatment plants, landfills or planned incinerators, gas pipelines, roads, electricity pylons, and mobile phone masts. As a member state of the European Union (EU), Ireland had begun to implement an environmental modernization policy regime. Nonetheless, grassroots campaigns emerged from communities concerned about the health risks associated with accelerated development which threatened natural resources such as water supplies (Leonard 2005, 2006).

This chapter will be divided into various sections, reflecting the extent of Ireland's water crisis. The island of Ireland is divided into two jurisdictions: Northern Ireland, which is part of the United Kingdom, and the Republic of Ireland, which has its capital in the city of Dublin. The chapter will present a case study on the water crisis that affected the western city of Galway in the Republic of Ireland in 2007. It will also briefly look at the problems of the cold snap that caused frozen municipal pipes in Northern Ireland in December 2010. At various points since the 1990s, both human-made crises emerging from poor planning and rapid development, and flooding or freezing temperatures that can be linked to climatic change, have impacted water supplies in all parts of Ireland. The relevant issues surrounding these cases will be discussed here, along with a discussion of the environmental regulatory framework surrounding water provision in the Republic of Ireland.

Methods

The subject matter in this chapter has emanated from a combination of research processes. Having developed an early understanding of the significance of ecological modernization as part of my masters research, my initial doctoral fieldwork was based on the mobilization processes surrounding community groups that resisted incinerators on health grounds (Leonard 2005, 2006). Through participant-observation, I was able to establish a wider understanding of the resource mobilization and framing processes of those campaigns that emerged in the period associated with accelerated growth in Ireland. Such growth threatened the environment in rural and suburban Ireland, and was resisted by regional social movements. I followed up this work with a research-based event analysis of the history of the environmental movement in Ireland (Leonard 2006, 2008). This was the first study of this kind on the Irish situation that located disparate elements of the Irish environmental movement within a wider process of regional sociopolitical community resistance to elements of central state policies. Subsequently, I covered the water crisis in Ireland through an analysis of media coverage of key events, based in part on my experience working on such issues as a journalist. This gave me access to key participants on the political scene, and I was able to chart their progression from activists through to electoral candidates, and then city councilors and mayors, which provided me with a closer understanding of the problems that emerged as part of these transformations.

Water supply in Ireland: a contested issue

For reasons I outline in the following section, water modernization initiatives in Ireland often generate intense public debate. Domestic water use charges, for instance, have been the subject of intense debate since first approved in 2009. Plans to pipe water from Ireland's largest river, the Shannon, to alleviate Dublin's ongoing shortages, have been resisted since 2007 by conservationists and farmers who fear long term damage to the Shannon's water quality as the result of the government's plan to extract over 300 million liters a day from Lough Ree, a special conservation area. According to conservationists, the planned 170 kilometer pipeline would lead to dramatically reduced water levels and would damage vital tourist interests. One report by environmental scientists drew comparisons with the damage caused by water extraction at the Dead Sea in Israel:

> Many experts believe that a proposed rescue plan for the Dead Sea—pumping water from the Gulf of Aqaba in the Red Sea to replenish it—will only make matters worse. The rate of abstraction in the Red Sea project could mean that 1 percent of the sea's volume would be abstracted every 1,200 years. In the case of Lough Ree, the abstraction rate proposed by Dublin City Council would mean that 100 percent of the volume of the lake would be removed every five years. This can only lead to damaging and irreversible consequences at an alarmingly fast rate. (Curtin 2007)

Ecological modernization and Irish environmental governance

After decades of economic stagnation, the Irish state embraced the dual agendas of neoliberalism and neocorporatist social partnership to create the economic basis for the multinational-led development that came to be synonymous with the boom decade of the Celtic Tiger. In addition, the state prioritized unregulated infrastructural development as a response to the social needs of communities traditionally marginalized by high levels of unemployment and emigration. In many cases, planning was dealt with in a haphazard manner by the state and the industrial sector. As a result, vital infrastructure that had been neglected by the state over the decades before the Celtic Tiger era of growth in the late 1990s was unable to cope with the post boom increase in demand, which has created many problems, including the problem of water provision (Leonard 2008).

The resultant tension between environmental and developmental considerations creates a conflict where the state should mediate. However, in the Irish case, the demands of globalized industry have gained prevalence over that of the community due to neocorporatist arrangements, with a series of regional crises occurring as a result in recent years. This transition from policies aimed at providing community services to policies favoring industrial development is reflected in the state's environmental policy agendas. These policy frameworks had originally been devised in order to prevent pollution and regulate industry, but have subsequently come to facilitate corporate and economic growth. Much of this regulatory transition has been facilitated by the governmental or state interpretations of the concept of ecological modernization (Janicke 1999; Leonard 2005; Weale 1992), which underpins much of the environmental legislation in the Irish case. Ecological modernization can be understood through the following quote:

> All commentators agree that the core message of the theory of EM [ecological modernization] is that economic development and environmental policy can be

reconciled, synergy can be established, so-called "win-win" opportunities can be seized upon. In one sense, this reconciliation is inevitably happening since advanced industrial societies have reached a stage where ecological resources emerged as scarce, or being bottlenecks to further economic growth. Therefore, affluent market societies now tend to economize upon ecological resources, attributing them increasing economic value, as well as investing in them, mainly through cleaner technologies. However, in another sense, government policy intervention is called for in order to smooth out, or accelerate, the transformation towards an ecologically sound society and economy. For that reason, innovative, market oriented policy instruments (such as eco-taxes, etc.) and a thorough, as well as widespread, policy integration are recommended. (Pataki 2005)

Perceived threats to environmental purity, as a feature of development in rural areas, have been a significant issue in contemporary Irish society. In the era of globalization, Ireland's relationship with the EU and the World Trade Agreement (WTA) has opened domestic markets through deregulation. Agricultural production has became more expensive as globalized competition increases. This in turn caused an erosion of the traditional rural lifestyle, which included clean water supplies. New forms of globalized trade, tourism, and development have created problems around the conservation and provision of resources like water, which is an increasingly important national commodity. Affluence has also placed a strain on newly urbanized areas, as landfills and sewage treatment plants reach capacity in the wake of an economy driven by construction. Conservationism has become marginalized in this development-led economy, with resultant impacts on the infrastructure that provides resources to communities.

Conservation issues in Ireland can be seen as part of a contest between two competing understandings of ecological modernization: one is a technocratic form of ecologically focused neoliberalism, while the other is expressed by a grassroots engagement with green political economy. The most prevalent form of ecological modernization is that which combines environmental and industrial concerns as a form of sustainable development (Pataki 2005). The ramifications for those wishing to adhere to green political economic thinking become clear in this process, as ecological modernization creates a proindustry rationale that then impacts on the direction of parties in Ireland's recent coalition governments.[1] The influence of grassroots conservationists on maintaining an adherence to green political economic tenets is of considerable importance in areas such as water provision.

The response of the environmental movement sector resulted from the Irish state's prodevelopment, neoliberal interpretation of the tenets of ecological modernization, demonstrating the extent to which the state and industrial partners used their alliances in order to overcome local concerns about resources and commonage. The shift in the state's policy direction undertaken by successive neoliberal Irish governments has led to increased conflict between communities and the state in relation to infrastructure projects and development (Leonard 2008). While the Irish state has attempted to follow the ecological modernization policy frameworks set out by the EU, there was an overemphasis on economic growth over environmental protection or concerns for community health issues (Leonard 2008). Ultimately, the costs of economic growth have led to greater environmental degradation and an undermining of significant aspects of the nation's resources, including its water supply.

What becomes clear is the degree to which local factors in Ireland such as alliances between the political sector and the property speculators who fund these parties lead to an ideological shift in the state's responses to EU-derived policy or directives. In this way, the Irish state can be seen to lack core green values within its wider policy framework, unlike other countries. Environmental values and state imperatives reached congruence at various stages of each state's development of environmental policy frameworks and movement activism over recent decades, with ecological modernization either becoming central to wider acceptance of environmental initiatives in the case of Germany or becoming part of the ecomodernist versus ecopopulist divide in the case of the United States, where an old-fashioned stand-off between economy and environment still exists. To an extent, this dichotomy between the economically derived imperative of the state and the subpolitics of conservationist movements exists in Ireland. While undoubtedly ecomodernism has provided an outlet for many competing elements in the Irish environmental arena and has crystallized current understanding of what environmental protection means from an industry perspective, it can be criticized. Although it facilitates policy making at an administrative level and brings diverse elements of pollution control together under holistic regulatory frameworks, ecomodernism has been criticized for being too inclined toward industry for an ecologically minded concept. On the other hand, proponents argue that ecomodernism might bridge the gaps between state, industry, and the environment and help overcome the regulatory failures of the past.

The impact of neoliberalism on ecological modernization in the Irish case becomes evident when we note the manner in which successive Irish governments have attracted multinationals for the purpose of job creation. When environmental regulations constrain job creation the regulations may often be altered or overlooked and the public finds itself caught in a propaganda war over the merits of industrial progress versus environmental protection. A feature of this debate is the conflict between the public relations section of the polluting industry, which is given practically limitless funding to give their viewpoint, and the oppositional efforts of concerned citizens and groups who have little or no previous expertise and no funding beyond public contributions. Ireland has followed a neoliberal plan for growth that has overridden environmental concerns and subsequently impacted on water supplies.

Drinking Water Legislation in Ireland

Ireland's drinking water is protected under both EU and national legislation. The Water Services Act 2007 prioritizes the protection of human and public health. The European Union Drinking Water Directive (98/83/EC of 3 November 1998) sets standards for the most common substances in Irish drinking water.[2] Where the water quality does not meet the specified standards, government regulations require remedial measures. These regulations do not change the monitoring requirements or water quality standards but they do change the enforcement of the standards, assigning new powers to the Ireland Environmental Protection Agency (EPA) and local authorities, and emphasizing the importance of the health service executive (HSE) in relation to both the water supplier and the supervisory authority. Water suppliers must consult with and reach agreement with the HSE when there are implications for public health. Failure of the water supplier to comply with specific aspects of the regulations is an offense; the supervisory authority may prosecute. The main provisions of these public health regulations emphasize the monitoring and protection of human health, notification of consumers when health risks occur, and the exercise of supervisory authority.

The local water authority works with the HSE during any crisis to provide actions or notice in the following manner:

a) The supply of such water is prohibited, or the use of such water is restricted, or such other action is taken as is necessary to protect human health

b) Consumers are informed promptly thereof and given the necessary advice, and

c) In the case of a public water supply, the Agency (EPA) is informed promptly. (Ireland HSE 2008: 5)

The responsibility for the provision of safe drinking water rests with the Local Authorities, while other agencies that are involved include the Ministry of Environment, Heritage and Local Government; the EPA; and the HSE. Respective roles and responsibilities are described below. Water services authorities made up of twenty-nine county councils and five city councils are responsible for the quality of drinking water, and the councils must work with the health authority when either party has health concerns with the supply.

While the EPA has supervisory authority over the water services authorities concerning the quality of the drinking water, the water services authorities maintain direct supervisory authority over the providers of other water supplies, such as private group schemes in rural areas and other private supplies. The water services authorities also ensure the water supplies that come under the regulations are monitored at the frequency provided for in the Drinking Water Regulations (Department of Environment, Heritage and Local Government 2007). This monitoring service is provided jointly through the Environmental Health Service (HSE) and the HSE laboratories in conjunction with water services authorities laboratories with a combination of monitoring procedures in place in certain geographic regions. In 2007, one such region, Galway city and county, suffered a major water crisis, with an ensuing health crisis due to a parasite in municipal supplies.

The Galway Water Crisis

In recent years, water legislation has been accompanied by the threat of water charges being levied on Irish householders; a number of community groups have protested against these charges (Leonard 2010). Nonetheless, Irish drinking water had been supplied with relatively few problems until the property construction boom of the late 2000s. European and national legislation was implemented regionally by the relevant local councils. However, in the case of the cryptosporidium outbreak in the western Irish city of Galway in the spring of 2007, a blame game of accusation and counteraccusation broke out between the various layers of institutional politics involved in the provision of municipal water to the city and its environs (Leonard 2010).

The issue of water pollution in the Irish case can be traced back some time before the current crisis emerged. The water crisis issue can be understood as a contestation between various forms of expertise. In the Galway case, local and national experts from a diverse range of fields made contributions to the water quality debate. These groups comprised scientific experts, politicians, planners, and civic groups.[3] Local fishers provide an example of one such group who provided an element of local expertise to the wider contested debate. Anglers on Lough Corrib have had long-standing concerns about water quality; scientists attempted to raise the issue as far back as 1997 when the *Geological Survey of Ireland* first highlighted a cryptosporidium risk for the Galway region due to the absence of soil layers prevalent in other parts of the country (Daly and Ball 1997). In the west, the indigenous karst limestone topography provided little in the way of the filtering effect necessary to break down microorganisms such as cryptosporidium, increasing the risk of water contamination (Leonard 2010).

The central issue of the water crisis was manifested through the infestation of local water supplies with the cryptosporidium parasite in the wake of a series of floods, a crisis compounded by years of poor urban and rural infrastructural development. The cryptosporidium protozoa is a microscopic parasite that attaches itself to the human intestine when ingested. It results in cramps and diarrhea for most people, with severe cases of infection in very young or old people leading to severe illness and, in a few reported cases, even death. The key source of the parasite is waste, both animal and human. As McNamara wrote in the *Irish Times* on April 14, 2007 ("Council Told of Threat to Water Last Year"), any infestation of water supplies is usually attributed to waste appearing in the water supply at some point.

In the case of Lough Corrib, the largest lake in the Republic of Ireland, effluence from large tracts of counties Mayo and Galway is washed through the lake and into the water supply for most of the local population, including the seventy thousand inhabitants of Galway City. Towns surrounding Lough Corrib and its tributaries have become part of the bourgeoning commuter belt around the prosperous city of Galway. Developers are enticed by the promise of profit from house prices that are second only to Dublin. As a result, sprawling housing developments have sprung up around all of the towns near Galway City, contributing to the pattern of national property market inflation that has been synonymous with the second phase of the Celtic Tiger phenomenon.

The issue of rapid development and subsequent social harms and environmental degradation can be applied to the Irish case through understandings of environmental sociology (Hannigan 2006) and risk society (Beck 1992; Giddens 1999). During the construction boom of the 1990s Irish society became preoccupied with risks and harms such as flooding as a result of rapid development on flood plains. The onset of risk society in the Republic of Ireland has thus been a characteristic of the Celtic Tiger years of growth, as a predominately rural society was transformed into a modern European state that urbanized rapidly in two decades (Leonard 2008, 2009). The manufactured risk stemming from human activity such as poor planning and development practices was compounded by a political system with shadowy links between property developers and mainstream political parties. These relationships have come under the scrutiny of successive tribunals of inquiry, and demonstrate a culture of corruption in the Irish planning process. According to Kirby, this culture of corruption is representative of a corrosive link between vested interests and political elites, alienating the wider population and leading to "the enrichment of an elite and the marginalization of ... citizens from the benefits of economic growth" (Kirby 2006: 181; cf. Leonard 2009: 280).

The fact that Galway had maintained its multinational presence through large multinational employers meant that a steady flow of people continued to be attracted to this western city. Its two college-level academic institutions and culturally rich reputation provided Galway with the reputation of a dynamic young and growing city. In March 2007, news began to spread of a contamination threat to the city's water supply. The HSE conducted tests across the city and county as the extent of the problem began to emerge. Residents in the city and surrounding areas were promptly advised to boil all drinking water until notified otherwise. Panic-buying of bottled water was reported over the bank holiday weekend as hotels, bars, and restaurants struggled to accommodate the influx of tourists for the traditional St. Patrick's Day festival (Leonard 2010).

The boil notice was announced as a precaution while the authorities searched for the source of the outbreak. Testing was initially concentrated at the city's aging waterworks, first constructed in the late 1940s with a second facility built in 1972.

Most municipal waterworks contain the cryptosporidium parasite through means of a filter system. However, the older Galway facility had a simple mesh to remove large pieces of waste, followed by a process of chlorination that in itself does not kill the cryptosporidium. The local council had been aware of a significant risk of cryptosporidium from 2005 (McNamara 2007). This news was enough to engender an immediate response from the government, particularly because it was an election year. The government was concerned about damage to tourism as much as it was about the health of its citizens, and tried to pass off the blame for the issue on the elected representatives of Galway's local authority.[4]

One issue that was clarified by the tests conducted by public health scientists was that the source of the crisis was human rather than animal waste. This confirmed the fact that unregulated development had been the cause of the water contamination due to the insufficient waste treatment plants in areas that had been overdeveloped along Lough Corrib. Lough Corrib is the largest lake in the republic; it flows via the Corrib River into Galway Bay and then into the Atlantic Ocean. However, for the beleaguered citizens of Galway, the cause of the crisis was of much less significance than the solution. Within weeks of the crisis, a populist community movement was mobilized in response to the issue. The Galway Water Crisis campaign was made up of ordinary citizens who had lost faith in the political system's ability to deliver on health-related issues, and included campaigners from previous local protests on incineration, municipal dumps, or the siting of the water treatment plant (Leonard 2008).

Representatives of a local group that was protesting the health risks of the Galway Water Crisis stated that the campaigners as a group felt let down by both local and national politicians who "seemed more interested in passing the blame than actually doing anything about this crisis" (Leonard 2010: 37). A letter from the group continued by stating that the state was taking €1.5 million a week in taxes for water sales, and that the group wanted that money used in the efforts to solve the crisis. Some local authority councilors criticized the proposed use of ultraviolet (UV) technology, stating that UV tubes were less efficient in dark water which is what the Lough Corrib had become due to pollution (Leonard 2010).

In addition, local politicians claimed that while UV light worked with most microbes, it was not effective on protozoan parasites such as the cryptosporidium. The councilors suggested a Russian method of electrochemically activated water or anolyte, which was a nontoxic way of destroying all microbes effectively.

Environmental social movements have benefited mainstream politics in Ireland as witnessed by the rise of the Green Party in local, national, and European elections over recent decades. One distinctive outcome of environmental conflict has been the establishment of the EPA in 1992. The EPA was originally criticized for its location of a regulatory framework within the context of the state's industrial development policies. However, some distance remains between the EPA's regulatory performance and the concerns of grassroots campaigners (such as those campaigning for clean water); the bridging of this gap remains an outstanding issue for all levels of environmentalism in the Irish case.

The response of the political sector to the cryptosporidium outbreak was one of passing blame for the crisis. Local politicians claimed the crisis arose as a result of a government which was seen to be overtly focused on speculative development. The political sector was already being blamed for poor planning in relation to Galway's sewage treatment plant, which was built in Galway Bay with a large investment from the European Union in the mid 1990s, only to reach capacity within a couple of years due to the explosion in home and industrial construction in a city sometimes called "the fasting growing city in Europe" (Leonard 2010: 29).

Others believed that culpability rested with a number of groups such as the HSE, the city and county managers, the elected councilors, the Department of the Environment, and the EPA. Commentators stated that risk assessment procedures were not carried out properly over many years, and that the exceptionally heavy rainfall and subsequent flooding that had occurred in Galway that winter should have alerted the authorities to the need to monitor the water supply for problems (Leonard 2010). One editorial claimed that gallons of raw sewage were being pumped into the rivers and lakes of Galway, as well as into Galway Bay, due to new housing developments.

For their part, Galway city managers claimed that the water crisis would end on June 15 with the increase in supply of clean water from a nearby plant. The council also had plans to repair water pipes in the city. With the general election taking place at the end of May, however, the local media began to speculate as to which candidate would be most affected by the water crisis. The issue had relegated all others to secondary importance, becoming a major issue in the Galway West electoral constituency during the 2007 national elections.[5] The Galway Water Crisis campaign had mobilized large levels of public anger that was now being directed at public officials due to the crisis. Local politicians also had to contend with the protest campaign of the Galway Water Crisis group. The group claimed that public faith in the political system had been irreparably damaged by the water crisis. Many believed that the risk of an outbreak of cryptosporidium even after a new plant was built would remain, due to poor monitoring and unregulated development. According to the group,

> The feedback we are getting from the public is that they would be afraid to drink the water from the city system again. Unless we tackle the major problem of cleaning Lough Corrib, then this problem will simply come back again and again. (*Galway Advertiser* April 26, 2007)

The campaign claimed that the problem was one that affected all the towns around Lough Corrib, not just the city areas reliant on the old Terryland sewage plant that was the source of the outbreak. These areas had unreliable sewerage systems that pumped effluent into the Lough. For the campaigners, "[T]he major problem lies with the cleaning of the Corrib. There will always be a problem as long as raw sewage is coming into the Lough" (*Galway Advertiser* April 26, 2007). The water campaign brought its protest to a meeting of Galway city council, calling for "clean water for Galway." Protesters brought their empty water bottles to deposit at city hall as part of their protest, to highlight the rising cost for citizens having to buy water and recycle bottles, in a city that ironically had led the way when recycling pilot schemes were first introduced (Leonard 2005). The water campaign's webpage had nearly one thousand hits in its first two weeks, indicating the lack of information available from official sources for citizens concerned about the health risks of the water crisis in Galway.

Local politicians called for localized filtering systems so people could collect free water. They claimed that the government had plans to privatize water supplies across Ireland, something that led to the delay in "the allocation of EU and state funds for water and sewage services to local authorities" (*Irish Times*, April 24, 2007). These elected representatives also pointed out that while citizens faced payment of large fines from the EU due to the nonimplementation of EU directives, they were also being denied clean water due to "government failure" (*Galway Independent*, April 25, 2007). The mayor did receive some support from the Heritage Group An Taisce who claimed that the construction of new housing should be frozen until the water crisis was sorted out. This moratorium should remain in place until "the essentials of life" such as good drinking water, sewage treatment, and waste disposal systems were sorted out (*Galway Independent*, April 25, 2007).

The outcome of this issue can be understood when viewing the water crisis as an example of societal risk born from poor planning practices. From a social movement theory perspective, the crisis is indicative of the difficulties faced by campaigns and key activists alike when electoral success deprives movements of their potential leadership group. The experience of activists who became part of the city council in Galway demonstrates the difficulties for campaigners who become "cut off from the activist base" (Leonard 2008: 57).[6] This was witnessed in the difficulties faced by some advocates who ran for city council and the position of city mayor, who were unable to use the mobilizing processes of advocacy when they had become part of the establishment. Once an issue such as the water crisis emerged, former advocates who had protested about issues were unable to deliver on changes that were outside the remit, presenting their response in a poor light to the electorate (Leonard 2010).[7] The problem of cooptation for activists who embrace the political mainstream remains an outstanding obstacle for those environmental advocates now faced with negotiating with a government that included a Green Party minister for the environment.

The beleaguered citizens of Galway had little respite through the economically important summer festival season. The HSE announced that up to 238 cases of cryptosporidiosis had been recorded in Galway in 2007, with nonreported cases estimated in much higher numbers. The boil notice was lifted in east Galway in July, due to the extension of services from the revitalized water plant in the city (*Galway Advertiser*, July 21, 2007: 1, 5). The famous Galway Race Festival was declared cryptosporidium free due to a €100,000 investment in water filters at the local race course. Commercial interests, which had complained of losses during the tourist season, were offered a 10 percent tax rate rebate.

In 2007, nearly a quarter of a million lost water supplies in the southwestern region of England due to extreme flooding, while other regional Irish towns such as Ennis and Clonmel were also hit with boil notices due to flooding. For Galway's local authorities, the loss of confidence from the public in its services was equaled by costs of over €400,000 in weekly laboratory tests at its water plants. After a series of delays, water supplies to the city and its western environs were returned to Galway City as the last obstacles to providing acceptable levels of clean water from the revitalized local plant to the city were overcome (with supply rising from fourteen thousand cubic meters of water to seventeen thousand cubic meters), according to Galway's city manager ("Chamber of Commerce Says Water Crisis Not Affecting Tourism," *Galway City Tribune* July13, 2007: 3). With cryptosporidiosis cases dwindling to precrisis levels, the Galway water debacle was coming to an end. Water supplies were restored to all parts of the city by August 20. The final bill for the Galway crisis was set at €15 million, including the costs to industry and private households (Hickey 2007).

In 2008, Galway would be beset by further water problems as lead corrosion began to affect local water supplies with a new risk, leaving householders reaching for their water drums again.

The Big Freeze

Ireland's water crisis has emerged as a serious threat to health across the island. In December 2010, thousands of residents in Northern Ireland were left without water due to record freezing temperatures, which caused freezing to municipal and domestic pipelines in areas where winter cold snaps rarely went below minus five degrees Centigrade. In the winter of 2010, the lowest overnight temperature reached minus twenty Centigrade, as Arctic-like conditions gripped the island. Water supplies had to be sourced from neighboring Scotland as pipes froze. The economic downturn had led to public sector water utility workers being laid off, and repairs to burst mains were severely hampered as a result. Officials from the government-owned water supply company in Northern Ireland received six hundred thousand phone calls and ten thousand emails while its website was bombarded by half a million hits during the worst week of the freeze. The Northern Ireland Water chief executive subsequently resigned due to protests about the poor response to the crisis (MacDonald 2011).

A similar situation faced citizens to the south in the Republic of Ireland, with water shortages becoming an annual problem in Dublin and surrounding counties, which are home to one-third of the nation's population. Record low temperatures created significant water shortages across the Republic of Ireland throughout December 2010 and early January 2011. Dublin City Council put water restrictions in place throughout the holiday season, with major implications for householders and businesses in the capital. A spokesperson for the council said only half the amount of water necessary (about twenty million liters of water) was being saved each night during the freeze and subsequent thaw.

The daily demand at the Dublin reservoirs was around 507 million liters, 36 million liters fewer than was produced by the city's water treatment plants during the cold weather. Local authorities in the worst-affected areas provided standpipes or tankers for residents, so people could collect water. Residents were reminded to take their own containers to water collection sites, and to boil all water collected before drinking. The Department of the Environment said it was satisfied with investment in water infrastructure, and claimed a water mains rehabilitation program could be completed in ten years if the current rate of investment continued (*Irish Times*, January 4, 2011).

Response to the water crisis

Ireland's water crisis has led to a crisis of what Foucault (1991) has called governmentality, as various state and regional bodies compete for ascendancy over the nation's water supply. The Department of the Environment's response was to ensure water bodies in the country met strict criteria to be set out soon in the EU's Water Framework Directive (WFD). Local authorities have huge challenges ahead if Ireland is to avoid fines and worse, with 85 percent of water bodies at risk of not meeting the objectives. EPA experts had long harbored concerns about the type of water crisis experienced in Galway, but their concerns were ignored during the property development–led economic boom from 1998 through 2006.

The cryptosporidium issue was believed to have arisen because one of Galway's two water treatment plants was not capable of removing the feces-borne parasite and other pathogens from the water supply, according to the EPA. The EPA also feared many of Ireland's other regional water supplies and many treatment plants serving the nine hundred public water supplies across the country are under threat from a similar crisis to Galway, particularly in an era of climate change–related flooding. An EPA report found that one in five plants were either high risk or very high risk for outbreaks of cryptosporidium or other water borne diseases out of a survey of half of the nation's facilities.

Water shortages are another concern, due to a rapidly aging supply system. The greater Dublin area system, which delivers 520 million liters of drinking water every day to the city and surrounding counties, faces serious stresses from a grow-ing demand that was never planned for in the boom years. Pipe leakage increased by thirty million liters a night at the height of 2010's big freeze. Daily demand in the summer can see increases of up to forty million liters. While plans have been set for an improvement to the municipal pipe system, the fact remains that Dublin runs the serious risk of having demand outstrip supply.

MacDonald (1989) has argued that the blame for much of Dublin's ills could be laid at the feet of rural and suburban developers, managers, and planners who lacked sufficient empathy with their surroundings. Conversely, the same lack of empathy could be said to be characteristic of many planners and developers when dealing with rural conservation sites. Essentially, he has articulated an understand-ing of the need to see cities, towns, and villages as living entities rather than as commercial zones. In his books and editorials, MacDonald has consistently argued that planning and development needs to be undertaken with a sympathetic under-standing of both place and hinterland, within an overall context that incorporates the history and ecology of an area.

MacDonald presents a view of planning as an ideological process based on deterministic concepts of spatial ordering and functionalist utility. From this per-spective, planning decisions are determined mainly by the concerns of the ruling social elite and the elite's derivative interest groups and supporters. At the heart of this equation is a capitalism that promotes private property and free market indus-try as two of its main tenets, forgoing concerns about equity of distribution.

Ireland has a natural imbalance in its distribution of water. According to the EPA, there are about seventy-six thousand liters of water per capita per day avail-able in the west of Ireland and about one-tenth of that (seventy-six hundred liters of water per capita per day) in the greater Dublin region. Climate change has caused further problems for the supply of water on the island: the EPA reported in 2009 that the implications of climate change will have serious results on both the availability and subsequently the quality of Ireland's domestic water supply. The state will be forced to arbitrate between competing industrial, agricultural, and householder demands for water as a result.

Water supplies will be reduced in the east by up to 40 percent, with most water emanating in low-lying areas that provide surface water that is more exposed to climatic related issues such as flooding. Further concerns have arisen due to the ongoing rise in global temperature, which will affect nations' water supplies glob-ally. Even a rise of 1 percent can lead to significant reductions of rainfall in regions subject to variable distribution of rainfall or drought. The supply to both new housing developments and industries in the east has reached its limits because the same bodies of water are used for the cooling of high tech equipment, domestic drinking, and effluence, a contest that seems to have evaded planners during the era of rapid development. These issues have been exacerbated by the problem of increased pollution.

The Irish state has included plans to increase water infrastructure spending by €5 billion. These plans are now under threat due to the nation's economic collapse and subsequent EU/International Monetary Fund fiscal bailout, however. Many large-scale development projects have been put on hold as a result. The state plans to introduce water meters, but needs to survey and repair the country's entire supply infrastructure before this can be achieved, a large-scale project that is also facing delays due to fiscal constraints.

Governance Implications: The Water Framework Directive

One area of ecological modernization that emerged within the EU and subsequently Irish policy was the idea of a holistic water framework that established processes for the provision and protection of clean water through member states. The EU WFD is an important piece of EU environmental legislation that aims at improving our water environment. It requires governments to take a new holistic approach to managing their waters. It applies to rivers, lakes, groundwater, estuaries, and coastal waters. Member states must aim to achieve good status in all waters by 2015 and must ensure that status does not deteriorate in any waters. National regulations implementing the national WFD were introduced in 2003 and, following the adoption of the WFD, water policy and protection have been devolved from European or national governance levels down to the community level. Ireland began implementing new national regulations in 2003. These regulations included the establishment of seven water advisory councils around the country, one in each river basin district, as part of an integrated approach to water management, providing communities with a role in the management of local water supplies and amenities. The Environmental Protection Agency (EPA), the Department of the Environment, Heritage and Local Government, local authorities, and associated consultancies all worked together to prepare the Article 5 Characterization Report for the Irish River Basin Districts. This was submitted to the EU in Brussels in November 2004.

Conclusion

Ireland is coping, albeit with some difficulties, to provide municipal water supplies to its population. The National Urban Wastewater Study concluded that only 48 percent of the wastewater treatment plants in Ireland will be able to adequately treat the projected loadings in 2022 (Department of the Environment, Heritage and Local Government 2005: 42). Eighty-five percent of the wastewater treatment plants studied in 2002 could not maintain required environmental standards for receiving waters when operating at full capacity (2005: 30).

In the bourgeoning city of Galway, the local population felt the results of poor planning. Drinking water supplies was contaminated with cryptosporidium. Lax regulation of the construction industry by the state had created a growing strain on the city's water and sewerage systems. After heavy flooding in the winter of 2007, Lough Corrib, which supplies Galway City with water, became contaminated. Hundreds of people fell ill , and a boil water notice was called by the local authority. Panic buying of bottled water ensued, and the boil water notice lasted from March until August 2007, costing retailers and hoteliers millions of euros in revenue. The water crisis also became an issue during the general election called in May 2007, and was said by some commentators to have cost the city's Green Party mayor a seat in the Dublin parliament. Despite this crisis, Ireland was again plunged into crisis in the winter of 2010, when freezing temperatures led to tens of thousands of residents losing water supplies over the holidays.

With reports of increased instances of extreme weather forecast by the United Nations World Meteorological Organisation, and after a decade when extreme weather events occurred in a "range which is well outside the historical norm" (*Irish Times* August 9, 2007), the threat of instances of flood-related water contamination will remain for communities in Ireland and across the planet. With poor infrastructure and increased fiscal constraints in the aftermath of the Ireland's economic bailout by the EU and International Monetary Fund, citizens across Ireland will be hoping for a sustained and effective response from their political leaders to address this issue before another water crisis occurs, particularly in the heavily populated east coast.

Notes

1. The system of proportional representation through the single transferable vote (PRSTV) used in the Republic of Ireland has created a series of coalition governments, which are usually center left or center right in orientation. Parties of the right include Fianna Fáil and Fine Gael, with smaller parties of the left including Labour and the Greens, both of which have been part of recent coalition governments with Fine Gael (current since 2011) and Fianna Fáil (previous to 2011), respectively.
2. These standards were translated into Irish law by the European Communities (Drinking Water) (No. 2) Regulations 2007 (Department of Environment, Heritage and Local Government 2007).
3. For a detailed account of civil society mobilizations on ecological issues in Galway, see Leonard (2005).
4. For a more in-depth account of the issue of contested notions of sustainable rural tourism in the Irish case, see McAreavey, McDonagh, and Heneghan (2009).
5. One election candidate stated that the issue was being raised on "at least 50 percent of all doors we canvass" (*Galway Advertiser* 2007a).

6. For those candidates who had cut their political teeth as environmental advocates (both anti-landfill and anti-incinerator), their new roles as city councilors or mayors created constraints because they could not campaign against the council that they now represented. This issue of cooptation of former advocates would become apparent during the 2007 general election campaign. For its part, the Galway Water Crisis group continued with its campaign of public meetings, protests, and petition gathering, tactics that had served these previous advocates well during their time as social movement campaigners (Leonard 2005, 2006).

7. Charged with solving the problems of the bourgeoning "City of the Tribes," the former advocates who were now in city hall could not distance themselves from the public's own desire to play the blame game in the aftermath of the water crisis. In many ways, this predicament displays the dilemma facing activists who become coopted by the mainstream, where the potency of collective action is sometimes traded for the robes of office (Leonard 2010).

References

Beck, Ulrich. 1992. *Risk Society: Towards a New Modernity.* London: SAGE.

Curtin, Joseph. 2007. Ireland's Water Crisis. *Business and Finance.* http://www.businessandfinance.ie/index.jsp?p=286&n=288&a=1021.

Daly, Donal, and David Ball. 1997. Protection of Wells: Where Dilution Is Not the Solution to Point Source Pollution! *Geological Survey of Ireland* 32 (November): 6–8. http://www.gsi.ie/NR/rdonlyres/E136C6F1-97A4-41A6-B52F-6B848F6D03EE/0/No32.pdf

Department of the Environment, Heritage and Local Government (Ireland). 2005. National Urban Waste Water Study, Volume 1, National Report. http://www.environ.ie/en/Environment/Water/WaterServices/NationalUrbanWasteWaterStudy/PublicationsDocuments/FileDownLoad,1533,en.pdf

Department of Environment, Heritage and Local Government (Ireland). 2007. European Communities (Drinking Water) (No. 2) Regulations 2007. Statutory Instruments, SI 278. http://www.environ.ie/en/Environment/Water/WaterServices/RHLegislation/FileDownLoad,14547,en.pdf

Foucault, Michel. 1991. Governmentality. In *The Foucault Effect: Studies in Governmentality*, edited by G. Burchell, C. Gordon and P. Miller. London: Harvester Wheatsheaf.

Giddens, Anthony. 1999. Risk and Responsibility. *Modern Law Review* 62 (1): 1–10.

Hannigan, John. 2006. *Environmental Sociology.* London: Routledge.

———. 2008. *Drinking Water and Health: A Review and Guide for Population Health, December 2008.* http://www.hse.ie/eng/services/Publications/services/Environmentalhealth/HSE_Drinking_Water_and_Health_Review_and_Guide_2008.pdf

Hickey, D. 2007. Vintners Calculate Cost of Water Crisis. *Connacht Sentinel* August 21: 1, 3.

Janicke, Martin. 1999. Democracy as a Condition for Environmental Policy Success: The Importance of Non-institutional Factors. In *Democracy and the Environment: Problems and Prospects*, edited by William M. Lafferty and James Meadowcroft. Brookfield: Elgar.

Kirby, Peadar. 2006. Bringing Social Inclusion to the Centre: Towards a Project of Active Citizenship. In *Taming the Celtic Tiger: Social Exclusion in a Globalised Ireland*, edited by D. Jacobsen, P. Kirby and D. Ó Broin, 180–199. Dublin: New Ireland.

Leonard, Liam. 2005. *Politics Inflamed: GSE and the Campaign against Incineration in Ireland.* Greenhouse Press Ecopolitics Series, vol. 1. Galway: Greenhouse/Choice.

———. 2006. *Green Nation: the Irish Environmental movement from Carnsore Point to the Rossport 5.* Greenhouse Press Ecopolitics Series, vol. 2. Galway: Greenhouse/Choice.

———. 2008. *The Environmental Movement in Ireland.* Dordrecht: Springer.

————. 2009. Social Partnership's Boiling Point: Environmental Issues and Social Responses to Neo-liberal Policy in Ireland. *Critical Social Policy* 29 (2).

————. 2010. The Galway Water Crisis. In *Ireland of the Illusions: A Sociological Chronicle 2007–2008*, edited by P. Share and M. Corcoran. Dublin: IPA.

McArevey, Ruth, John McDonagh, and Maria Heneghan. 2009. Conflict to Consensus: Contested Notions of Sustainable Rural Tourism on the Island of Ireland. In *A Living Countryside? The Politics of Sustainable Development in Ireland*, edited by J. McDonagh, T. Varley, and S. Shortall. Aldershot: Ashgate.

McDonald, Frank. 1989. *Saving Our City: How to Halt the Destruction of Dublin.* Dublin: Tomar.

McNamara, Denise. 2007. Council Told of Threat to Water Last Year. In *The Irish Times,* April 14.

Pataki, György. 2005. The Theory of Ecological Modernisation from a Critical Organisation Theory Perspective. Paper for the Seventh Conference of the European Sociological Association, Rethinking Inequalities, September 9–12, Torun, Poland.

Weale, A. 1992. *The New Politics of Pollution.* Manchester: Manchester University Press.

Part IV

∞

GOVERNANCE

The term "governance" is now often used to indicate the entire network of governmental and nongovernmental organizations involved in decision-making processes. Its meaning is thus distinct from "management," which refers to the specific day-to-day actions of organizations, such as water utilities, that are formally mandated to implement management decisions. "Governance" is also distinguished from "government" since the former refers to the actions and interrelationships of a much broader set of institutions and interest groups than government agencies (de Loë et al. 2009). It can also be an ideological concept that signals, for some, a radical shift away from the conventional control-and-command role of government toward more decentralized approaches. The governance concept became prominent, first, during the 1980s, within a neoliberal discourse that advocated for less-regulated marketplaces and a stronger role for corporations (De Angelis 2003), but over the past two decades political scientists and others have developed the concept of distributed, multilevel governance to refer to approaches in which the decision-making authority of state agencies is shared with local and regional institutions and is informed by local as well as scientific knowledge (Bache and Flinders 2004; Hooghe and Marks 2003; Wagner and White 2009).

Some researchers argue that distributed, multilevel governance approaches are inherently more effective than conventional approaches, but the strongest support for this argument may be ideological rather than empirical. Multilevel approaches require more public and local involvement in decision making and stress the importance of fairness, equity, and inclusivity. Effectiveness, from this perspective, cannot be measured solely in terms of efficient economic delivery of services, but must also include the assessment of social outcomes. Support for distributed, multilevel approaches is thus best understood as support for the democratization of decision-making processes.

This section begins, therefore, with chapters by Wutich and colleagues (chapter 11) and Sam and Armstrong (chapter 12) that focus on issues of fairness and human rights. Amber Wutich and her coauthors begin their chapter with a discussion of the idea of water as a human right. They describe the progress made by the United Nations on this issue over the past decade but point out that many countries have not agreed to recognize water as a human right, while others are unsure of how to implement the UN resolutions they have ratified. Based on a cross-cultural study of socially and ecologically distinct sites in Bolivia, Fiji, New Zealand, and the United States, the authors seek to understand whether there are "general principles of water distribution that people understand as fundamentally fair or unfair" (this volume p. 221), and how "key differences might be grounded in local cultures, ecologies, and water governance systems" (this volume p. 235). They conclude that perceptions of fairness do have much in common across all their sites, but that some concerns are not adequately addressed in existing international agreements. Equity of access to water is found to be especially at threat in situations where water is scarce and where governance processes are not transparent and democratic. Under these circumstances, the authors conclude, there is "there is an unmet need for institutions that establish equity" (this volume p. 236).

In chapter 12, Marlowe Sam and Jeannette Armstrong address the issue of indigenous rights to water, noting the continuity of contemporary water privatization schemes with older forms of colonialism. The Cochabamba protest in Bolivia, they remind us, was not just a popular uprising against an abusive privatization program—it was an indigenous uprising. Most people living in the Okanagan Valley of British Columbia understand water scarcity as a recent threat, but indigenous Okanagan (Syilx) communities have been experiencing water scarcities for a century and a half, since the arrival of European settlers and the creation of water laws that denied indigenous rights in favor of water licensing schemes that use water as an instrument of development and exclusion. Though written from the perspective of resistance, and set in the context of legal battles over water, Sam and Armstrong emphasize in their conclusion that the formal recognition of indigenous water rights by states such as Canada would necessarily bring with it recognition of a different understanding of water. This different understanding, in turn, could provide a foundation for a more equitable and sustainable approach to water for all people, both indigenous and nonindigenous.

Bryan Bruns, on the basis of insights gathered over many years of work as a water management consultant in Aceh, Indonesia, and in other settings, points out in chapter 13 that there are no governance panaceas. Strongly influenced by the collective action theory of Elinor Ostrom, he recognizes the weaknesses of both conventional, top-down approaches and recent decentralization initiatives that tend to idealize local-level management capacity. Bruns' approach, which he describes as "bureaucratic bricolage," is grounded and balanced and provides an essential foundation for strategic thinking about how policy interventions in any given setting can be developed.

The volume concludes with a chapter by John Donahue that illustrates the type of work that needs to be done in watersheds around the world to resolve or mitigate water crises. Echoing the foundational theory articulated by Bruns concerning the polycentric nature of water governance, he describes the successful resolution of a governance crisis in central Texas that was caused by the overexploitation of the Edwards Aquifer. In the midst of court battles and political conflict among diverse interest groups, a new institution was created and given a mandate to resolve outstanding issues. The result is a transformational learning process in which multiple stakeholders gradually develop a sense of collective responsibility for the watershed as a whole, and generate a new language and new cultural values consistent with that sense of collective responsibility. While the image of a boardroom full of stakeholders may not be the most striking image with which to end this volume, I believe it does honestly portray the real work that must be done around the world to resolve water crises. Solutions everywhere require that people with diverse interests come together to develop a renewed and collective vision of their responsibilities toward water, one another, and the complex web of socioecological relations that bind us all together.

References

Bache, Ian, and Matthew Flinders, eds. 2004. *Multi-Level Governance*. Oxford: Oxford University Press.

De Angelis, Massimo. 2003. Neoliberal Governance, Reproduction and Accumulation. *The Commoner* 7 (Spring/Summer): 1–28.

de Loë, R.C., D. Armitage, R. Plummer, S. Davidson, and L. Moraru. 2009. *From Government to Governance: A State-of-the-Art Review of Environmental Governance*. Final Report. Prepared for Alberta Environment, Environmental Stewardship, Environmental Relations. Guelph, ON: Rob de Loë Consulting Services. http://environment.gov.ab.ca/info/library/8187.pdf

Hooghe, Lisbet, and Gary Marks. 2003. Unraveling the Central State but How? Types of Multi-level Governance. *American Political Science Review* 97 (2): 233–243.

Wagner, John R., and Kasondra White. 2009. Water and Development in the Okanagan Valley of British Columbia. *Journal of Entreprising Communities* 3 (4): 378–392.

Chapter 11

FAIRNESS AND THE HUMAN RIGHT TO WATER
A Preliminary Cross-Cultural Theory

Amber Wutich, Alexandra Brewis, Sveinn Sigurdsson,
Rhian Stotts, and Abigail York

∝

Over the past decade, a global consensus has emerged around the idea of water as a fundamental human right. The United Nations (UN) Committee on Economic, Social and Cultural Rights (CESCR) General Comment 15 requires national governments to make progress toward providing "sufficient, safe, acceptable, physically accessible and affordable water" for all citizens (UN Committee on Economic, Social and Cultural Rights 2003: 1). The UN Human Rights Council Resolution 7/22 in 2008, UN Human Rights Council Resolution 12/8 in 2009, and UN General Assembly Resolution 64/292 in 2010 have all expanded basic human rights to water. This trend is not without political dissent. Forty-one countries abstained from the 2010 vote on UN Resolution 64/292 because, they argued, it confused the Human Rights Council's ongoing efforts to clarify national governments' legal obligations to guarantee access to water (International Service for Human Rights [ISHR] 2010). In arguing in favor of a human right to water, proponents often appeal to a basic sense of fairness and justice (Louka 2006). This speaks to a key tension in current global debates around the human right to water: How do we determine what is fair? Specifically, what should people's entitlements and expectations be with regard to water access and availability?

The problem is not just confined to forging international agreements. National governments struggle to translate global guidelines into locally appropriate and effective policies. Middle-income developing countries, such as South Africa, India, and Argentina, that adopted a constitutional right to water have found achieving all of the UN goals to be a practical impossibility (Bluemel 2004) because these principles fail to account for complex on-the-ground, local realities. This basic problem of how to define fair rights to water is also critical to untangling the complex debates around how water law and policy are framed and implemented. Considerations of how principles of fairness around water vary from place to place generally fail to inform the policy debate because even the most basic comparative research on this point is lacking (cf. Syme et al. 2000).

Our challenge in this chapter is to develop a preliminary cross-cultural theory of conceptions of fairness around the right to water. We use an innovative process of theme and metatheme analysis to examine, contrast, and integrate local beliefs in this domain using interview data collected in four ecologically and culturally different sites—squatter settlements in the Bolivian highlands, an indigenous coastal Fijian village, urban and rural communities in central New Zealand, and a desert city in the southwestern United States. To develop such an elemental theory, our analysis of people's ideas in these varied places focuses on three key questions: (1) How are conceptions of fairness in water grounded in local cultures, ecologies, and governance systems? (2) What general factors or conditions might best explain variation in ideas around specific dimensions of fairness in water distribution? (3) Are there general principles of water distribution that people understand as fundamentally fair or unfair cross-culturally?

A distinction between equity and equality is often made by scholars interested in the social dimensions of water management (e.g., Boelens 1998; Ingram, Whiteley, and Perry 2008); that distinction is important to addressing these questions. Equal water distribution assumes uniform needs and rights to water. By contrast, equitable water distribution is based on fairness in terms of local histories, norms, and beliefs rather than equal allocations alone. Growing evidence suggests that locally sustainable and stable water distribution systems should create and maintain equity, rather than equality (Boelens 1998). Yet most water research focuses on equality, leaving our understanding of equity relatively undeveloped. There are several reasons for this. First, equality is easier to define, design, and achieve in water systems than equity (Murray-Rust, Lashari, and Memon 2000). Second, policy makers generally tend to emphasize equality, which is more consistent with the project of nation-making, than equity, which may undermine states by promoting self-governance and self-determination in ways that deviate from national interests (Boelens 1998). Third, the scant research on equity overemphasizes idealized and philosophical representations and fails to provide detailed descriptions of locally embedded rules, norms, and values (Lauderdale 1998). Achieving equity involves consideration of the cultural value of water, community welfare, the protection of vulnerable groups, transparency and participation in governance, and long-term environmental sustainability (Ingram, Whiteley, and Perry 2008). By contrast, attaining equality involves imposing one-size-fits-all water rules on groups with diverse needs, potentially depriving them of essential water resources and forcing them to change their lifeways to accommodate these water allocations (Oliverio 1998).

Regardless of whether the goal is equity or equality, the sad reality of water distribution today is that neither is close to being realized. Over the past thirty years, water sector reforms, for the most part, have ignored or even undermined both equity and equality while focusing instead on economic efficiency (Araral 2010). The efficiency approach treats water as a commodity, favors economic incentives over rules and regulations, and assumes that markets more efficiently allocate water than do public enterprises (Ingram, Whiteley, and Perry 2008). Proponents of the efficiency approach argue that it represents our best hope to conserve limited freshwater supplies and extend underfunded municipal water systems, while critics contend the efficiency approach is a thinly veiled attempt to privatize the world's water resources—while excluding the poorest and most vulnerable—for the profit of a few (Budds and McGranahan 2003). Indeed, recent efforts to establish a human right to water are a response to the redefinition of water as an economic good after the 1992 Dublin Statement on Water and Sustainable Development (Bluemel 2004). In accordance with these previous debates, future attempts to guarantee a human right to water will most likely consist of both public and private approaches to water provision. These varied approaches also frame our understanding that it might be important to distinguish market, commons, and mixed water institutions in our considerations of how fairness might be differently perceived by actors on the ground.

Research Settings

The interview data that form the basis for our analysis were collected as one component of the Global Ethnohydrology Study, a multiyear, multisited study examining comparative cultural knowledge of water. Interviews regarding notions of fairness in water institutions were collected with 219 adults in four countries in 2007–08. The specific countries and study locations were selected to facilitate two-way comparisons on both economic development (developing sites were Bolivia and Fiji; developed sites were New Zealand and the United States) and water availability (semi/arid sites were Bolivia and the United States; water-rich sites were Fiji and New Zealand). While this research design allows us to make broad comparisons across contexts, the inclusion of only four research sites and two dimensions of comparison limits our ability to generalize our findings to other communities globally.

The Bolivia data were collected with women from extremely water-scarce squatter settlements (N=41) in Cochabamba, Bolivia's third-largest city. Over the past thirty years, migrants relocated from the impoverished highlands to densely populate a former greenbelt in a temperate semiarid valley on the city's south side.

These squatter settlements are unable to obtain water from SEMAPA, the municipal water utility. Since 2000, Cochabamba has been recognized as the site of the Water War, a months-long conflict in which locals protested the privatization of scarce local water resources and delivery infrastructure. The privatization deal was ultimately annulled, leaving vast swaths of the city where squatters reside with no municipal water services. Some of these communities run their own small-scale commons water systems, although groundwater scarcity in this region often means that these systems run dry or provide insufficient water allotments. In many of Cochabamba's squatter settlements, however, the lack of public water delivery forces people to obtain water from private water-vending trucks. Because water is so scarce, average water consumption in Cochabamba squatter settlements falls well below the established international standard of fifty liters per person per day (Wutich 2009).

The Fiji data were collected with adults in one indigenous Fijian village (N=37) on the semirural south coast of the main island of Viti Levu. To protect the confidentiality of respondents within this small community, we have chosen not to publish the village name. Despite some rain-short months that affect agriculture, water in Viti Levu is readily available via the regular rains and large volcanic island system of rivers. These villagers access most of their water from a dammed river uphill from the village. A main social unit in indigenous Fijian society is *vanua*, meaning the people tied to shared ownership of *vanua* (land) and *wai* (water) resources in a defined territory. Thus, collective interests around water are implicit in the social and economic organization of much rural Fijian village life. The national government in Fiji has only recently considered policies regarding consolidation of water use. Currently, however, there is no central governmental management of Fijian water resources. The village has been engaged in local water system upgrades over the past several years (2007–09), and this filtered water is now their main domestic supply. They also developed fairly effective grey and black water systems to recycle water for agricultural use in concert with expanding and upgrading the household water supply. Despite a major Viti Levu aquifer being the source of exported Fiji Water, surface water quality is often an issue in Fiji because of contamination with pig or human waste.

The New Zealand data were collected in a rural (N=29 in Piopio) and urban setting (N=52 in Wellington, the capital) on the North Island. Wellington residents access city water mains, while those residing in Piopio are among the roughly 15 percent of New Zealanders who collect and use water directly on their properties (such as from captured rainfall or springs). By global standards, water quality and quantity in New Zealand is very high. New Zealand is a freshwater-abundant country with manifold rivers, lakes, glaciers, and aquifers. Droughts occur periodically and local floods are common, but freshwater for domestic use is safe, cheap, and almost always available. Most New Zealanders drink water straight from the tap without concern and many take unfettered water access completely for granted in their day-to-day lives. Dairy farming is a major user of freshwater resources on the North Island, and historically a major contaminator through waste and chemical runoff into waterways. Because some 60 percent of New Zealand's water supply is from surface sources, pathogens affecting untreated water quality include giardia, E. coli, and campylobacter (Weinstein, Russell, and Woodward 2000). The Resource Management Act 1991 is the key parliamentary legislation governing freshwater resource management and makes local and regional authorities responsible for decisions on water allocation, use, quality, and charges. Wellington residents on mains supply (the piped, government supplied water system) do pay modestly for water, but costs have been hidden within local taxes that reflect property valuation rather than actual water use. Water is billed separately to residents in the largest city of Auckland based on volume used. Wellington is now beginning pilot implementations of such metered charges for use of city water, but at the time of research this was not the case.

The United States data were collected in the desert city of Phoenix, Arizona, from residents of the Willo (N=30) and Laveen (N=30) neighborhoods. The Seven States agreement determines water rights for the Colorado River, a major source of water for western states. According to this interstate compact, Arizona has junior rights to the Colorado River and its allocation will be the first cut under shortage-sharing agreements. In 2010, it was expected that shortage-sharing would take effect the next year, 2011, unless a ten-year drought came to an end. Within Arizona, water rights are allocated according to Prior Appropriation law under the "first in time, first in right" principle that affords more water to established communities and water users. As the original residents, indigenous groups (known as native nations) have the most senior water rights, but this has only recently been recognized and water reallocations are contentious and ongoing. There are twenty-one federally recognized Native American tribes in Arizona, some of which have pending water rights adjudication. Phoenix, as one of the oldest nonindigenous settlements in the region, holds senior water rights, which enables residents to enjoy a desert oasis lifestyle exemplified by golf courses, swimming pools, and other water-intensive amenities. In contrast, many of the twenty-two satellite cities located in the metropolitan Phoenix region with junior water rights are dependent on less-abundant or less-predictable water sources. Households pay the city for water based on the volume they use for domestic consumption.

Methods

Data were collected in face-to-face interviews at each of the sites using the same survey tool. Interviewer training included consistent methods of recruiting, elicitation, and prompting for use at all the sites. The original protocol was developed and pretested in English for use in Fiji, New Zealand, and the United States.

Interpreters assisted as needed in Fiji, but since many people speak the colonial language of English, we had few issues. A Spanish version was translated for use in Bolivia. Participants were recruited conveniently in public spaces. The theoretical rationale for this approach is that if agreement exists in local cultural knowledge, this should be represented within any public space sampling. The protocol posed general and more-detailed questions about whether the water situation in their community was fair or unfair, and why, and was designed to tap locally appropriate concerns in each site (i.e., water rights in Phoenix, water access in Fiji/New Zealand, and justice in Bolivia). Interviewers recorded each of the open-ended responses by hand, and were instructed to capture verbatim responses as closely as possible. Research assistants then entered these responses into electronic formats, and translated the Bolivia data into English.

The results were analyzed using a two-stage method for analyzing themes and metathemes. First, the theme analysis was designed to examine inductively how fairness in water availability is conceptualized at a local level in Bolivia, Fiji, New Zealand, and the United States. Themes are defined as underlying dimensions of meaning that cut across a variety of texts (Wutich and Gravlee 2010). We identified themes on a country-by-country basis to ensure that none of the cultural knowledge specific to each site was lost in the initial coding process. Once we had elicited all possible themes, we selected three themes per country based on frequency of mention as the focus of our further analysis (Ryan and Bernard 2003). We then created code definitions that rigorously define the inclusion and exclusion criteria for identifying the presence of a theme in a text (MacQueen et al. 1998). After pretesting the codebook, we assessed inter-rater reliability using a kappa coefficient (Cohen 1960) to establish the validity and reliability of each code used in text analysis. To do so, we selected random 20 percent subsamples of texts from each site (between thirteen to thirty-four texts per subsample). For all twenty of the theme codes tested, final kappa scores were considered to have good reliability or better (≥ .60) (Landis and Koch 1977). With the codebook finalized, we coded each of the four country datasets. Finally, we (1) ranked themes by frequency in each site and (2) calculated the percentage of respondents who mentioned each theme by site.

Second, we designed the metatheme analysis to identify conceptualizations of fairness in water availability that are shared cross-culturally and to examine differences in how these ideas are expressed in Bolivia, Fiji, New Zealand, and the United States. Metathemes are defined as broad or overarching dimensions of meaning that cut across many smaller themes (Ryan and Bernard 2003). To find metathemes, two coders independently generated a list of metacodes based on the 240 themes in the initial elicitation of themes from the four field sites. The coders identified nine identical items; three additional items were generated by only one coder (75 percent agreement). We proceeded with the analysis based on the nine key metathemes, and collapsed the remaining three items into those themes.

Following the methods outlined above, we defined codes for each metatheme and tested kappas with a random 20 percent subsample of themes (forty-eight themes total). For all of the nine metatheme codes, final kappa scores were good or better ($\geq$ .60). We then coded the themes derived from the four field sites for the presence of the nine metathemes and used this coding to link individual respondents to metathemes. Finally, we (1) ranked metathemes by frequency in each site and (2) calculated the percentage of respondents who mentioned each metatheme by site in order to determine which metathemes were most relevant at each field site as well as across sites.

Results

Local Conceptualizations of Water-related Fairness: Core Thematic Differences

Table 11.1. Top three most frequently mentioned themes in Cochabamba, Bolivia

Theme	Respondents reporting (%)
Unreliable water vendors	46.3
Water scarcity	41.5
Water is essential	17.0

Bolivia

The most common theme to emerge from our Cochabamba interviews was that private water vendors made water delivery unfair (table 11.1). One issue was the water vendors' frequent failure to appear or provide service. Many respondents reported, "Sometimes the water vendors come, sometimes they don't." Even when the water vendor did appear, the timing of water delivery was a pervasive obstacle. Water vendors may visit a specific squatter settlement only before dawn (e.g., "They deliver the water at 5 a.m."), after most people need to leave for work (e.g., "At 10 a.m. we are still waiting"), or they deliver the water "only ... a very short time" before water runs out. Additionally, because water demand outstrips supply, water vendors are powerful and can behave capriciously. For example, people commonly reported that the water vendors "just don't stop" and "we have to run behind the water truck" and "we always have to beg" to buy water. Many others said "the water vendor is very arrogant" because "if we buy from one ... the other one refuses to bring us water." In general, people felt that being forced to buy water from private trucks was unfair because the experience is arduous and often humiliating.

The second-most-common theme to emerge from interviews in Cochabamba was the insufficiency of water to meet people's basic needs. One respondent put it simply: "There is no water." When households are unable to obtain water, several reported, household water stores run out and "sometimes we go two days without water." The scarcity of water is so severe that it restricts people's ability to complete the most basic tasks. For instance, people reported that "sometimes we don't have water to cook" or "water to drink." To meet these needs, household members seek help from family or neighbors. As one respondent reported, "I have to borrow water from my siblings." Others find that water is so scarce that "people don't even want to give each other one glass of water." In a number of ways, then, people explained that "we suffer greatly because of the water."

A third theme to emerge from our interviews was that the water situation in Cochabamba's squatter settlements was considered fundamentally unfair because "water is essential." This idea was also expressed in statements such as "water is life," "there must be water," "without water, we can't live," and "there ought to be water for everyone." Implied in these statements is the idea that people have a fundamental right to the water they need to survive. During the Water War of 2000, one of the most common slogans was, "Water is life." In 2004, Cochabamba's squatter settlements formed ASICASUR, an umbrella advocacy organization with the motto, "[W]ater is life: it is an inalienable right and not a merchandise" (UN Habitat 2010). Most recently, Bolivia was responsible for introducing UN Resolution 64/292 on the right to water and sanitation. Thus the squatters' belief that water is an essential, life-giving, inalienable right is deeply tied to the politics of water in Cochabamba.

Table 11.2. Top three most frequently mentioned themes in Viti Levu, Fiji

Theme	Respondents reporting (%)
Improved infrastructure	32.4
Inadequacy of past water system	18.9
Everyone has access to water	18.9

Fiji

Overall, the consensus regarding the water situation was high among respondents in Viti Levu, Fiji (table 11.2). The most common theme to emerge in the Fiji interviews was the recent improvement in water infrastructure. The infrastructural upgrades included a "new tank," "bigger pipes," and "better water dams." These improvements were credited with "improving water system for reliability and security." Respondents told us, "Thanks to the new project, everyone has [water] pressure" and as a result, "We can use the sink and shower at the same time." The water improvement scheme had been a focus of community development efforts in the prior two years and most male village members were directly involved in different elements of its construction, such as helping to lay the pipes. This community-based project was a source of shared pride and symbolizes some core Fijian ideals: the importance of sharing and allocating resources fairly and the importance of contributing to communal causes (e.g., Brison 2007).

The contrast between the current and past situation is also made clear in the second-most-common theme, which dealt with the disrepair and unfairness inherent in the prior water system. Some respondents mentioned that the old water system had technical problems such as "only one pipe" and "low pressure." More often, respondents emphasized how inadequate infrastructure created unfairness and social tensions. People disliked sharing water when "three houses only had one tap." In some cases, water would run out because "if one house was using water, other houses could not get it." Inequities were built into the system because households "on the end of the water line ... would not get water early in the mornings." To many, this uneven water delivery system ran counter to the Fijian ideal of "sharing and caring" (see Brison 2007: 13; Katz 1993: 27) and the importance of fair distribution of resources within communities.

Relatedly, the third key theme identified from respondents in the Fiji site was the idea that everyone now has access to water. Among respondents in Viti Levu, universal water access across the village was clearly identified as a characteristic of fairness. Respondents repeatedly said—with very little variation—that the water situation was fair because "each house has access to water." According to the respondents, the new water system "allows for fair access" that was not possible when households were sharing tapstands, ensuring "through the village all houses receive water access." Currently, "everyone has equal distribution" to their homes and the water system provides enough water so that no household runs out. In statements in this domain, respondents underscored the importance of equality, saying that the situation is fair because "everyone in each home is getting their supply."

Table 11.3. Top three most frequently mentioned themes in Wellington and Piopio, New Zealand

Theme	Respondents reporting (%)
Everyone has access to water	28.9
Easy access to water	28.4
Ample water	27.2

New Zealand

Our respondents in Wellington and Piopio, New Zealand, living in both rural and urban areas, understood water to be always available, accessible, and abundant—simply a matter of turning on a tap (table 11.3). Because of consensus around these ideas, the theme analysis was less analytically complicated than the other sites. The primary theme to emerge focused on the concept that water is universally available to everyone, commonly expressed as, "Everyone has running water." One respondent explained that the local council government "takes care of everyone and creates equal opportunities for everyone." People said that water is "piped into all houses," "Water is available in every home," or "Everyone has access to springs." As a result, there is "very fair access." That these notions of fairness around water are connected to its universal availability ties well to the deep, pervasive, and historical New Zealand cultural ideals of the need to seek and maintain both equality and social harmony (Baehler 2007).

The second key theme identified in responses from Piopio and Wellington concerned the ease of acquiring water. Informants described water access as close by and readily accessible, with descriptions of "good," "easy," "stable," "readily available," and "available everywhere." Responses underscored water availability at home (e.g., "Tap water is easy to access") and in public (e.g., "There is water at most parks and other public areas"). In New Zealand, people access water through city or local supply taken from major rivers or dammed catchment areas, through rain capture off their own house roofs, or in rural areas by tapping into local rivers or springs; thus, informant discussions of easy access to water are consistent with low-cost water systems.

The third-most-prevalent theme concerned the volume of water available. Despite the sources of water being different in urban (city mains supply) and rural (springs, rain tanks) settings, people living in both settings perceived water as always copious, far outstripping demand. Water resources were repeatedly described as "ample," "plentiful," "a large supply," and "quite a lot of water." One respondent said, "There is free water everywhere and not many people." In the cities, people said that there was a "plentiful town supply." In the farming community interviews, people said, "Living rurally means that water is readily available," and "Fresh and clean water is always available." Some respondents mentioned that there are "no shortages" and "no droughts in most places." Notably, this common view of abundant water in Piopio and Wellington does not match the current governmental positions on freshwater supply in New Zealand, which increasingly emphasizes water resource limits being met in many parts of the country. For example, the current New Zealand government strategy entailed in the New Start for Fresh Water program managed by the Ministry for the Environment (2009: 14) explains the need for more systematic planning for water use in all parts of the country, noting that New Zealand's water resources are finite and that "many New Zealanders do not understand the long-term economic and environmental risks posed by the status quo in water management."

Table 11.4. Top three most frequently mentioned themes in Phoenix, United States

Theme	Respondents reporting (%)
Differential allotment/distribution	16.7
Water pricing	15.0
Water restrictions	11.7

United States

The most common theme to emerge in Phoenix was water pricing (table 11.4). Respondents said that Phoenix residents' water rights are fair because "we pay for our usage." According to some respondents, the role of government is to supply water and the role of citizens is to pay for the water they consume, and it is fair that "we are supplied with water and we pay for it." Though one respondent described the water price as "outrageous," most respondents highlighted the affordability of water, noting the "affordable prices" and "low water bills," and that "irrigation is $250 a year." In Phoenix, utilities price water on a uniform or seasonally varied flat rate, as opposed to the block rate pricing used to discourage high-volume water use in other regions. Recognizing this, some respondents think the price of water is too low and "We should be paying dearly for it." Notably, none of the Phoenix respondents said that people have a basic right to water regardless of ability to pay or that water should be free. However, one respondent observed that Phoenix residents' water rights are fair only "as long as [people] have enough money."

The second-most-common theme dealt with differential water allotments or distributions. Several respondents felt that water allotments under the prior appropriation rule are unfair to new communities because water is "not fairly proportioned" and "Phoenix metro probably gets more than its fair share to detriment of other outlying areas." Additionally, some felt that native nations living on reservations have unfair water allotments. One respondent said, "Tribes don't have access to water running through their land," and another noted, "Why should we get water and the Hopi … reservation doesn't have water?" However, other respondents, referencing pending water adjudication, thought native nations were being awarded too much water. Beyond these controversies, people indicated, "A lot of people use more water than others," which they indicated was unfair. On the whole, however, most concerns were focused on inequities in water rights across communities.

The third-most-common theme in Phoenix was the lack of restrictions on water use. In other parts of the United States, there are restrictions on the timing and frequency of water-intensive activities such as watering lawns and washing cars. In Phoenix, however, "there are no hard-and-fast rules" and "there have never been any restrictions" on water use. Several statements indicated that this signaled fairness. Respondents said, "I can consume as much as I want to" or "I've never been limited." Yet lack of restrictions runs counter to some people's expectations for living in a water-scarce environment. One respondent said that water use rules in Phoenix are "way too loose." Some hinted that they were in support of stricter water restrictions or "would not be opposed to restrictions." Others went farther, stating water "misuse should not be rewarded, it should be stopped." People favoring more water restrictions generally felt water must be conserved in order to maintain the desert oasis lifestyle in the future.

Cross-Cultural Conceptualizations of Water-related Fairness: Nine Key Emergent Metathemes

Table 11.5. Percentage of respondents reporting nine metathemes in four international sites

Metatheme	Bolivia	Fiji	NZ	US
Water access	63.4	56.8	76.5	36.7
Water quantity	53.7	13.5	30.9	38.3
Equality and equity	17.1	10.8	37.0	40.0
Government	26.8	27.0	23.5	23.3
Infrastructure	7.3	45.9	17.3	5.0
Water cost	7.3	2.7	29.6	20.0
Water quality	7.3	8.1	19.8	1.7
Water rights	19.5	8.1	9.9	23.3
Water source	7.3	10.8	39.5	10.0

Water Access

The access metatheme deals with the extent to which water access is convenient, consistent, easily attained, and available to all (table 11.5). Across sites, respondents had different perspectives on fairness in water access. In the Fiji and New Zealand sites, respondents underscored its adequacy and availability. Several Fijians commented, "Each household has access." Respondents in New Zealand similarly noted, "Everyone has access" because water is "easy to access" and "readily available." In contrast, respondents in the Bolivia site almost universally agreed that access was inadequate, highlighting difficulties in accessing water from vendors because the trucks "never arrive" and "we have to wait" to buy water. In the U.S. site, some respondents emphasized inequities in water availability such as those across landholdings of different sizes, while others underscored the universality of access at the household level. Access to water appears to be an important part of conceptualizations of fairness cross-culturally, regardless of the ease of access in each field site.

Water Quantity

The water quantity metatheme deals with the amount of water available, water conservation, and restrictions, and the idea that adequate water is essential to survival. Respondents in the Fiji, New Zealand, and U.S. sites perceived adequate quantities of water and that this strongly indicated fairness. "Ample water" emerged from the New Zealand interviews as one of the most salient themes. In the Fiji site, too, people said, "We get enough water," but noted that would last only "as long as there is still rain." In the U.S. site, there was "plenty of water," but a few mentioned the need to conserve water. In contrast, the most salient indicator of unfairness in the Bolivia site was the inadequate quantity of water. People were concerned that "there is no water" and expressed fears such as, "I could die because of the lack of water." Water quantity was identified as a key indicator of fairness in all four field sites, but it was particularly salient in the Bolivia site due to concerns over water scarcity.

Equality and Equity

This metatheme deals with equality and equity in water provision. In sites with universal water distribution, such as the Fiji and New Zealand sites, comments emphasized equality. In the New Zealand site, respondents said, "Water is available to every class level." In the Fiji site, the new water system has "fair and equal water pressure and amount." In the Bolivia and U.S. sites, more respondents focused on inequities in water allotment, usage, pricing, and restrictions. In the U.S. site, many people felt native nations or cities received more or less than their "fair share" of water. In the Bolivia site, concerns were focused on the denial of access to water because "renters do not have water" and, with water vendors, "only some people get to buy water." This metatheme was common in all four field sites. However, equality was more salient in sites with universal water access, while equity was more salient in sites in which water access is not evenly proportioned across households and communities.

Government

The government metatheme deals with local and national water regulation, taxation, and political leadership, revealing government's complex role in perceived fairness of local water distribution. In the New Zealand site, government is perceived to play an appropriate role in taxing citizens and providing high-quality water service, but disagreement emerged between those who believe "the government does not control the water supply" and those who believe government "control[s] availability." In the U.S. site, concern with the need for more restrictions on water use and approval of the lack of restriction were common governance-related responses. In the Bolivia site, people frequently opined, "The leaders do not *caminar* (walk)," indicating that ineffective community-level leadership, via a failure to secure water rights for the squatter settlements, was to blame for water scarcity. A related perspective was corruption: "The leaders are lying sometimes; they are just swindling us." In the Fiji site, respondents credited the village government for the new water project "because the water committee ensures there is water" and "the water committee wants everyone to have fair access to water in the village." To accomplish this, however, the "village made a deal" with a variety of external groups, including regional universities and foreign government aid agencies, and the project was "funded by the New Zealand government." The role of governments emerged as a common theme cross-culturally, regardless of the abundance or scarcity of local water resources, though sites with the greatest proportions of displeasure toward the water system (Bolivia and the United States) held correspondingly critical views toward government's role.

Infrastructure

The infrastructure metatheme deals with the physical water delivery system including taps, pipes, trucks, tanks, and water pressure. This metatheme was most prominent in the Fiji site, where recent and extensive improvements, such as dams, tanks, water pipes, and in-house taps, are credited with "improving the water system for reliability and security" and greatly enhancing fairness. In the New Zealand site, where infrastructure is universally considered to be satisfactory, a standard comment was, "Water is available in every home and public place through the tap." In contrast, in the Bolivia site few respondents mentioned infrastructure, but some noted that there is virtually no infrastructure for water delivery or sewerage in squatter settlements. Finally, water infrastructure in the U.S. site was considered adequate for household use but inadequate for some urban agricultural needs, as exemplified by one comment: "The flow of irrigation is not constant every month." Across the sites, infrastructure was not a highly salient indicator of fairness unless the site had recent changes in local water infrastructure.

Water Cost

The water cost metatheme focuses on the need to pay for water and the affordability of water. Cross-culturally, the responses showed large disparities in the ways people perceive fairness in water costs. Respondents in the New Zealand and U.S. sites expressed the idea that water pricing was fair—but for different reasons. In the New Zealand site water was "free" with "no dollar figure attached" or "included in city council rates" at a very low cost. Respondents in the U.S. site approved of the water affordability, but some favored pricing that recognized the value of water as a scarce resource. In the Bolivia site, water pricing increased water insecurity for impoverished households because it costs "money, money, and we never have enough money." In the Fiji site, pricing was not frequently mentioned because people "never pay water bills," as tap water is seen as free from cost. Respondents in the wealthy sites, in New Zealand and the United States, made specific comments about water cost more frequently than in the less wealthy sites, in Bolivia and Fiji.

Water Quality

The water quality metatheme deals with the purity, taste, and cleanliness of water. In the New Zealand site, the water is "clean and pure," "the best," and "amazing," although some are "just hoping it stays clean and free of contamination" as there have been "boil water advisories" and contamination related to "dairy farming." In the Fiji site, a few people mentioned that water can cause "stomach ache," "diarrhea," and "skin diseases." In the Bolivia site, respondents mentioned varied concerns such as, "We do not know if the water is contaminated" or "dirty." Others mentioned that their water is "very salty." In the U.S. site, although frequent complaints about poor taste have previously been recorded (Gartin et al. 2010), in this study only one respondent raised a quality comment dealing with the cleanliness of canals used to transport water through the city. Thus, cross-culturally, water quality was not a highly salient indicator of fairness except where people were focused on either excellent or poor water quality.

Water Rights

The water rights metatheme deals with water ownership, control, entitlement, and allotment. It addresses public and private ownership of water, how rights are tied to land or location, and the ownership by specific populations (especially indigenous groups). In the Fiji site, water rights were fair because the community members "own access to water" and "manage [their] own water." In the New Zealand site, some people felt they had "full control" because they have "access to [their] own water supplies" such as springs or streams. In the U.S. site, diverse opinions were raised about "property ownership," "tribes" and "reservation land," "economic factors," "longtime law," "state control," and "fairly proportioned" rights. In the Bolivia site, "there ought to be water but there isn't any." Cross-culturally, water rights are a much more salient issue where water is scarce and ownership is disputed, as in the U.S. and Bolivia sites, than where water is abundant and such disputes are largely unnecessary, as in the Fiji and New Zealand sites.

Water Source

The water source metatheme describes natural sources such as rivers, streams, or lakes and market sources such as bottled water or water vendors. Across the four sites, the sources shaped perceptions of fairness. In the New Zealand site, water access was fair and secure because of the many and rich natural water sources available including "plenty of rainfall," an "abundance of stream water," and "many lakes and rivers." In the Fiji site, several people noted, "We drink from our own natural source," but this theme was not especially salient. In the U.S. site, a handful of responses mentioned the Colorado River and related reservoirs. In the Bolivia site, a couple of people noted, "We are in a dry environment" or "There is no river or well," but relatively few identified water sources as a main reason for water scarcity. As with water quality, water sources did not appear to be a highly salient indicator of fairness except for the New Zealand site, where water is ubiquitous and the landscape contains a multiplicity of water sources.

Discussion and Conclusions

Here we have presented a very basic model of local perceptions of fairness around water in four different field sites to determine (a) if there are any shared ideas across socially and ecologically distinctive sites, as well as (b) how any key differences might be grounded in local cultures, ecologies, and water governance systems. The goal is to explicate a preliminary theory of concepts of fairness around water that might be applicable cross-culturally. So, what do we find is agreed by all to be fair? In our meta-analysis, we traced commonalities across the sites, finding nine cross-culturally shared, core concerns that people recognized as fundamental to evaluate the fairness of water issues: water access, water quantity, equality or equity, the role of government, infrastructure, water cost, water quality, water rights, and water sources. Based on the metatheme analysis, we find four key domains—water access, water quantity, the role of government, and equity and equality—around which there is consistent concern.

By contrast, water quality, water cost, water source, water rights, and infrastructure turn out to be relevant to people only in some sites. The distinction between water-rich and water-scarce sites and wealthier versus poorer economies seems to provide much of the contextual explanation of this variation. For example, in semi-arid and impoverished Bolivian shantytowns, the most common theme expressed by people is the unreliability of water vendors, who are the main source of water access in squatter settlements. Other common themes are related to the scarcity of water and, to a lesser extent, the fundamental human right to water. This highlights the importance of local water procedures and institutions in shaping perceptions of fairness, particularly when water is scarce. By contrast, in Phoenix, in the United States, key themes deal with the fairness of rules for water allotment, pricing, and restrictions. In this desert city, local governance developed extensive institutions to secure and protect water availability in the long term. Yet our analyses suggest that, even with democratically developed institutional mechanisms, water scarcity itself engenders divergent opinions over what is fair. In the Fiji site, common themes deal primarily with the adequacy of water infrastructure in the local village, highlighting the social value of collective, community-based action to solve local problems, but also suggesting the importance and cultural saliency of engineered solutions to water problems. Even when rules and institutions reflect local values, engineering failures can undermine the fairness of such systems. In Piopio and Wellington, New Zealand, responses reflect the universality, ease of access, and abundance of water. This critically shows that where water is plentiful and highly affordable we may expect little or no discord around issues of fairness. And, as scarcity and costs rise, we would predict that discord around key notions of fairness will increase.

Importantly, our findings can be directly related to the global movement toward defining water as a basic human right discussed at the opening of this chapter. Particularly, seven of the nine identified metathemes that comprise our preliminary cross-cultural theory—water access, water quantity, equality or equity, infrastructure, water cost, water quality, water rights—are referenced in some manner in the UN CESCR General Comment 15 where the human right to water is defined as "sufficient, safe, acceptable, physically accessible and affordable water" (UN Committee on Economic, Social and Cultural Rights 2003: 1). The similarities between the results of our metatheme analysis and General Comment 15 suggest that the evolving global definition of the human right to water, to a large extent, reasonably reflects views on the ground.

However, our findings also show that there is another set of shared concerns that are not well developed or well represented in the current international agreements. A major concern is the role of national governments. Fairness in this context depends on people's perception that their interests are represented, and yet water governance and management is not always democratic and transparent. Particularly, we observe that where water is scarce, there is an unmet need for institutions that establish equity. Such institutions should establish what base water rights are afforded to individuals and communities and make clear which water uses and needs are prioritized and protected. Water rights at the national level should be more clearly defined, be they the fundamental human right to water claimed by people in Bolivia or community rights to water in the western United States. The World Health Organization's (WHO; 2003: 31) review of General Comment 15 does recognize this need for national governments to more clearly define water rights at individual and community levels, and to enhance the transparency of development and enforcement. To succeed, national governments need to engage in this process in a way that is sensitive to local variations in perceived needs and values.

Our analysis also suggests some pathways forward for future research and ultimately improved policy development and implementation. Certainly, collecting comparable data from a wider range of socioecological settings will expand our ability to identify which aspects of fairness are the most salient when viewed on a larger, global scale, and hence which could have more utility to setting shared policies around fair water access and distribution. In terms of building a more sophisticated theory of fairness related to the human right to water, we also need to develop and test core hypotheses around why notions of fairness might vary from place to place. If the differences prove to be mostly tied to ecological factors (e.g., water-poor or water-rich), such as we observed in the patterns of concern over water rights in this analysis, rather than—say—sociocultural factors (e.g., collectivistic or individualistic cultural beliefs), then this has implications for how we can conceptualize and implement the human right to water in a meaningful and sustainable way. For instance, in water-rich sites in Fiji and New Zealand, respondents generally pointed to equality in water distribution as evidence of fairness. In water-poor sites in Bolivia and the United States, comments were much more focused on the need for equity across social groups. If this pattern holds cross-culturally, we may expect countries and communities to become increasingly concerned with issues of equity as freshwater resources shrink locally and globally and costs to individuals go up. Equity—a relatively neglected concept in water research—may out of necessity come to the forefront of both global and local water planning and provision.

Notes

We thank the in-country research directors, student researchers, and study participants who contributed with their time to this research. The work was locally seeded by the National Science Foundation (NSF) Grant No. SES-0345945, Decision Center for a Desert City (DCDC), and NSF grant number DEB-0423704, Central Arizona–Phoenix Long-Term Ecological Research. We received funding supporting the international research from the Arizona State University Late Lessons from Early History program. Any opinions, findings and conclusions, or recommendations expressed in this material are those of the author(s) and do not necessarily reflect the views of the funding agencies.

References

Araral, Eduardo. 2010. Improving Effectiveness and Efficiency in the Water Sector: Institutions, Infrastructure and Indicators. *Water Policy* 12 (S1): 1–7.

Baehler, Karen. 2007. Social Sustainability: New Zealand's Solution for Tocqueville's Problem. *Social Policy Journal of New Zealand* 31: 22–40.

Bluemel, Erik. 2004. The Implications of Formulating a Human Right to Water. *Ecology Law Quarterly* 31: 957–977.

Boelens, Rutgerd. 1998. Equity and Rule-Making. In *Searching for Equity: Conceptions of Justice and Equity in Peasant Irrigation*, edited by Rutgerd Boelens and Gloria Dávila. Assen: Van Gorcum.

Brison, Karen. 2007. *Our Wealth is Loving Each Other: Self and Society in Fiji*. Plymouth: Lexington Books.

Budds, Jessica, and Gordon McGranahan. 2003. Are the Debates on Water Privatization Missing the Point? Experiences from Africa, Asia and Latin America. *Environment and Urbanization* 15 (2): 87–113.

Cohen, Jacob. 1960. A Coefficient of Agreement for Nominal Scales. *Educational and Psychological Measurement* 20 (1): 37–46.

Gartin, Meredith, Beatrice Crona, Amber Wutich, and Paul Westerhoff. 2010. Urban Ethnohydrology: Cultural Knowledge of Water Quality and Water Management in a Desert City. *Ecology and Society* 15 (4): 36.

International Service for Human Rights (ISHR). 2010. GA Pre-empts Council and Recognises the Human Right to Water and Sanitation. http://www.ishr.ch/ga-64th-session-2009/858-ga-pre-empts-council-and-recognises-the-human-right-to-water-and-sanitation

Ingram, Helen, John M. Whiteley, and Richard Perry. 2008. The Importance of Equity and the Limits of Efficiency in Water Resources. In *Water, Place, and Equity*, edited by John M. Whiteley, Helen Ingram and Richard Warren Perry. Cambridge: MIT Press.

Katz, Richard. 1993. *The Straight Path of the Spirit: Ancestral Wisdom and Healing Traditions in Fiji.* Rochester, NY: Park Street Press.

Landis, J. Richard, and Gary G. Koch. 1977. The Measurement of Observer Agreement for Categorical Data. *Biometrics* 33 (1): 159–174.

Lauderdale, Pat 1998. Justice and Equity: a Critical Perspective. In *Searching for Equity: Conceptions of Justice and Equity in Peasant Irrigation*, edited by Rutgerd Boelens and Gloria Dávila. Assen: Van Gorcum.

Louka, E. 2006. *International Environmental Law: Fairness, Effectiveness, and World Order.* Cambridge: Cambridge University Press.

MacQueen, Kathleen M., Eleanor McLellan, Kelly Kay, and Bobby Milstein. 1998. Codebook Development for Team-Based Qualitative Analysis. *Cultural Anthropology Methods Journal* 10: 31–36.

Ministry for the Environment. 2009. New Start for Fresh Water. Cabinet paper issued by office of the minister. http://www.mfe.govt.nz/issues/water/freshwater/new-start-for-fresh-water-paper .html

Murray-Rust, Hammond, Bakhshai Lashari, and Yameen Memon. 2000. Water Distribution Equity in Sindh Province, Pakistan. Working Paper 9. International Water Management Institute, Lahore, Pakistan.

Oliverio, Annamarie. 1998. Reclaiming Equality, Equity and Diversity. In *Searching for Equity: Conceptions of Justice and Equity in Peasant Irrigation*, edited by Rutgerd Boelens and Gloria Dávila. Assen: Van Gorcum.

Ryan, Gerry, and H. Russell Bernard. 2003. Techniques to Identify Themes. *Field Methods* 15 (1): 85–109.

Syme, Geoffrey J., Elisabeth Kals, Blair E. Nancarrow, and Leo Montada. 2000. Ecological Risks and Community Perceptions of Fairness and Justice: A Cross-Cultural Model. *Risk Analysis* 20 (6): 905–916.

UN Committee on Economic, Social and Cultural Rights (CESCR). 2003. *General Comment No. 15: The Right to Water (Arts. 11 and 12 of the Covenant)*, 20 January 2003, E/C.12/2002/11, available at: http://www.refworld.org/docid/4538838d11.html.

UN Habitat. 2010. Strengthening the National Service for the Sustainability of Basic Sanitation Services (SENASBA) for the joint work with water community systems associations in peri-urban areas of Cochabamba, Bolivia. http://www.unhabitat.org/content.asp?cid=8376&catid=577& typeid=61&subMenuId=0

Weinstein, Philip, Nina Russell, and Alistair Woodward. 2000. Drinking Water, Ecology, and Gastroenteritis in New Zealand. In *Interdisciplinary Perspectives in Drinking Water Assessment and Management*, edited by Eric G. Reichar, Fred S. Hauchman, and Ana Maria Sancha. Proceedings of the Second International Symposium on Assessing and Managing Health Risks from Drinking Water Contamination: Approaches and Applications, held at Santiago, Chile, in September 1998. IAHS Publication no. 260. Wallingford, Oxfordshire, UK.:IAHS Press.

World Health Organization (WHO). 2003. *The Right to Water.* Health and Human Rights Publication Series 3. http://www2.ohchr.org/english/issues/water/docs/Right_to_Water.pdf/

Wutich, Amber. 2009. Estimating Household Water Use: A Comparison of Diary, Prompted Recall, and Free Recall Methods. *Field Methods* 21 (1): 49–68.

Wutich, Amber, and Clarence Gravlee. 2010. Water Decision-Makers in a Desert City: Text Analysis and Environmental Social Science. In *Environmental Social Sciences: Methods and Research Design*, edited by Ismael Vaccaro, Eric Alden Smith, and Shankar Aswani. Cambridge: Cambridge University Press.

Chapter 12

Indigenous Water Governance and Resistance

A Syilx Perspective

Marlowe Sam and Jeannette Armstrong

∞

The social life of water related to water governance, for indigenous peoples, represents struggle against colonial exclusions under water law and injustices in water governance as one aspect of the much larger injustice of the forces of globalization and the overarching resistance the world over by indigenous peoples to political annihilation.[1]

Modern resistance to the privatization of water was brought to the world's attention in January 2000 when indigenous peoples shut down the entire city of Cochabamba in Bolivia to protest the economic impositions created by Bechtel, a U.S. corporation that had assumed private control of water distribution in the city (Barlow and Clarke 2002: 138, 155).

According to Oscar Olivera, a leader of the resistance movement, the water distribution system in Cochabamba became privatized in 1999 under a condition imposed by a World Bank loan that stipulated the elimination of public subsidies that governed drinking water and sanitation. Compounding the situation was Bolivia's Law 2029 that eliminated any and all guarantees of water distribution to the rural areas of Bolivia and demanded that autonomous water systems be handed over without compensation (Olivera 2004: 8–9). Rocio Bustamante, senior researcher with the Water Law and Indigenous Rights Project contends that ownership of water had been a critical issue in Bolivia since 1879. Bustamante adds that continued resistance spearheaded by the indigenous intelligentsia of Bolivia had led to autonomous control over large territories in Bolivia and also had promoted the growth of its national advocacy movement (Bustamante 2006: 125). According to Bustamante, the principles and values of the indigenous peoples strongly influenced the outcome of struggles over proposed changes in laws regarding ownership and use by the people of Bolivia (Bustamante 2006: 120).

Despite launching complaints to the Bolivian government and global lending institutions, the foreign corporations were ousted from Bolivia, allowing Bolivians to reassume control of their water resources, an action inspiring indigenous peoples and others the world over. According to applied ethics author Michael K. Green (1995: 2), the Zapatista's uprising in the Mexican state of Chiapas was triggered by the signing of NAFTA but was influenced by preceding collective action and solidarity against water privatization by the indigenous people throughout Central and South America.

Water, Globalization, and the Indigenous Rights Movement

In 2003, the authors attended the World Water Forum in Kyoto, Japan, which included an official session on Water and Indigenous People. The Forum included a section focused on the struggles for recognition, a section on worldviews and water management, and a section on water rights and national legislation, as well as an indigenous peoples caucus to create an Indigenous Peoples Kyoto Water Declaration (UNESCO 2006: 1–16, 174–179).

The indigenous peoples caucus met on a collective position to deliver an indigenous declaration on water to the Forum. Indigenous representatives relayed stories of how their governments and foreign corporations were imposing restrictions that denied access to critically needed water resources. At one of the strategy sessions, a discussion arose over the use of the word "consultation"; the discussion illustrated a critical point in the complexities indigenous peoples face under colonizing forces and in particular critical differences in water laws imposed by colonizing governments. The South American delegation, which included the Bolivian grassroots organization led by Evo Morales who subsequently became the president of Bolivia, repeatedly reminded other delegates that indigenous peoples were never consulted in any decision-making processes that involved resource development and resource extraction in their traditional territories, and so consultation needed to be a critical point. The Canadian delegation, however, explained that the Canadian government uses the term "consultation" to tout obligatory transparency to stakeholders, but that Canada continues to effectively block any real participation by indigenous people in decision making, and by doing so masks ongoing appropriations of unceded natural resources. Indigenous leaders from the Asian delegations and leaders of the Tebtebba Foundation, Joji Carino and Victoria Tauli-Corpuz, requested the North American and South American delegates to discuss the problem among themselves with instructions to return only when a consensus had been reached.

After a great deal of deliberation and explanation through interpreters that the term "consultation" presumes government possession, the indigenous peoples caucus agreed that the word "consultation" would be left out of the indigenous declaration. In the end, however, the term found its way into the final draft of the Indigenous Peoples Kyoto Water Declaration.

In 2006 the authors attended the "Indigenous Peoples' Resistance to Economic Globalization" conference hosted in New York City by the International Forum on Globalization. Among the other invited participants were frontline indigenous and nonindigenous activists who recounted their efforts to resist the advancement of economic globalization within indigenous territories around the world. The many stories of resistance actions going on in indigenous peoples' traditional territories brought back memories to the authors of a large tent meeting on the plains of South Dakota and the heart-wrenching testimonials delivered at the International Indian Treaty Council gathering held at Wakpala, South Dakota, during the summer of 1976. Participants repeated the same stories almost word for word during the 2006 conference.

The International Indian Treaty Council evolved as an international voice of indigenous resistance out of the American Indian Movement (AIM). AIM, in its earliest form, was mostly made up of urban-based chapters that spread across the United States from Cleveland, Ohio, to the West Coast (Smith and Warrior 1996: 127–148). According to historian Peter Mathiessen (1991: 34), in 1968 a small group of American Indians in Minneapolis, Minnesota, formed AIM with the initial intent to heighten awareness about civil rights violations that included police abuse and violence against Indian populations in that Midwestern city.[2] Throughout the 1970s, the leadership of AIM organized direct actions to counteract the aggressiveness and violence associated with the colonization policies of the European settler society. During an AIM organized gathering on the Standing Rock Indian Reservation in South Dakota in 1974, more than five thousand representatives of ninety-eight indigenous nations formed the International Indian Treaty Council (2011). In November 1977, indigenous representatives from the United States and Canada traveled to various European countries to make it known that an indigenous liberation movement existed in North America. According to AIM spokesperson William (Bill) Means (1981), the establishment of the Document Center for Indigenous Peoples (http://www.docip.org/) was a result of this action and led to the indigenous peoples being given a place on the agenda at the United Nations in 1978. In the same year, the International Indian Treaty Council became the first organization of indigenous peoples to be recognized as a nongovernmental organization (NGO) with consultative status to the United Nations Economic and Social Council (Means 1981).

The 2006 indigenous peoples conference on economic globalization occurred thirty years after the International Indian Treaty Council gathering at Wakpala, South Dakota, and yet indigenous peoples in the Amazon basin were still being hunted and forced from their traditional territories to accommodate resource development schemes intended for commoditization in the global economy. Vilson Benedito de Oliveria, traditional chief of the Tupininkin tribe whose traditional homelands are located in the rainforest jungles of Brazil, related how his people were forced to resist and defend themselves with traditional hunting bows and arrows while development company mercenaries used modern armaments and weaponry. Traditional leaders reported many unjust social and environmental issues from Canada, Central America, Mexico, South America, the Philippines, and the United States; delegated spokespersons and leaders from these countries sought assistance to resist and combat corporate development in their respective countries. Leaders from Brazil, Ecuador, and Peru related stories about how resource development corporations had forcibly displaced indigenous peoples of their countries from their villages as a result of exploratory oil drillings; leaders from the Philippines related stories about how large water and mining development projects were reported to be causing massive destruction to the environment, while displacing upwards to twelve million indigenous peoples from valuable agricultural lands.

Indigenous activists from Canada and the United States shared similar stories of rapid environmental changes affecting the plants and animals in wilderness areas of Canada, which they continue to utilize as traditional subsistence sources in territories that have not been legally ceded. Foreign investment companies are allowed to harvest large amounts of timber from many forests unceded by treaty with indigenous peoples of Canada and to export raw and finished products to overseas markets. At the same time, the economic development projects and the consequent environmental degradation have negative consequences within the indigenous communities and are reflected in the social, cultural, and health problems rampant in those communities.

It is increasingly clear that the phenomenon termed "globalization" is not new and has impacted indigenous peoples of the world for many centuries. The advancement in military technologies and the ability to cross oceans allowed Western Europeans to establish nationalized colonial authority over indigenous peoples and the lands they own and occupy; there is no question that the process of colonization is directly based on opening up the ability for trade and commerce back to colonial empires. Madame Daes, UN special rapporteur on indigenous issues, has characterized globalization as an extension and assertion of "military power and systems of trade to other continents" (Daes 2008: 77–78). The authors concur with Daes. From the indigenous standpoint, the colonial history of North America is based on overt militarism as its governments control and influence dispossessions through military threat masked under the rubric of upholding domestic laws. Many of these laws are designed to protect trade and commerce dependent on economic benefits arising out of an antagonistic and unjust commoditization of the natural resources of the Americas. Indigenous peoples are forced into collaborative actions dependent on political relationships to curtail the unjustifiable acts that threaten their lives within the particular ecosystems that are their homelands. It is clear that the foundational principles of globalization impose and maintain the economic structures and its legal underpinnings that are held in place through military dominance. Globalization, from the indigenous standpoint, can be thought of as subsuming the phenomenon of colonization, engulfing indigenous peoples and their lands in domestic law as part of an economic order revolving around and centered on the systematic global movement of natural resources. This movement chiefly benefits foreign investment interests through trade agreement lawmaking designed to exclude, dispossess, and disengage indigenous and local autonomies from decision making.

Historic Patterns of Indigenous Resistance in North America

Whether overt or indirect, the threat of militarism defines the political approaches taken by indigenous peoples in North America. Resistance movements by indigenous peoples in North America who were continuing to practice their inherent rights to fish for salmon in the 1960s and 1970s were a response to the aggressive, historical, and political development actions of special interest groups that had the backing of local, state, and federal governments. Fishing disputes in northern California and in the Pacific Northwest in both Canada and the United States offer many examples of violence by state and provincial game officials and police departments who engaged in violent acts against the indigenous peoples under the rubric of enforcing laws protecting settler interests. The very existence of a tribe has come under legal challenge to protect settler commercial and sport fishing interests.[3] An example is in the legal court action by the state of Washington in the 1964 superior court finding against the Puyallup and Nisqully Tribes of Oregon and Washington State that used the argument, there was "no Puyallup tribe." The court, citing the Puyallup reservation allotment acts of 1893 and 1904, said the Acts, in effect, had abolished the reservation and the fishing rights that had been accorded under the Treaty of Medicine Creek of 1854. The judge issued an immediate injunction against all net fishing by the Puyallups that was followed by resistance and ensuing extreme violent confrontations on the river as the Puyallups and Nisquallys continued to exercise their rights, until the 1968 appeal of *the Puyallup Tribe v. Department of Game* confirmed their treaty protected rights to fishing off and on the reserve (American Friends Service Committee 1970: 94–97).

In British Columbia where treaties had not been concluded, the historical process of the legal capture of salmon was drawn from centuries-old English law as a common law doctrine that constructed fish as an open-access resource and therefore accorded public rights to fish and constrained the Crown's ability to allocate exclusive fisheries. Government encouragement of the industrial commercialization of fishing provided "cannery control of the resource" (D. Harris 2001: 55–78), erasing the long history of indigenous laws residing in preexisting exclusive conservation management and control of each fishery. Creation of the Federal Department of Marine and Fisheries in 1868 and application of the Marine and Fisheries Act in British Columbia in 1877 legally circumscribed indigenous peoples' rights to net and weir fishing, instituting a distinction between what was called "Indian food fishing" and commercial fishing. The Act allowed fisheries' agents to open and close fishing at will, favoring commercial access and limiting indigenous harvesting to food fishing, not as a right but simply as a privilege, and dispossessing them of any economic benefits (D. Harris 2001: 55–78). Intensive resistance to the cruelties that took place as a result of opening aboriginal fishing for extremely short periods, in poor or absent salmon run times, led to violence and confrontation and subsequent prosecutions throughout British Columbia in the ensuing years. After decades of legal and political struggle, the 1990 *Sparrow* decision of the Supreme Court of Canada determined a priority of rights in fisheries, in which conservation came first, followed by the rights of indigenous peoples to take fish for food, social, and ceremonial purposes. In most cases protected rights to commercial harvests are in limbo, however, pending hoped-for treaty-negotiated "harvest agreements". An interim and controversial aboriginal pilot sales program was terminated in 2003 as a result of a lower court ruling (*R. v. Kapp 2003*) in which the judge determined that the sales program was "racially discriminatory" (Wright 2004: 169–70). The lower court ruling was overturned in a series of appeals and in 2008 the Supreme Court of Canada (*R. v. Kapp 2008*) ruled that the constitutional guarantees in the *Charter of Rights*, specifically under s. 25, protected the rights and freedoms of natives in this case. While R. v. Kapp was connected to natural resource rights pertaining to the salmon fishery, the legal determination in the appeal could have immense ramifications on other natural resources within unceded indigenous territories in the Canadian Province of British Columbia.[4]

In terms of the historical circumstances that unraveled here in North America, greater clarity needs to be brought to bear on the legal theory related to indigenous title and rights, especially as that theory pertains to the international agreements that underpin the mechanisms of globalization. A response to the effects of globalization is the ongoing resistance to the continuation of the forces of colonization unjustly dispossessing indigenous peoples of their lands and resources. The aboriginal or native societies of the North American continent are considered indigenous peoples based on their long-term occupancy of their ancestral domain on this continental space. The ability to have survived and thrived in the local natural world environment of their ancestral domain is considered the eminent requirement to be considered indigenous to a specific place. According to Madam Chairperson of the UN Permanent Forum on Indigenous Issues, Victoria Tauli-Corpuz, and Tebebba Foundation organizer Erlyn Ruth Alcantara, a basic ambiguity lies in the interpretation of "ancestral domain" (Tauli-Corpuz and Alcantara 2004: 69). Tauli-Corpuz and Alcantara (2004: 69) argue that ancestral domain refers to "all the land and resources collectively held by a distinct people as their inheritance from ancestors in stewardship for the generations yet to come. It is, in short, a peoples' patrimony. It covers both water and land rights, surface and subsurface rights."

In the discussion on ancestral domain, Tauli-Corpuz and Alcantara argue for a peoples' patrimony over water and land rights, surface and subsurface rights. It is a well-documented common practice, however, for colonizing nations to create and enforce a political environment that deliberately diminishes indigenous rights to water, land, and other natural resources. The argument of ancestral domain clearly dispels claims for indigenous rights for settler families whose ancestors have cultivated lands or lived in a specific place for several generations. The legal distinctions between indigenous or aboriginal rights and the rights of the settler over water and land is made more complex, however, as a result of domestic lawmaking that historically has favored settler rights over indigenous or aboriginal inherent rights. European colonization inaugurated radical transformations that altered the long-term coexistence between indigenous peoples and their natural world. Land tenures by settler populations led to the creation of new geopolitical boundaries that set in place processes that severed indigenous peoples from their natural world. Forced removal and displacement from land imposed restrictions on water and other resources that disallowed freedom of movement to vital subsistence procurement sites and inhibited the ability of indigenous peoples to continue ancient customary relationships and responsibilities within their ecosystems. In that way, the creation of reservations added another layer to the question of water rights situated in treaty recognition and reservation-based rights. As indigenous peoples were severed from customary resources, their fight for survival turned to agricultural alternatives on the lands to which they were relegated and the struggle for survival became increasingly one of struggle for access to water for subsistence agriculture.

In 1908, the U.S. Supreme Court in *Winters v. United States* decided that the indigenous peoples on the Ft. Belknap Indian Reservation in Montana held preexisting water rights. Without going into details on the precedent established in *Winters v. United States* (1908: 564), the ruling established a doctrine of reserved water rights for indigenous peoples that transformed the manner in which water was to be appropriated under the doctrine of prior appropriation. Reisner and Bates (1990: 62–63) contend that the colonial powers employed a doctrine of prior appropriation that bestowed superior rights to the first licensed user of waters in a stream, and in that way based exploitation of the resource on a concept of beneficial use rather than on land ownership.

Despite the *Winters'* decision, U.S. government agencies routinely ignored the prior appropriation rights of indigenous peoples, as did Canadian government agencies. In 1877, prior to the *Winters'* decision, the Joint Indian Reserve Commission of British Columbia officially recognized the water rights of indigenous communities, but the provincial government promptly disavowed these rights. According to historian Olive Dickason (2002: 303–304), the Joint Indian Reserve Commission was formed in 1876 to resolve a deadlock between the province of British Columbia and the federal government in Ottawa concerning the land allocation for the Indians of British Columbia. The Commission was continuously active from 1876 to 1878 and intermittently convened until 1910 (Dickason 2002: 304). Despite the Commission's recognition of existing indigenous water rights, settler water licenses were given priority on streams that flowed through reserve lands. A local example can be found in the way the Joint Indian Reserve Commission acknowledged that the Penticton Indian Band held indigenous priority rights to water resources (Jolly 1999: 13). Almost immediately following this ruling, the province of British Columbia disclaimed the Commission's acknowledgement of priority rights to water on the Penticton Indian Reserve and granted to settlers hundreds of water licenses on streams that flow across reserve boundaries. These licenses were granted without regard for the indigenous water use required for supporting the natural habitats that supported their food subsistence supply or the newly imposed need for acculturation to agricultural water use.

In 1909, the province of British Columbia created its own water act to accommodate the needs of growing agricultural water use by settler populations. It was a time in which the water resources were more than plentiful to accommodate British Columbia's growing population. For indigenous peoples in British Columbia, historical land tenure under the Department of Indian Affairs policy determined whether individuals, families, or bands would be able to engage in agricultural practices. Water allocations for their agricultural needs depended on the availability of remaining water resources after settler needs had been accommodated.

The Joint Indian Reserve Commission granted the Penticton Indian Reserve, along with several other reserves in the Okanagan region, in 1877 to avoid the likelihood of Okanagan warriors joining a war effort to the south, across the Canada–United States border, or of those warriors joining in a war alliance with the Shuswap to drive out the settlers (C. Harris 2002: 29,31). Soon after the establishment of the Penticton Indian Reserve, many of the Penticton Indian Reserve families became engaged in the ranching industry. Employing natural flood irrigation methods on riparian bottomlands adjacent to the Okanagan River, families converted lowland meadow into hay fields to produce adequate amounts of winter feed for the large herds of cattle and horses they were raising on reserve lands.

In 1951, the Joint Board of Engineers, appointed by the government of Canada and the province of British Columbia, proposed a flood-control project that would remove the oxbows of the meandering Okanagan River that flowed between Okanagan and Skaha Lakes. Despite warnings of the possible environmental consequences, the project proceeded unabated and was completed in 1953 (Clark 1956: 10; Symonds 2000: 5). The riparian and wetlands habitats adjacent to the Okanagan River were essentially destroyed within a few months following the completion of the flood-control project.

At that time, a majority of the members of the Penticton Indian Reserve were living in the lower village area west of the Okanagan River on and at the periphery of the floodplain area, and they experienced the full force of these changes. The ecological disturbances created a cascading effect that extended beyond the riparian areas and had immediate and long-lasting negative effects on the social, cultural, economic, and political structures of the Penticton Indian Band membership.

During a 2004 research project, interviews of Penticton Indian Band elders revealed that the once-prolific producing hayfields suddenly had dried up once the flood control project had been completed, forcing many Penticton Indian Band members to sell their cattle and horse herds. Many of these families who were dependent on income from the ranching industry were forced to relocate to the United States to find labor jobs in orcharding to support themselves. Many of these families never returned to the Penticton Indian Reserve and remained in the United States (Sam 2008: 69).

Other water diversion allocations supported by provincial water acts gave up-stream users unrestricted access to the limited water supplies on Shingle Creek as noted during the late nineteenth century (Jolly 1999: 21). All the creeks running through the reserve, including Shingle, Shatford, and Trout Creeks, originate out-side the Penticton Indian Reserve boundaries. Members of a former ranching-dependent family living on the reserve at the confluence of Shingle and Shatford Creeks contend that overuse of waters by upstream users in the early 1970s did not leave adequate amounts of water to continue flood irrigation of their hayfields next to the creeks (Sam 2008: 69). As a result, the hayfields dried out, making the raising of cattle and horses a nonviable economic venture.

Modernization of the British Columbia Water Act

After a century of population growth, the competition over water resources has prompted a new need for radical changes in the management and allocation of this natural resource, and the province of British Columbia is engaged in modernizing the province's water legislation. One of the principles embedded within the mod-ernization process is the accommodation of aboriginal interests and needs related to water as a result of the 1997 *Delgamuukw v. British Columbia* (1997) decision that upheld the concept of aboriginal rights to traditional resource use. This decision stated, however, that the aboriginal right is a shared right that must be reconciled with the interests of the broader society and that puts the onus on indigenous peo-ples of British Columbia to prove the existence of inherent entitlements to land and resource use. Although many aboriginal groups consider the *Delgamuukw* deci-sion a victory, it remained silent on the issue of aboriginal title, opening instead a wide path for entitlements and rights to lands and resources to be extinguished through a negotiated consent process. The public discussion on water regarding

Canada's policy differs from U.S. policy after the *Winters* decision in that Canada requires indigenous peoples to either "(1) negotiate and enter a succession treaty in exchange for limited water and land rights which will remain subject to federal and provincial laws; or (2) prove the existence of aboriginal (original) title, or aboriginal rights in water (e.g., to fish, protection of watersheds, or actual use of water) in a costly and time consuming court process" (Walkem 2004: 2–3).

The Water Act Modernization (WAM) initiative calls on numerous water-dependent users such as cattle ranchers, mining and forestry interests, agriculturists, municipalities and First Nations to contribute to potential alterations to the 1909 Water Act.[5] The aforementioned groups are being petitioned to engage in the discussions that will reform the current Water Act to meet current and future water needs of British Columbia's population. The strategy and ongoing discussions are inclusive of First Nations traditional knowledge in water stewardship and decision making, yet there is no mention of the inherent or priority rights to water held by the indigenous peoples of British Columbia.

Watershed management within the WAM proposal is based in a new approach of sustainability, with the intent of meeting the consumption objectives of the populations of British Columbia. The challenges confronting all water-dependent users go beyond assessing the water quality and the current needs distribution of this resource in order to ensure sustainable water use practices based in a concept of long-term responsibility rather than short-term economic development strategies.

Complicating the WAM process are the many diverse laws, regulations, and international agreements, and the unanswered questions on what exactly the aboriginal water rights are in the province of British Columbia. Throughout British Columbia's colonial history, aboriginal water rights have been severely compromised to accommodate settler interests. Any new revisions to the Water Act must clearly define and state that participation of aboriginal peoples in no manner compromises future claims to this natural resource. From the indigenous perspective, participation in the framing of the WAM initiative must not be used to position aboriginal peoples as consenting to give away inherent rights to this natural resource.

A danger to the inherent and priority rights of the aboriginal peoples of British Columbia is that within the WAM proposal aboriginal peoples are considered to be stakeholders, giving British Columbia's Indian bands equal—or, more accurately, lesser—rights equivalent to the rights accorded cattlemen's associations; municipalities; or forestry, mining, and agricultural interests. An underlying problem is that Canadian courts interpret aboriginal water rights as implying rights of occupation and use but disallow the ownership of the water itself (Notzke 1994: 12).

It is clear that the province of British Columbia is attempting to settle unresolved water issues with as little confrontation as possible. The WAM process relies on the success of provincial stakeholders to consult with aboriginal peoples to accommodate water use needs of British Columbia's aboriginal peoples from within a narrow definition of water use rather than the wider definition of aboriginal water use tied to the wider and underlying question of unceded land and water. Participation in the consultation process as water use stakeholders by aboriginal peoples of British Columbia diminishes and silences the injustice of the subsuming dispossession of ancestral domain taking place. Aboriginal leaders should make it abundantly clear from the outset that indigenous attendance and participation at these meetings does not constitute consent or recognition of the alienation of aboriginal rights and title.

The question of aboriginal water rights looms large in British Columbia in areas where neither treaties nor land claims are in progress, and where instead a process of assertion of aboriginal rights and title is, in most circumstances, in various stages of progress through the courts and in fewer cases is being negotiated into comanagement frameworks. Water use and therefore water governance in contemporary indigenous perspective is intricately bound with water rights and water law. Nowlan in "Customary Water Laws and Practices in Canada," and others, have described customary water rights as the "sleeping giant of water in western and northern Canada," in that, where treaties and land claims exist, they do not address water rights (Nowlan 2004: 1–3).

Indigenous Customary Rights vs. Canadian Statutory Rights

The fact that water rights are not addressed in treaties triggers questions in the legal gray area between the customary rights of indigenous peoples and statutory rights in Canada. Without entanglement in legal questions, for the purpose of this chapter suffice it to say that the main considerations relate to several looming areas of incongruity within the government framework of water governance. Questions arise from divergent perspectives related to aboriginal rights as protected by the constitution, aboriginal rights arising from aboriginal title being defined through the courts, and the access and use of water by aboriginal people on lands that have not been ceded and are not within a reserve. The key question revolves around the difference in meaning between aboriginal water use in indigenous water laws and the provincial and federal governments definitions of water use as constructed in water legislation. Water law and therefore water governance in Canada is limited to addressing the government definition of water use. For the purpose of differentiating water use from aboriginal water use the former would be better defined as water exploitations, in that water governance addresses water extractions, diversions, and manipulations for domestic, recreational, agricultural, industrial, and megadevelopment purposes such as hydroelectric damming and big mining.

The concept of water use from an indigenous perspective relates directly to the social life of water as experienced in all the customs and traditions that are dependent on water. From this perspective, the use of water is tied to the relationship that indigenous peoples have with water, in that water is the source of all of the life forms of each specific habitat that water creates by its presence, and is the basis of their livelihood. In that way, the indigenous social relationship to water is specific and unique to each place, and the benefits of water and the surroundings it creates are the source of and are foundational to aboriginal rights. Aboriginal water use is thus not limited to the access right in fishing that is defined in law as an aboriginal right to fish; rather, it is an aboriginal water use right. The consequence is that the aboriginal right to fish can be and has been quantified and limited as a shared right to be recognized and accommodated or compensated in relation to its loss as advised in *Delgamuukw*. In other words, the aboriginal right to fish can be removed from the water place and exchanged for compensation or for access to other sites in order to make way for other fishing interests or for alteration or destruction of the water place and its surroundings for other purposes and interests. In that way, the aboriginal water use right can be seen as a right separate from an aboriginal right to fish.

Aboriginal use of water includes customary water governance methods such as maintaining fish-ceremonial activities that are foundational to traditional in- tertribal law observance for harvest access and distribution at fishing locations, fishing stations, and tribal boundaries. The concept of aboriginal use of water, from that perspective, relates to the aboriginal rights involved in social mechanisms to maintain the water system to support a healthy watershed sustaining quantities of food, medicine, and materials customarily utilized, traded, or sold as direct economic benefit. Aboriginal water use arises in social customs as a system of water governance historically held in place through a network of localized opera- tional, indigenous political jurisdictions that determined access and limitations in aboriginal water use related to each place. Harris contends, the "laws of Native peoples governed the use of local fisheries performing the same functions as Cana- dian fisheries, law-defining and determining access to and ownership of particular fisheries." He goes on to explain, "Native resistance to Dominion regulation was born of Native systems of resource allocation that were produced and reproduced in Native legal cultures" (C. Harris 2001: 208–212).

The recognition of existing aboriginal rights as required by Section 35 of the Canadian constitution, when situated in regard to aboriginal water use, requires, at the minimum, a recognition of the right to exercise traditional and customary so- cial practices and, at the optimum, the recognition of aboriginal law in its entirety. The aboriginal use of water from that perspective is an existing aboriginal right. For example, the recognition of aboriginal water use would be compromised by a lake being filled with mine tailings or a fishing station inundated by a dam, or a riparian forest drained off, or a meadow land deprived by channeling, or a water- shed destroyed by a clear cut. The governments of British Columbia and Canada would then be required, from that perspective, to recognize aboriginal water use as well as water use as currently defined by legislation. The rights to aboriginal water use, when thus defined, become bound into the legal struggles related to aboriginal title in unceded territories because modern water law only recognizes water users as stakeholders if their requirement falls into one of the exploitative areas of water use. The consequence of defining water use simply from the negative end of entitlement to water exploitation is to silence, trivialize, and patronize aboriginal water use in the eyes of the current Canadian legal system. The governments of British Columbia and Canada continue a highly unjust and systematized racism in structuring water exploitation as the basis in law of water use.

Conclusion

With respect to the recognition of existing aboriginal rights, as required by Canadian law, recognition of aboriginal water use means that Canada must formally recognize the right of aboriginal people to exercise traditional and customary social practices. Recognition would include mechanisms to allow managing, maintaining, and protecting aboriginal water use within modern water use laws, and would be implemented in modern federal and provincial water use governance. Aboriginal water use in aboriginal social custom is a system of water governance that is held in place through historical, aboriginal, and local jurisdictions determining limitations to use related to each place, and based on sustainability of use. These practices have continued to be exercised through political and legal avenues of resistance.

The inability to implement and exercise aboriginal water governance is simply the result of the modern structuring of water jurisdictions, notwithstanding the countless ongoing interventions of indigenous peoples to continue to protect their water governance practices to maintain their responsibility to sustainable local governance. The reorganization of water law to accommodate aboriginal water use, in Canada for example, could put into practice community cooperation with the goals of equity and sustainability at its core, likely to the betterment of water governance locally as well as the protection and recovery of the environment. The province of British Columbia could engage in meetings with the aboriginal peoples to determine aboriginal water use priorities that could be broken down by watershed areas because each differs in the needs of humans, the land, and its animals depending on environmental conditions and geographic variances. A blanket policy that accommodates aboriginal interests cannot take into consideration the local scope of aboriginal water use in relation to the supply and demand of all water-dependent users in a given area. The inclusion of indigenous rights in the development of water management strategies directed at protecting water bodies and stream health and the surrounding aquatic and land environments in indigenous territorial water systems would form a critical foundation for a collaborative approach. Indigenous knowledge related to aquatic and land environments and aboriginal water use could then be given equal weight in defining a beneficial water management code in British Columbia. Following that logic, a process providing for a specific dialogue for each indigenous group in recognition of British Columbia's indigenous peoples' sui generis status of rights would be appropriate. Wright has outlined the growing body of scholarship on traditional ecological knowledge and the relationship of local communities, in particular indigenous communities, to their environments, which emphasizes that local communities, working within communal property regimes, sustainably manage common-property resources such as fish (Wright 2004: 173).

Managing, maintaining, and protecting aboriginal water use practices in contemporary legal systems could also provide a unique source of protection for water sovereignty in Canada from jurisdictional erosion as a result of overriding international water treaties and trade agreements favoring interests in the United States. Indigenous legal scholar Ardith Walkem has pointed out that as the demand for continental harmonization of economic and political systems continues, waters will be increasingly viewed as "a shared North American resource" with pressure for large-scale diversion or export of water from Canada to the United States (Walkem 2004: 3–4). Groenfeldt, in his UNESCO essay on water development and spiritual values in Western and indigenous societies, argues that although differences between Western views on water and indigenous values about water suggest great potential for conflict, with education there could be needed cooperation: "The ethical perspective embedded in indigenous views about nature and water is largely missing from the Western toolkit on water management" (Groenfeldt 2006: 114–115). For indigenous peoples, the social life of water related to water governance provides security and sustainability. In the future, sustainability and security for local communities, in relation to water use, may be best preserved through reconciliation of water governance rights with the rights and practices of indigenous peoples.

Notes

1. Where used in this chapter, the term *indigenous* corresponds to the definition in the United Nations Declaration on the Rights of Indigenous Peoples.
2. The term "American Indian" is commonly interchangeable with the term "Native American" to refer to indigenous peoples of the United States.
3. The term "tribe" refers to groups recognized under U.S. reservation policy.
4. The term "aboriginal" refers to the three groups with defined legal rights in Canada: the Inuit, Metis, and Indians under the Indian Act.
5. The term "First Nation" is used informally in Canada to replace a less-politically correct although legal definition of Indian bands under the Indian Act. The Indian Act is a federal statute that defines and constrains the rights of Indians within Canada and defines the responsibilities of the federal government in respect to the management of Indian reserves.

References

American Friends Service Committee. 1970. *Uncommon Controversy: Fishing Rights of the Muckleshoot, Puyallup, and Nisqually Indians.* Seattle: University of Washington Press.

Barlow, Maude, and Tony Clarke. 2002. *Blue Gold: The Battle against Corporate Theft of the World's Water.* Toronto: Stoddard Publishing.

Bustamante, Rocio. 2006. Pluri-, Multi-Issues in the Reform Process: Toward New Water Legislation in Bolivia. In *Water and Indigenous Peoples,* edited by Rutgerd Boelens, Moe Chiba, and Douglas Nakashima, 118–129. Paris: UNESCO.

Clark, F.J. 1956. *Okanagan Flood Control Project: Field Trip Tour.* Penticton, British Columbia: Columbia River Basin Water Forecast Committee.

Daes, Erica-Irene A. 2008. *Indigenous Peoples: Keepers of our Past—Custodians of our Future.* Copenhagen, Denmark: International Work Group for Indigenous Affairs.

Delgamuukw v. British Columbia. 1997. 3 S.C.R. 1010.

Dickason, Patricia Olive. 2002. *Canada's First Nations: A History of Founding Peoples from Earliest Times.* Don Mills, ON: Oxford University Press.

Green, Michael. 1995. Cultural Identities: Challenges for the Twenty-First Century. In *Issues in Native American Cultural Identity,* edited by Michael K. Green, 1–38. New York: Peter Lang.

Groenfeldt, David. 2006. Water Development and Spiritual Values in Western Indigenous Societies. In *Water and Indigenous Peoples,* edited by R. Bolens, M. Chiba, D. Nakashima, 108–115. Paris: UNESCO.

Harris, Cole. 2002. *Making Native Space: Colonialism, Resistance, and Reserves in British Columbia.* Vancouver: University of British Columbia Press.

Harris, Douglas C. 2001. *Fish, Law, and Colonialism: The Legal Capture of Salmon in British Columbia.* Toronto: University of Toronto Press.

International Indian Treaty Council. 2011. Working for the Rights of Indigenous Peoples. http://www.treatycouncil.org//home.htm

Jolly, Diana. 1999. *First Nations Water Rights in British Columbia: A Historical Summary of the Rights of the Penticton First Nation,* edited by Daniela Mogus and Miranda Griffith. Victoria, BC: Water Management Branch.

Matthiessen, Peter. 1991. *In the Spirit of Crazy Horse.* New York: Penguin Books.

Means, Bill. 1981. Presentation at the Okanagan Indian Historical Symposium, sponsored by the Okanagan Tribal Council and Curriculum Project. July 28–30. Penticton Indian Reserve, Penticton, British Columbia.

Notzke, Claudia. 1994. *Aboriginal Peoples and Natural Resources in Canada.* North York, ON: Captus University Publications.

Nowlan, Linda. 2004. Customary Water Laws and Practices in Canada. A paper commissioned by FAO under a joint FAO/IUCN research project. http://www.fao.org/Legal/advserv/FAOIUCNcs/Canada.pdf

Olivera, Oscar. 2004. *¡Cochabamba!: Water War in Bolivia.* Cambridge, MA: South End Press.

Puyallap Tribe v. Department of Game. 1968. 391 U.S. 392.

R. v. Kapp. 2003. BCPC 279.

R. v. Kapp. 2008. 2 S.C.R. 483, 2008 SCC 41.

Reisner, Marc, and Sarah Bates. 1990. *Overtapped Oasis: Reform or Revolution for Western Water.* Washington, DC: Island Press.

Sam, Marlowe. 2008. Okanagan Water Systems: An Historical Retrospect of Control, Domination, and Change. Unpublished master's thesis.

Smith, Paul Chaat, and Robert Allen Warrior. 1996. *Like a Hurricane: The Indian Movement from Alcatraz to Wounded Knee.* New York: The New Press.

Sparrow v Her Majesty the Queen. 1990. I. S.C.R. 1075.

Symonds, B.J. 2000. Background and History of Water Management of Okanagan Lake and River. In *Water Management*, 1–8. Penticton, BC: Ministry of Environment, Lands and Parks.

Tauli-Corpuz, Victoria, and Erlyn Ruth Alcantara. 2004. *Engaging the UN Special Rapporteur on Indigenous People: Opportunities and Challenges.* Baguio City, Philippines: Tebetebba Foundation.

UNESCO. 2006. *Water and Indigenous Peoples.* Edited by R. Bolens, M. Chiba, and D. Nakashima. Knowledge of Nature 2. Paris: Author.

Walkem, Ardith. 2004. Indigenous Peoples' Water Rights: Challenges and Opportunities in an Era of Increased North American Integration. Canada and the New American Empire. Centre for Global Studies, University of Victoria, BC. November.

Winters v. United States. (1908). 207 U.S.

Wright, Guy. 2004. Aboriginal Fishing Rights in Practice: Australia and Canada. In *Water and Fishing: Aboriginal Rights in Australia and Canada*, edited by Paul Kaufman, 165–175. Woden Act, Australia: Aboriginal and Torres Strait Islander Commission.

Chapter 13

BUREAUCRATIC BRICOLAGE AND ADAPTIVE COMANAGEMENT IN INDONESIAN IRRIGATION

Bryan Bruns

∞

Introduction

In 1963, the Acehnese religious and political leader, Daud Beureueh, led the digging of a seventeen-kilometer-long irrigation canal in Pidie District, resolving a long-standing conflict created by a previous canal (Siegel 2000, 60–67). This was one of a variety of projects Daud Beureueh led to build mosques, roads, and bridges; Beureueh had earlier headed the All-Atjeh Union of Religious Scholars, been military governor of Aceh during the Indonesian revolution in 1945–46, and led a rebellion against the central government from 1953 until a settlement was agreed to in 1962. While farmers whose lands would benefit formed the core of the workforce to build the canal, it was only possible through the assistance of many others, who, Siegel (2000) argues, were inspired to carry out the work as part of a religious duty, something set apart from everyday life, something that would be rewarded in the hereafter.

The role of earlier Acehnese leaders in irrigation development was noted by Snouck Hurgronje (the ethnographer who helped mastermind the Dutch conquest of Aceh), who said that an irrigation tax was collected "in districts where irrigation canals had been constructed at the behest of rulers in ancient times" (Hurgronje 1906: 272).

The influential role of political leaders in organizing multivillage efforts to build and improve irrigation, in Aceh and elsewhere, is an example of the persistent importance of politics and collective action in irrigation development. While the smallest irrigation systems may be built and managed by a few neighbors or a single hamlet, many irrigation systems require larger-scale cooperation.

Top-down state control of large irrigation works was central to Karl Wittfogel's concept of hydraulic bureaucracies, which he contrasted with the hydroagriculture of small-scale irrigation (Wittfogel 1957: 3, 18). Locally initiated irrigation development may also build extensive infrastructure with large-scale impacts. Edmund Leach ([1959] 1981) critiqued Wittfogel by pointing out how reengineering of landscapes by apparently massive hydraulic works, as in Sri Lanka, could result from cumulative construction over decades and centuries, rather than necessarily being the product of a single despotic mastermind. More recently, analysis of Balinese irrigation by Stephen Lansing (1991, 2006) has clarified how irrigated landscapes may evolve through an emergent self-organizing process of complex adaptive systems, without a single mastermind or master plan, and nevertheless may create an elegantly ordered socioecological system.

The differences between top-down bureaucratic construction and bottom-up local irrigation organization have been a central theme in social science literature on irrigation. This concern has continued in the context of recent reforms aimed at decentralization through participatory irrigation management and irrigation management transfer (IMT).[1] The need for irrigation rehabilitation has often been framed in terms of a slow crisis of irrigation degradation, of unsustainable policies of external investment perpetuating a vicious cycle of poor performance, neglected maintenance, and pulses of inefficient rehabilitation (see, among others, Araral 2005; Rap 2006). More recently, arguments for irrigation reform have been linked with the broader rhetoric of water crisis, growing scarcity, and the need to reduce inefficiency in irrigation. In many areas of natural resources management, the need to understand the interplay between institutions at different levels of resource governance, for example irrigation systems, river basins, and nation-states, has emerged as a major concern (Young 2002). The variety of attempts at decentralizing governance and empowering community-based natural resources management undertaken in many sectors in the past few decades has led to a process of rethinking the limits and lessons of decentralization (see, e.g., Agrawal and Gibson 2001).

This chapter looks at some interactions between state agency intervention and local collective action in irrigation, and seeks to highlight the potential for comanagement. It draws on my experience as a practitioner and on some particularly relevant literature, with examples from Aceh and other parts of Indonesia, and seeks to address wider questions about the potential and problems of developing irrigation comanagement. In the next section, I provide some personal context for the material presented in this chapter, based on my experience as a consultant on irrigation projects. I then outline patterns of episodic mobilization, suggesting that these may characterize local irrigation management in many contexts. Bureaucratic intervention often overlays new institutional arrangements, and, as the third section discusses, may yield, not the obliteration of customary practices or imposition of standard procedures, but instead a bricolage that recombines institutional concepts and practices from diverse sources. Attempts at devolving irrigation management have met with mixed results, and the potential exists for more-focused attempts to improve cooperation between irrigation communities and government agencies in carrying out key tasks in irrigation construction and operation. The fourth section looks at opportunities for joint problem-solving and adaptation, particularly in participatory design, wider networking, and joint problem-solving to improve irrigation operations and water delivery. The final section of the chapter offers conclusions about how adaptive comanagement may weave a polycentric bricolage of diverse institutions and understandings.

A Practitioner's Perspective

I illustrate many of these issues with examples from Aceh, where I worked on a series of consulting assignments in 2005–07 for an irrigation rehabilitation project, part of the Earthquake and Tsunami Emergency Support Project (ETESP) funded by a grant from the Asian Development Bank (ADB). This chapter represents part of an effort to learn from earlier attempts to promote participation in irrigation development and conceptualize better ways of supporting communities of irrigators in acting collectively, in cooperation with government where relevant, to improve their livelihoods.

My first working experience in Southeast Asia came after studying anthropology as an undergraduate at Beloit College in Wisconsin, with an interest in applied work. I went to northeast Thailand as a Peace Corps volunteer where I was placed with a large Thai nongovernment organization, the Population and Community Development Association, led by Meechai Virivdaidya. That NGO was trying to branch out from its earlier work in family planning. I ended up helping villagers build household rainwater storage tanks. The Peace Corps offered me a chance to see whether living and working overseas was something I wanted to pursue as a career, and I decided to go on to study for a Ph.D., which at the time seemed to be the necessary qualification for professional work. When I inquired in 1978, the Cornell University Anthropology Department said they had no interest in students who wanted to do applied work, and instead encouraged me to look at the development of sociology program, based in the Department of Rural Sociology, which I did. I chose as an advisor Walt Coward, who had recently returned from Indonesia helping the Ford Foundation with a program of applied research on irrigation. I later included two anthropologists, Milton Barnett and A. Thomas Kirsch, as members of my Ph.D. committee. During the 1980s, much of the conceptualization of irrigation development was framed by the pioneering transformation of the Philippines National Irrigation Agency (Korten and Siy 1988), and the interdisciplinary experience of Cornell faculty, including Walt Coward, Milton Barnett, Randy Barker, Gil Levine, and Norman Uphoff, in supporting participatory rehabilitation of the Gal Oya irrigation system in Sri Lanka. (For one perspective on the Sri Lankan experience, see Uphoff 1991.)

In 1985, I returned to northeast Thailand for research, deciding to build on my previous experience in the area and knowledge of Thai and Lao. After the usual adjustments needed when research plans encounter field conditions, I ended up combining part-time residence in a village near a complex set of interrelated small irrigation systems with participation in applied research projects at Khonkaen University; these projects were exploring ways to improve participation in the development of small-scale irrigation. My research compared a New Zealand–funded project that worked with communities to upgrade existing small earthen weirs with the Royal Irrigation Department's much more technically oriented, and much less successful, approach to intervention (Bruns 1990, 1991).

While back at Cornell in 1988, I was invited to work as an institutional advisor with an Indonesian NGO, the Institute for Social and Economic Research, Education, and Information (LP3ES), and began working there in late 1988. LP3ES was providing training and support for a project that built on the earlier Ford Foundation–funded work to introduce participatory approaches within the irrigation management turnover component of the World Bank Irrigation Sector Support Project (Bruns and Atmanto 1995; Bruns and Soelaiman 1992).

After living in Jakarta for about two and a half years, I moved to northern Thailand in 1991 but returned to Indonesia frequently during the 1990s, particularly as a short-term consultant on irrigation projects related to small-scale irrigation and later as part of the World Bank's attempt to promote reform in the water resources sector (Bruns 2004; Suhardiman 2008). These projects were part of a generation of international projects aimed at promoting participatory irrigation management IMT (Rap 2006; Vermillion 1991, 2006). After becoming somewhat burned-out, for personal and professional reasons, on the reform efforts in Indonesia, and feeling that some kind of different approach was needed, I moved back to the United States, and shifted to work elsewhere, including Vietnam, China, and the Philippines, and continued to pursue a more academic interest in water rights (Bruns and Meinzen-Dick 2000; Bruns, Ringler, and Meinzen-Dick 2005).

The ETESP project in Aceh offered a chance to work again somewhat closer to the field, rather than in the policy stratosphere, in a part of Indonesia that was new to me. As an international expert coming in for a few weeks or months at a time over a period of two years, my experience was very much mediated through my colleagues, particularly the late Totok Hartono, who led the institutional side of the irrigation work, and the set of district coordinators who oversaw the institutional work of community facilitators. An Indonesia NGO, Bina Swadaya, trained and managed the work of facilitators, nominated by local communities, to work on the development of water users' associations (WUAs) and local involvement in project planning and implementation.

The discussion here draws not from prolonged residence in a single village, but instead from experience and participant-observation as a practitioner during a series of consulting assignments over a total of eight months during 2005–07. Field experience usually took the form of short visits, including formal meetings, walking through irrigation schemes together with farmers and engineers and informal discussions. Discussions with district coordinators and facilitators, in regular meetings and workshops and on other occasions, also were a rich source of information on local institutions, irrigation and other aspects of rural life, as was interaction with provincial and district government officials. Work on the project included areas like those surrounding Pidie, which form the heartland of irrigation in Aceh, as well as communities along the west coast and in other parts of the province. The discussion here attempts to suggest some general patterns, while remaining aware of the great diversity of local conditions and practices that prevail in irrigation in Aceh, other parts of Indonesia, and elsewhere.

During my earlier work in Indonesia, mostly dealing with Java, Western Sumatra, and South Sulawesi, I had become increasingly concerned about the misfit between the formal and relatively rigid models for irrigation organization that were being propagated by internationally funded projects and national policies, and the much more diverse and flexible ways in which local communities organized themselves to manage irrigation (Bruns 1992). The problems faced by many efforts to organize formal water users' associations (WUAs) confirmed the earlier warning by Robert Hunt (1989) that it was not simply a matter of scaling up principles that had contributed to success in small, locally organized irrigation systems. In Aceh, the disjunction between the formal models of organization supported by national policy, and the actual practices on the ground was particularly prominent.

Episodic Mobilization

In Aceh, as in many monsoon climates, local collective action to irrigate rice typically follows an episodic pattern of mobilization, focused on preseason canal cleaning and other key events. Aceh's oldest irrigation systems were built in the area around Pidie, where large rivers and gentle topography make conditions more favorable for irrigation, in contrast to the steeper topography and narrow strip of lowland along the west coast. In other areas, however, similar patterns of rice cultivation and social organization were used for irrigated agriculture. Those who share water usually join together in *kenduri blang* (agricultural feasts), before and after the cropping season. Such mobilization responds to annual cycles of monsoon rains, shorter-term fluctuations between rainfall and dry spells, and longer-term extreme events of drought and flood. Traditionally, the need for large amounts of labor to repair or rebuild diversion works one or more times per season acted to unite those who shared water from the same stream or river. The ability to rapidly mobilize large amounts of labor and materials was crucial to dealing with weir washouts, canal breaches, and landslides that threatened access to irrigation water. Rights to water were linked with obligations to contribute to irrigation construction and maintenance. The application of rules was often flexibly adapted to the fluctuating abundance and shortage of water, however, through shifts between different sets of rules, such as switching to rotational distribution of irrigation water during periods of shortage, with more-relaxed and more-tolerant sharing during periods of plenty.

Narratives of indigenous irrigation have often concentrated on areas with intricate organization and autonomous village-level irrigation organizations, known as *subak* in Bali (Geertz 1980), *muang fai* in northern Thailand (Potter 1976; Sirivongs Na Ayudhaya 1983; Tan-Kim-Yong 1995), and *zanjera* in the northern Philippines (Lewis 1980). Often these systems carefully allocate water in proportional shares, with requirements for labor contributions matched to benefits.

Such accounts sometimes omit or gloss over the complications, conflicts, and improvisation that lie behind the performance of such traditional irrigation systems. Thus, for example, the identity and affiliation of a *subak* could be a complex and highly contested matter, reflecting struggles within and between local communities (Spiertz 2000). In many irrigated areas, patterns of organization are more informal, ad hoc, and improvisational than suggested by accounts of well-organized systems. Among other things, the extent of organization may reflect factors such as differences between whether water flows are stable or fluctuate greatly, and the level of dependence on irrigation rather than rainfall, something that had distinguished the more informal irrigation systems of northeast Thailand from the more elaborate and formal institutions of communal irrigation systems in northern Thailand (Bruns 1991).

In contrast to famous examples of irrigation systems in Bali organized in ways that were relatively autonomous from local government, irrigation systems in Aceh have a long history of close linkage with village government (Coward 1991). Irrigation leaders were usually appointed by *kampong* (village heads), typically reflecting a local consensus among farmers, with the *keujreun blang* (field leaders) having a recognized role in local leadership for irrigation and other matters in the set of *blang* (fields) for which they were responsible. The formal status of these positions changed with the reorganization of village government resulting from Indonesia's 1979 Law on Village Government. This new law established a local government structure with no specific place for Aceh's *keujreun blang*. Traditional *ulu-ulu* (watermasters) were similarly displaced in Java. What had been an acknowledged position in village government was no longer recognized by law. In Aceh as in Java and other parts of Indonesia, however, communities have frequently perpetuated the institution of *keujreun blang* as part of sustaining patterns of social organization that are responsive to both water users and village leaders, while dealing with the challenges and opportunities posed by government agency actions.

The strong linkages with village government contrast with the tendency in some literature on irrigation, and other commons, to idealize autonomous specialized institutions, organized along the "hydraulic" lines of those whose lands were watered by the same canal. Empirically, there are areas such as the *subaks* studied by Geertz (1980), where irrigation is organized distinctly from other local governance. The topography of irrigation flows often crosscuts the boundaries of hamlets, villages, and larger administrative jurisdictions. However, irrigators often find pragmatic means to cope with such complications, often coordinating flows through one community with the largest irrigated area taking the leading role in irrigation management. In a process of sociotechnical coevolution, the boundaries of irrigation systems also are often shaped to fit the social and administrative networks, as has occurred in the coevolution of Javanese *desa* (villages) and their associated irrigation systems. Attempts to divorce irrigation governance from other local governance systems cut across the grain of local social relationships. Irrigation often benefits from its integration, or embeddedness, in other social linkages (R. Hunt and E. Hunt 1976), and from forms of social capital that can help generate trust, encourage leadership, monitor behavior and enforce a range of subtle rewards and sanctions. Government agencies may desire to have their own specialized local organization, whether WUAs, farmer groups, health groups, and so on. Communities often find this insistence on separate organizations confusing and burdensome; if the rewards are compelling enough, however, they are more than willing to comply, at least superficially, with the mandated formalities.

Success in episodic mobilization, including forming WUAs and using their legal status, was particularly notable in how communities in Pidie and other parts of Aceh responded to opportunities to participate in construction works funded by the Earthquake and Tsunami Emergency Support Project. WUAs were formally established and engaged in contracts for minor works, mostly canal lining. These used a new technology for the area of wire-mesh reinforced concrete (Hartono 2007). Many government engineers were skeptical about how well villagers could build using this technique. As it turned out, farmers were highly successful in learning to use the methods, and achieved much-higher quality standards than did contractors doing similar work. Neighboring WUAs joined together to purchase the wire mesh, which was not available in local markets; WUAs had to purchase it and truck it in from larger towns. While the sustainability of formal WUAs in terms of indicators such as formal meetings, records, and routine fees is in doubt, participation in construction demonstrated the resilience of patterns for episodic mobilization that combined customary and government-introduced forms of organization.

Bureaucratic Bricolage

Bureaucratic attempts to develop irrigation in Aceh, as in many other parts of the world, have often sought to impose new organizational forms, a form of social engineering. During the 1980s, the Indonesian government invested a portion of its revenues from oil and other natural resources in irrigation and other forms of rural development.[2] In Aceh, government projects built permanent diversion weirs and extensive canal systems on many of the rivers that flow to the north and east coasts. These projects often incorporated irrigation canals that had previously been built and rebuilt by smaller communities. Along with new physical structures, the government sought to establish new organizational arrangements. These were supposed to follow national doctrines for technical irrigation management. Irrigation system designs were based on standard procedures for operation and maintenance, specifying how water was supposed to be distributed by dam and gate operators. Irrigation management took place within the top-down apparatus for controlling cropping patterns that had been imposed during the 1970s as part of the green revolution expansion of high-yielding varieties through centralized programs. The role of farmers in irrigation management, at least in theory, was limited to the lowest level, below tertiary outlets, which typically served from 50 to 150 hectares.

Indonesian national standards for irrigation operation and maintenance are primarily based on irrigation management in Java. Many of the standards derive from systems dating back to nineteenth century East Java where bureaucracies had relatively high levels of staff by comparison to Aceh and more experience managing scarce water supplies during long dry seasons. Standard procedures assume adequate funding for operation and maintenance. In practice, the condition of infrastructure, hydrology of water flows, cropping patterns, and other conditions frequently mean that it is difficult or impossible to actually operate the irrigation system as designed. In Aceh and in many other parts of Indonesia, it is tempting to blame poor irrigation performance on insufficient staff, limited budgets, inadequate coordination, and other institutional factors. Perceptions of inadequate institutional capacity have underlain approaches to irrigation governance reform, such as participatory irrigation management and IMT.

The pattern of irrigation organization promulgated by the Indonesian government fit the New Order regime's strategy of political demobilization of rural communities, restricting organization above the village level and subordinating local institutions to the hierarchical control of national organizations. The politics of power were practiced through manipulation to produce a show of deliberation, *musyawarah dan mufakaat* (consensus) and compliance with government policies and commands (Anderson 1990). In practice, what happened was not a genuine consensus but more often a matter of imposing officially prescribed policies, underlain by intimidation of potential opponents and behind-the-scenes brokering of conflicts.

The general pattern of restricting farmer activity to the tertiary level, below the secondary canal outlet, reflects common policies in many countries. Management at the secondary canal level and higher is often ostensibly monopolized by agency staff. Field observation and research (for India, see Chambers 1988; Wade 1987) have revealed the many ways in which local farmers may influence higher-level water distribution through direct action to operate gates and take water, as well as lobbying, bribes, sabotage, and intimidation.

Historically in Aceh, as in many other parts of Indonesia, the lack of staff and budget often meant that in practice farmers took a much larger role in management, often working in cooperation with the available government staff. Although the budget year officially began in January, agencies often had no funds for maintenance until July or even later in the year, necessitating mobilization of users if urgent repairs had to be carried out before then. In practice, budgets were often late, inadequate, or diverted to other activities such as minor construction contracts that better fit the interests of officials with discretion over their disbursement. Staffing levels were low, with meager wages, so staff were not motivated to carry out their responsibilities. During periods of violent conflict, security fears further reduced government presence in the field. All these limitations on government agency action created a situation where farmers were often left to their own devices, to do what they could, with only very limited support from government staff and budgets.

Bureaucratic efforts to organize formal WUAs continued through the 1990s and beyond. Nominally, the new policies recognized the traditional *keujreun blang* irrigation institutions and it was recommended that the former *keujreun blang* become WUA heads or *pelaksana teknis* (technical staff). However, the internal structure was a standardized arrangement with a head, deputies, secretary, and treasurer, and a *pengurus* (board of leaders) who were often chosen to represent different parts of the irrigation system. WUA constitutions and by-laws were written in Indonesian, the national language, and not in the local language, and followed a standard example established by government decree.

Establishment of a formal WUA was often a condition for receiving government assistance. In practice, as elsewhere in Indonesia and many other countries, what often resulted was the establishment of paper organizations, while collective action in irrigation continued largely along previous lines. From the government perspective, organizations had been established. Meetings were held, officers selected, and training courses conducted. Such training programs sought to build a common understanding of the importance of WUAs and the proper procedures to be carried out. Subsequently, when necessary, WUAs could hold meetings and put on a show of following the prescribed patterns for WUA activities, especially if it would enhance the prospects of receiving government aid for canal repair and improvement. Some selected sites received special attention, as pilot sites entered into WUA *lomba* (competitions) at the district, provincial, and national level.

Much of what the polices expected of WUAs did not happen, however. Officials bemoaned the inability of most WUAs to keep proper records, collect *iuran* (regular monthly fees), and enforce government-mandated plans for water distribution. Evaluation of WUA development concentrated on conformity with formal requirements for preparing and implementing plans and filling in record books, highlighting the deficiencies of most WUA on such criteria. Thus, most WUAs were categorized as underdeveloped or not yet developed, while closer examination even of those that were categorized as developed usually revealed large gaps between what was on paper and what was done in practice. At the same time, farmers sharing a source of irrigation water typically continued to come together before each season to clean canals. Cropping patterns and schedules were confirmed and announced at *kenduri blang* feasts attended by local leaders and other farmers, although this traditional forum was incorporated into *kecamatan* (subdistrict) government procedures. *Keujreun blang* watermasters supervised water distribution, helped to resolve conflicts, and after harvest collected contributions from farmers, to pay for repairs and compensate their work.

Agency officials working in the field were often quite aware of the strength of traditional patterns of organization, and of the fragility of the new formal arrangements. However, the targets set for projects, were framed in terms of indicators such as formal establishment of WUAs, works carried out, and budgets disbursed. These policies and targets all encouraged government officials to go along with the official text of formal organization, although many officials acknowledged and sometimes bemoaned the limited extent to which it resembled how things were actually done.

It is tempting to portray this pattern of irrigation organization in terms of dualism, emphasizing the gaps between policy and practice, the discontinuities between bureaucratic prescription and local customs. Thus, the outcome could be depicted in terms of the strong disjunction between the formal procedures government sought to impose and the traditional patterns communities continued to practice. The story could also be narrated in terms of domination and resistance, since bureaucratic attempts to control resources are ignored, opposed, subverted, reinterpreted, and otherwise manipulated by the victims of state power. Such depictions do capture part of the story, of differences in perceptions, practices, and power.

A dualistic interpretation oversimplifies the extent to which government-introduced arrangements influenced local management practices, however, through various factors, including the attractiveness of government financial aid, the use of government terminology by farmers themselves to discuss and justify irrigation water distribution, and the influence on local leaders of training, government-mandated crop planning, and relationships with water resources agencies and other parts of local government. Government efforts to impose new organizational forms did not erase or eliminate previous concepts and practices, nor did they create a separate superficial layer of organization on top of local customs.

Instead, the pattern may better be characterized as a bricolage (Cleaver and Franks 2005; Levi-Strauss 1966), mixing old and new concepts and practices, with substantial variation among communities and with dynamics responding to availability of subsidies and political backing for those who could comply with government-mandated organizational forms. The dynamics of irrigation management form an improvisational combination of ideas and institutions in which both state officials and local people are active agents. Understanding the organization of collective action in irrigation as bricolage also better fits with the ways in which local actors opportunistically seek out state assistance while state agents pragmatically pursue accommodation between local traditions and new organizational forms. By highlighting agency and opportunity, such a perspective can help identify opportunities for creative comanagement rather than pretending that a solution might be found in either more complete implementation of national norms, or in state withdrawal.

From Devolution To Comanagement

During the 1980s and 1990s, various initiatives sought to introduce more participatory approaches to irrigation management in many countries. Many of these focused on reducing government roles in irrigation, often justified by hopes that costs of operation and maintenance could be shifted from government to water users. The emphasis and objective were often framed in terms of shifting tasks and responsibilities from government to farmers. In Indonesia, participatory initiatives included a program to turn over small irrigation systems to WUAs and the development of WUAs and WUA federations in larger irrigation systems to take part in collecting irrigation service fees and deciding how they would be used.[3] These represented an attempt to shift away from a highly hierarchical and technocratic approach where government controlled decision making, and restricted farmer roles to activities within tertiary areas. With the end of the Suharto regime in 1998, political reformation brought increased acceptance of principles of participation and democracy, opening new opportunities for participation, as well as hasty efforts to decentralize many roles and responsibilities, including transfer of irrigation systems to district governments.

At the national level in the late 1990s, the Indonesian government agreed to an ambitious program to reform irrigation and water management institutions, supported by the Water Sector Adjustment Loan from the World Bank. This program was supposed to include transfer of irrigation governance authority to WUA federations, replacing and going beyond activities under the earlier turnover and irrigation service fee programs. However, projects to expand implementation of the new program, with support from the World Bank and ADB, went through a process of prolonged preparation, extended negotiation, and delayed implementation.

In 2004, Indonesia enacted a new water law, and in 2006 a new government regulation on irrigation was issued to follow up on the water law. These regulations nominally endorsed local level participation in irrigation management, but, in contrast to earlier reform commitments, they did not proceed with any ambitious IMT programs. Draft regulations on irrigation management, prepared as part of the Water Sector Adjustment Loan program, underwent a long period of revision. Revised guidelines and manuals, which the government finally began to issue in 2007, largely reiterated past practices for technical irrigation management. This prolonged period of policy shift and uncertainty created somewhat of a policy vacuum for projects such as irrigation rehabilitation in Aceh.

In Aceh, during the 1990s and thereafter, attempts at reforming irrigation institutions had played out in the context of armed conflict and military repression. Spending on irrigation construction was one of the ways the government sought to offer benefits and align interests to obtain support in its conflict with the Gerakan Aceh Merdeka (Aceh independence movement, or GAM). However, the extent of the government's power to impose detailed changes in local organization of irrigation was quite limited, and the imperative was more on making conspicuous expenditures to show government's beneficence, accompanied by the personal interests of officials in profiting, in terms of careers and corruption, from the proceeds.

In the wake of the December 2004 earthquake and tsunami, armed conflict was finally halted by the August 2005 Memorandum of Agreement between the government and GAM. Post-tsunami national and international assistance brought funding for investment in irrigation reconstruction and rehabilitation. This assistance took place in the national policy context described above, where major efforts for more participatory irrigation governance had begun, but then faltered.

Post-Suharto decentralization reforms meant that, officially, authority over most irrigation systems had been transferred from national and provincial governments to *kabupaten* (districts). Irrigation infrastructure assets had been transferred to districts, and they became responsible for hiring and paying operations and maintenance staff. In Aceh, turnover of secondary canals within irrigation systems to WUA had never been implemented. Officially, WUA activities were still focused at the tertiary canal level and below. The pattern was thus one of devolution of irrigation management down to a lower level of government, but with little impetus for further devolution to user groups.

Projects funded by the national government and international aid continued to be implemented through special purpose project implementation units. These were separate from the routine administrative structure of district and provincial water resources agencies. The ADB's Earthquake and Tsunami Emergency Support Project offered grant funding for reconstruction and rehabilitation work in a range of sectors, including irrigation.[4] It sought to involve communities in the planning and construction of irrigation rehabilitation. Initially, irrigation rehabilitation in Aceh was arranged through a project management unit directly appointed by the national government's Directorate General of Water Resources Development. Subsequently, control over externally funded reconstruction and rehabilitation activities was transferred to the specially created Agency for Reconstruction and Rehabilitation in Aceh. Within the project, project managers controlled the flow of funds for civil works. District governments had a formal role in approving project proposals, and continued to be responsible for operation and maintenance of the improved schemes, but had relatively little influence over implementation.

Within government, there was significant ambiguity and uncertainty about responsibility for irrigation. The *penjelasan* (explanation) of the 2004 Water Law stated that responsibility for irrigation operation and maintenance would be divided in accordance with irrigation system size, with districts responsible for irrigation areas under one thousand hectares, provincial government responsible for areas of one thousand to three thousand hectares, and the national government responsible for areas larger than three thousand hectares. Similar language was included in the 2006 government regulation on irrigation. The ways in which this division of responsibility would be put into practice were not specified, and hypothetically could include various forms of delegation of funding and tasks to lower-level government units. In Aceh, schemes were not formally transferred back to national or provincial management, although the national government provided some funding for operations and maintenance, channeled through project activities under the supervision of the provincial irrigation office.

Implementation of reforms to devolve authority to WUA faces challenges since agencies tend to oppose reforms that would reduce their power. This factor affected attitudes toward reforms in irrigation management, such as proposals for IMT. As part of decentralization, however, the government implemented much more drastic shifts, most notably the transfer of irrigation systems to district governments and severe downsizing of central agency staff, including the transfer of national and provincial civil servants to become employees of district governments. This drastic reform refutes any hypothesis that major institutional changes in irrigation were politically impossible. Entrenched interests and institutional inertia were overcome. However, this reform did not take the form of IMT—currently popular in international discourse on irrigation reform. Instead, devolution of authority to WUAs was one part of a larger set of reforms in which governance authority was devolved downwards but without necessarily strengthening the role of individual water users in irrigation governance.

Water users have tended to be relatively unenthusiastic about taking over full control of irrigation systems or subsystems, and relatively accepting of continuing agency roles in water management. While farmers often complain about poor performance of irrigation systems, they still see state action in irrigation as legitimate, and attractive in terms of the benefits of state subsidies. Farmers' views contrast with the romantic, populist, or localist notions that sometimes seem to underlie policies for increasing participation and reducing the passivity attributed to state patronage. Communities are not pressing for the state to withdraw from irrigation or for the state to wither away. For these communities, the social life of water is one in which village governments and higher-level governments are expected to play a legitimate and continuing role.

Policies and programs for IMT internationally, including those in Indonesia discussed in this chapter, focus heavily on shifting tasks to farmers. Less attention is paid to the tasks that require continuing cooperation and interaction between government and farmers, such as distribution of water to secondary canals and construction of major repairs and improvements. As mentioned earlier, the rationale for IMT is often framed in terms of reducing government expenditures on operation and maintenance by shifting costs to farmers, rather than improving performance in delivering services. Development projects typically provide funding for a single round of rehabilitation, as part of turnover. Much less is done to develop capacity for future repairs and improvements. Funds are often used to restore irrigation systems to their earlier condition, continuing works that had not been completed earlier, and adding canal lining, without detailed analysis and development of the institutions and infrastructure needed for new management partnerships.

IMT projects are often conceived in terms of having farmers take over responsibility for management of the entire irrigation system, or a larger part of it, with an agency delivering water to a canal headgate. However, IMT reforms are often partial and incomplete (Vermillion 1996, 2006), resulting in continuing ambiguity and uncertainty about responsibilities.

Rather than creating a clean separation of roles, as sometimes suggested by the concept of wholesaling water, the result in Indonesia has typically continued to be a combination of cooperation and conflict between government agencies and communities in managing access to water in a system where both are still involved. Responsibility for maintenance is spread among farmers and different levels of government in ways that are uncertain and that may discourage preventive maintenance. Major repairs and improvements are still carried out through special-purpose project institutions that are relatively disconnected from farmers and line agencies. This situation creates both problems and opportunities, for agency officials and for farmers.

Despite clear doctrines for irrigation water distribution, actual practices commonly reflect a mix of past practices and accommodations to the capacities and constraints of new infrastructure. In particular, water distribution practices often diverge widely from the highly technical procedures for planning and measuring water distribution prescribed by official standards.

The failures of past efforts to transfer irrigation management, the existence of policies for participatory irrigation management, and the continuing need for cooperation between farmers and government in key tasks create conducive conditions for developing comanagement (Berkes 1994). In Aceh and elsewhere, for medium- and large-scale irrigation systems, the concept of comanagement could provide a useful framework for facilitating joint efforts to improve the flow of information about water needs and adequacy of service delivery, understanding of the potential and constraints for improving irrigation system performance, and organizing cooperative action to make improvements.

Comanagement is particularly crucial for situations such as management of larger irrigation systems and provision of government aid for irrigation system repair and improvement, where cooperation between users and government can be essential for obtaining good results. However, comanagement requires not just effective organization, but also adequate infrastructure and operational procedures.

Internationally, experience with participatory governance and comanagement, and the difficulties of implementing sharply contractual approaches to devolution have led to what might be called a second generation of ideas about how comanagement can be understood and practiced (Carlsson and Berkes 2005). Interactions between agencies and communities are not just a matter of giving stakeholders a voice in agency decisions or devolving management to the lowest possible level. Instead, many tasks require joint decision making and dealing with the interaction between different scales, in space and over time. Comanagement is a more contested and confused process than that suggested by policies for formal devolution of authority, wholesaling water, and service agreements that would attempt to sharply delineate agency and farmer roles.

A more flexible approach to comanagement may involve multiple and shifting sets of participants, adaptive decision making, and evolution in how problems are defined and solutions attempted (Carlsson and Berkes 2005). A comanagement approach would fit well with conditions in Aceh and many other parts of Indonesia. Comanagement is subject to continued contestation as multiple agencies and levels of government seek to assert or reassert authority, and communities pragmatically produce the appearance of compliance while perpetuating local practices. In Indonesia, the retreat from bolder IMT policy created the need to more carefully examine the options for crafting better cooperation between water users and irrigation agencies.[5] In Indonesia and elsewhere, a key challenge seems to lie in the development of more-effective approaches to comanagement of critical tasks in ways that support cooperation at multiple scales.

Polycentric Problem Solving

Lessons from attempts to improve participation in irrigation management help to clarify the potential and constraints for adaptive comanagement to solve problems in irrigation water management. The implication of a practical perspective on participation is the need to focus carefully on those forms of cooperation likely to be most feasible and worthwhile for all involved. A practical perspective does not rule out more drastic institutional restructuring such as IMT if suitable conditions exist to make it feasible, but is intended to highlight the opportunities that may exist where such shifts seem difficult or impossible. These opportunities seem likely to lie in three areas: (1) participatory design, (2) networking among water users, and (3) an adaptive problem-solving approach to improving irrigation systems. In Aceh, as elsewhere, experience has shown the feasibility of involving users in design and construction. Widening linkages between WUAs is also generally feasible, through the formation of WUA federations. In response to problems and crises in water distribution, comanagement may provide a framework for instituting improvements to adapt irrigation operation and maintenance to changing conditions.

A polycentric perspective (V. Ostrom 1999) goes beyond narrow conceptions of centralization and decentralization to look not only at multiple levels and scales of social organization, but also at the ways these may productively overlap and interact, including the formation of local public economies to provide goods and services. Nested organizational structures of smaller units aggregating into larger ones was one of the institutional design principles of robust common property governance identified by Elinor Ostrom (1990, 1992). A polycentric perspective perceives governance as a form of problem solving, where coalitions of those concerned use various institutional means to meet their needs. A polycentric perspective can thus sidestep some of the polarized debates on centralized power or decentralization, and public control or privatization, to look more flexibly at the potential for creative combinations of various kinds of institutions at multiple scales. Polycentricity does not mean that such institutions are without conflicts, achieve completely egalitarian results, or always survive, but that they greatly expand the diversity of institutional options that may be available for fitting institutions to the goals and circumstances of those involved, in contrast to the simplistic replication of uniform, one-size-fits-all models as panaceas (E. Ostrom 2007; Meinzen-Dick 2007).

Recent projects in Aceh, particularly the Northern Sumatra Irrigation and Agriculture Sector Project and the Earthquake and Tsunami Emergency Support Project, have sought to more thoroughly apply participatory approaches to the design of irrigation construction and rehabilitation. Participatory approaches benefit from users' knowledge of local conditions, and their interest in obtaining good performance. Participation has usually involved a package of institutional arrangements including joint irrigation system walkthroughs to discuss problems and priorities, a series of meetings to discuss draft designs, strengthening of irrigators' organizations, and involvement of users in carrying out construction, with the process aided by a facilitator of some sort. These approaches drew on ideas initially pioneered in the Philippines and Sri Lanka and subsequently widely emulated in Indonesia and elsewhere (Bruns 1993). While not always easy to introduce, they have generally proved feasible to apply. This experience with participatory design and construction provides a repertoire of ideas and procedures that agencies and farmers can draw on in cooperating to solve problems that require repair or improvement of irrigation systems.

Organization of WUAs on a wider scale, as federations at the secondary canal and irrigation system levels, has also proved feasible in many parts of Indonesia and in other countries. There is potential for wider networking within watersheds, districts, and provinces. While full takeover of management of secondary canals is problematic and may be unlikely to proceed, there is nevertheless much scope for improving cooperation and performance in main system management. In some cases, development of WUA federations has been linked with changes to improve water distribution within irrigation systems, particularly providing more-timely delivery, better coordination of cropping schedules, and more-equitable sharing with tail-end areas. The most important opportunities for wider-scale networking seem to lie in improving the equity of water distribution within irrigation systems and watersheds. The capacity of bureaucratic management to enforce compliance with water allocation is quite limited, so participatory forums open up the opportunity to mobilize more general support for changes in water distribution, based on consent from users and their representatives, for rules they are willing to help monitor and enforce.

Recent proponents of irrigation modernization (Renault, Facon, and Wahaj 2007; and of modernization more generally, see Burt and Styles 2004; Plusquellec 2002) argue that many irrigation systems are difficult or impossible to operate as designed, not only due to deficiencies in infrastructure as actually built and maintained, but also due to inherent complexity in how fluctuations in water flows propagate disturbances along canals in response to gate operations, rainfall, changes in water taken by farmers, and the nonlinear ways these factors interact. The argument is that irrigation systems must be understood as complex adaptive systems, which require incremental adjustment, learning and adaptation, for which there is no single or stable solution for water distribution.

Rather than being able to impose a standardized doctrine for water distribution, the need is for learning and adaptation customized to local conditions. Various adjustments in infrastructure and operating rules may help to reduce and stabilize fluctuations, but will still require continuing adaptation to changing and unpredictable conditions. This analysis argues that centralized command-and-control methods are inherently incapable of delivering good performance in irrigation management, particularly in run-of-the-river irrigation systems that lack storage reservoirs to buffer fluctuations in water availability and to stabilize flows. Analysis of irrigation modernization in terms of management of irrigation as a complex adaptive system clarifies the challenges involved in providing reliable delivery of water to secondary irrigation canals and fields, showing the need for continuing two-way flows of information, adjustment, and learning to adapt irrigation operations to changing conditions. Modernization concepts emphasize water delivery to secondary canals as a key area where continuing cooperation between farmers and agencies is essential for effective irrigation management.

Sophisticated approaches to irrigation modernization that actively involve users in diagnosis and testing of new operational arrangements have the potential to stimulate fertile forms of comanagement. There is also a risk, however, that irrigation modernization is misunderstood to mean simply the installation of specific infrastructure, particularly of long-crested weirs in canals and measuring flumes, along with standardized operational procedures. Furthermore, there is a crucial need not just to look at how to improve the reliability of water delivery to canal headgates, but also to improve delivery to tail-end areas. Perpetuating an exclusive division of labor that leaves distribution within tertiary canal areas purely to farmer responsibility would mean being blind to the actual impact on farmers. Without improvement in main canal management, improving water delivery in tertiary areas is often difficult or impossible. Improvement in delivery to canal headgates is often insufficient to lead to wider and more equitable distribution, however, and government may have a crucial role to stimulate attention to lower-level water distribution. Therefore, again, the need is for comanagement, rather than for an exclusive division of labor between agency and farmers.

Once current water distribution practices are seen as a pragmatic bricolage, blending top-down and bottom-up ideas and institutions, then it becomes easier to see how they provide space for adapting to changing conditions, and for continuing current practices if they do not pose critical problems for performance. Many irrigation systems can deliver water reasonably well most of the time to most users, although there may still be serious problems for those toward the tail-end. The question may not necessarily be one *of optimizing* performance, but rather of ensuring there is enough institutional capacity for problem solving in response to drought, tail-end deprivation, and other problems. In the context of state intervention, it also becomes important to ask to what extent such intervention tries to promote or require changes to pursue objectives such as increased equity, productivity, and environmental sustainability.

In Indonesia and elsewhere, it seems likely that future efforts at improving management institutions in larger irrigation systems are likely to be less ambitious and conceptualized in different terms from simply achieving decentralization, devolution, or transfer. A key question thus concerns what concepts may be used to frame future projects. Ideas about irrigation modernization offer one approach. Another option is more business-as-usual rehabilitation projects to restore infrastructure and perhaps upgrade lining and gates, possibly accompanied by a gloss of superficial institutional reforms. Another possibility may be relative neglect or abandonment of irrigation investments (as argued by Frederiksen 2005), with governments reducing budget allocations for irrigation construction and operation. There may be more attention to context specific approaches that seek to avoid the perils of simplistic panaceas (Lankford et al. 2007). The political appeal of aiding farmers and increasing shortages of water seem likely to drive continued government intervention in irrigation, in one form or another. In many countries, rising expectations for democracy, participation, and agency accountability are likely to encourage implementation of participatory approaches, creating conditions conducive for further development of various forms of comanagement and the potential for participatory problem solving.

Conclusions

Historically, irrigation governance has been organized at multiple levels, in Aceh as in many other areas. Irrigation construction and management often has involved cooperation across multiple communities and government agencies. Even in systems that appear highly centralized, substantial scope for local adjustment and contestation has been present. Conversely, even in systems that appear to be farmer-managed, a closer examination of history often reveals episodes of intervention for construction or reconstruction, as well as of a larger context of states, small or large, promoting cultivation of lowland *sawah* (rice fields) by settled populations. Patterns of polycentric irrigation governance are likely to persist so that comanagement, in one form or another, is likely to continue to be an important part of irrigation management.

Governments in Indonesia and elsewhere can be expected to continue to intervene in irrigation, to secure food supplies, aid rural areas, and respond to increasing water scarcity, a process now increasingly framed in the international discourse of climate change. Many irrigation systems encompass more than a single community, so some relatively formal governance structures are essential. In this context, decisions will be made about how to manage shared resources. Therefore, attempts at institutional crafting or social engineering may be inevitable (North 1990). Top-down visions of neatly hierarchical arrangements for water distribution may well continue to appeal. However, practice is likely to continue to be much more complex and contested, creating a polycentric bricolage of improvisation and adaptation. Rather than a triumph of centralization or decentralization, the future may well lie in the middle, in an improvised mixture of social engineering and participatory problem solving.

The 2006 law on regional government for Aceh stipulated that the government would legally recognize *keujreun blang*, as well as other positions such as *panglima laut* (fisheries leaders). However, the specific mechanisms through which this recognition may occur have not been formulated. While there is interest in and concern for encouraging the vitality of local institutions such as *keujreun blang*, there are not yet proposals for reforms as strong as those embodied in the attempts in West Sumatra to restore the control of traditional *nagari* (communities) over local land, water, and forest resources. Irrigation continues to be an important part of rural livelihoods in Aceh, driving action by farmers and government. Social engineering and participatory problem solving both seem likely to continue, playing out in the context of episodic mobilization by farmers, bureaucratic bricolage, and polycentrically contested comanagement.

Notes

Statements in this chapter are the views of the author and do not represent those of any organization with which he is or has been affiliated.

1. For an international review of experience and directions for irrigation development, see the chapters on institutions and irrigation in *Water for Food, Water for Life: A Comprehensive Assessment of Water Management in Agriculture* (Molden 2007); see especially Chapter 5 on institutions (Merrey et al. 2007) and Chapter 9 on irrigation (Faures, Svendsen, and Turral 2007). See also analyses in Mollinga and Bolding (2005) and Shivakoti et al. (2005).
2. What the government invested in Aceh was far less than what it gained in revenues from oil and gas from Aceh, a major point of conflict between those in Aceh and the national government in Jakarta.
3. For a discussion of the turnover project and subsequent attempts at participatory reform, see Bruns (2004).
4. General information on the project is available on ADB's website at http://www.adb.org/Projects/ETESP/default.asp.
5. The policy shift was most clearly marked by the 2004 water law; this law did not provide any explicit support for IMT, which had been enabled by an earlier government regulation on irrigation. Subsequently, the new 2006 government regulation on irrigation reaffirmed government responsibility for managing irrigation systems to the tertiary gate and fifty meters beyond it, while still calling for participatory irrigation management but with little detail as to how such management would occur.

References

Agrawal, Arun, and Clark C. Gibson. 2001. The Role of Community in Natural Resource Conservation. In *Communities and the Environment: Ethnicity, Gender, and the State in Community-Based Conservation,* edited by Arun Agrawal and Clark C. Gibson. New Brunswick, NJ: Rutgers University Press.

Anderson, Benedict R. O'G. 1990. *Language and Power: Exploring Political Cultures in Indonesia.* Ithaca, NY: Cornell University Press.

Araral, Eduardo. 2005. Bureaucratic Incentives, Path Dependence, and Foreign Aid: An Empirical Institutional Analysis of Irrigation in the Philippines. *Policy Sciences* 38: 131–157.

Berkes, Fikret. 1994. Co-management: Bridging the two Solitudes. *Northern Perspectives* 22: 2–3.

Bruns, Bryan. 1990. Design for Participation: Elephant Ears, Crocodile Teeth and Variable Crest Weirs in Northeast Thailand. In *Design Issues in Farmer-Managed Irrigation Systems. Proceedings of an International Workshop of the Farmer-Managed Irrigation Systems Network held at Chiang Mai, Thailand from 12 to 15 December 1989,* edited by Robert Yoder and Juanita Thurston, 107–119. Colombo, Sri Lanka: International Irrigation Management Institute.

———. 1991. The Stream the Tiger Leaped: A Study of Intervention and Innovation in Small Scale Irrigation Development in Northeast Thailand. Ph.D dissertation, Cornell University, Ithaca, NY.

———. 1992. Just Enough Organization: Water Users Associations and Episodic Mobilization. *Visi: Irigasi* (6): 33–41.

———. 1993. Promoting Participation in Irrigation: Reflections on Experience in Southeast Asia. *World Development* 21 (11, November): 1837–1849.

———. 2004. From Voice to Empowerment: Rerouting Irrigation Reform in Indonesia. In *The Politics of Irrigation Reform: Contested Policy Formulation and Implementation in Asia, Africa and Latin America.* Aldershot: Ashgate.

Bruns, Bryan, and Sudar Dwi Atmanto. 1995. How to Turn Over Irrigation Systems to Farmer? Questions and Decisions in Indonesia. In *Irrigation Management Transfer: Selected Papers from the International Conference on Irrigation Management Transfer*, edited by S.H. Johnson, D.L. Vermillion, and J.A. Sagardoy, 267–284. Rome: International Irrigation Management Institute Food and Agriculture Organization of the United Nations.

Bruns, Bryan, and Ruth Meinzen-Dick, eds. 2000. *Negotiating Water Rights.* New Delhi: Vistaar.

Bruns, Bryan, Claudia Ringler, and Ruth Meinzen-Dick, eds. 2005. *Water Rights Reform: Lessons for Institutional Design.* Washington, DC: International Food Policy Research Institute.

Bruns, Bryan, and Irchamni Soelaiman. 1992. From Practice to Policy: Agency and NGO in Indonesia's Program to Turn Over Small Irrigation Systems to Farmers. *Overseas Development Institute Irrigation Management Network Paper* (10e): 25–38.

Burt, Charles M., and Stuart W. Styles. 2004. Conceptualizing Irrigation Project Modernization through Benchmarking and the Rapid Appraisal Process. *Irrigation and Drainage* 53 (2): 145–154.

Carlsson, Lars, and Fikret Berkes. 2005. Co-management: Concepts and Methodological Implications. *Journal of Environmental Management* 75: 65–76.

Chambers, Robert. 1988. *Managing Canal Irrigation: Practical Analysis from South Asia.* New Delhi: Oxford and IBH.

Cleaver, Frances, and Tom Franks. 2005. How Institutions Elude Design: River Basin Management and Sustainable Livelihoods. Bradford, UK: Bradford Centre for International Development Research Paper No. 12.

Coward, E. Walter Jr. 1991. Planning Technical and Social Change in Irrigated Areas. In *Putting People First: Sociological Variables in Rural Development*, edited by M.M. Cernea. Oxford: Oxford University Press.

Faures, Jean-Marc, Mark Svendsen, and Hugh Turral. 2007. Reinventing Irrigation. In *Water for Food, Water for Life: A Comprehensive Assessment of Water Management in Agriculture*, edited by David Molden, 353–394. International Water Management Institute. London: Earthscan.

Frederiksen, Harald D. 2005. The Future of Rural Water Services in the Developing Countries and the Government's Role. *Irrigation and Drainage* 54: 501–508.

Shivakoti, Ganesh, Douglas Vermillion, Wai Fung Lam, Elinor Ostrom, Ujjwal Pradhan, and Robert Yoder, eds. 2005. *Asian Irrigation in Transition: Responding to Challenges.* New Delhi: SAGE.

Geertz, Clifford. 1980. Organization of the Balinese Subak. In *Irrigation and Agricultural Development in Asia*, edited by E. Walter Coward Jr., 70–90. Ithaca, NY: Cornell University Press.

Hartono, A. Totok. 2007. *Participatory Irrigation Approach in Nanggroe Aceh Darussalam: The Case of Introducing Concrete Canal Lining Through Participatory Construction.* Asian Development Bank, Manila. http://www.docstoc.com/docs/49642317/Participatory-Irrigation-Approach-in-Nanggroe-Aceh-Darussalam-%C3%A2%E2%82%AC%E2%80%9C-The.

Hunt, Robert. 1989. Appropriate Social Organization? Water User Associations in Bureaucratic Canal Irrigation Systems. *Human Organization* 48 (1): 79–90.

Hunt, Robert, and Eva Hunt. 1976. Canal Irrigation and Local Social Organization. *Current Anthropology* 17: 389–411.

Hurgronje, C. Snouk. 1906. *The Acehnese*, vol. I. Translated by A.W.S. O'Sullivan. Leiden: E.J. Brill.

Korten, Frances F., and Robert Y. Siy. 1988. *Transforming a Bureaucracy: The Experience of the Philippine National Irrigation Administration.* West Hartford, CT: Kumarian.

Lankford, Bruce A., Douglas J. Merrey, Julien Cour, and Nick Hepworth. 2007. *From Integrated to Expedient: An Adaptive Framework for River Basin Management in Developing Countries.* Colombo, Sri Lanka: International Water Management Institute.

Lansing, J. Stephen. 1991. *Priests and Programmers: Technologies of Power in the Engineered Landscape of Bali.* Princeton, NJ: Princeton Univesity Press.

———. 2006. *Perfect Order: Recognizing Complexity in Bali.* Princeton, NJ: Princeton University Press.

Leach, Edmund. 1981/1959. Hydraulic Society in Ceylon. In *The Asiatic Mode of Production*, edited by Anne M Bailey and Josep R Llobera. London: Routledge and Kegan Paul.

Lévi-Strauss, Claude. 1966. *The Savage Mind.* Chicago: University of Chicago Press.

Lewis, Henry T. 1980. Irrigation Societies in the Northern Philippines. In *Irrigation and Agricultural Development in Asia*, edited by E. Walter Coward. Ithaca, NY: Cornell University Press.

Meinzen-Dick, Ruth. 2007. Beyond Panaceas in Water Institutions. *Publications of the National Academy of Sciences* 104 (39): 15200–15205.

Merrey, D.J., R. Meinzen-Dick, P.P. Mollinga, and E. Karar. 2007. *Policy and Institutional Reform: The Art of the Possible.* In *Water for Food, Water for Life: A Comprehensive Assessment of Water Management in Agriculture*, edited by David Molden, 193–232. International Water Management Institute. London: Earthscan.

Molden, David, ed. 2007. *Water for Food, Water for Life: A Comprehensive Assessment of Water Management in Agriculture.* International Water Management Institute. London: Earthscan.

Mollinga, Peter P., and Alex Bolding. 2005. *The Politics of Irrigation Reform: Contested Policy Formulation and Implementation in Asia, Africa and Latin America.* Aldershot: Ashgate.

North, Douglass C. 1990. *Institutions, Institutional Change and Economic Performance.* New York: Cambridge University Press.

Ostrom, Elinor. 1990. *Governing the Commons: The Evolution of Institutions for Collective Action.* Cambridge: Cambridge University Press.

———. 1992. *Crafting Institutions for Self-Governing Irrigation Systems.* San Francisco: Institute for Contemporary Studies Press.

———. 2007. A diagnostic approach for going beyond panaceas. *Proceedings of the National Academy of Sciences* 104 (39): 15181–15187.

Ostrom, Vincent. 1999. Polycentricity. In *Polycentricity and Local Public Economies*, edited by M.D. McGinnis. Ann Arbor: University of Michigan Press.

Plusquellec, Herve. 2002. How Design, Management, and Policy Affect the Performance of Irrigation Projects: Emerging Modernization Procedures and Design Standards. Bangkok: Food and Agricultural Organization of the United Nations.

Potter, Jack M. 1976. *Thai Peasant Social Structure.* Chicago: University of Chicago Press.

Rap, Edwin. 2006. The Success of a Policy Model: Irrigation Management Transfer in Mexico. *Journal of Development Studies* 42 (8): 1301–1324.

Renault, Daniel, Thierry Facon, and Robina Wahaj. 2007. Modernizing Irrigation Management: The MASSCOTE Approach: Mapping System and Services for Canal Operation Techniques. Food and Agricultural Organization of the United Nations Irrigation and Drainage Paper 63. Rome: FAO.

Siegel, James T. 2000. *The Rope of God.* Ann Arbor: University of Michigan Press.

Sirivongs Na Ayudhaya, Abha. 1983. A Comparative Study of Traditional Irrigation Systems in Two Communities of Northern Thailand. Bangkok: Chulalongkorn University, Social Research Institute.

Spiertz, H.L. Joep. 2000. Water Rights and Legal Pluralism: Some Basics of a Legal Anthropological Approach. In *Negotiating Water Rights*, edited by Bryan Bruns and Ruth Meinzen-Dick. New Delhi: Vistaar.

Suhardiman, D. 2008. Bureaucratic Designs: The Paradox of Irrigation Management Transfer in Indonesia. Ph.D. thesis, Wageningen University, Netherlands.

Tan-Kim-Yong, Uraivan. 1995. *Muang-Fai Communities Are for People: Institutional Strength And Potentials.* Bangkok: Chulalongkorn University Social Reseach Institute.

Uphoff, Norman. 1991. *Learning from Gal Oya: Possibilities for Participatory Development and Post-Newtonian Social Science.* Ithaca, NY: Cornell University Press.

Vermillion, Douglas. 1991. The Turnover and Self Management of Irrigation Institutions in Developing Countries. Colombo, Sri Lanka: International Water Management Institute Discussion Paper. June.

———. 2006. *Lessons Learned and to be Learned about Irrigation Management Transfer.* Paper presented at Survival of the Commons: Mounting Challenges and New Realities: the Eleventh Conference of the Annual Association for the Study of the Commons, Ubud, Bali, Indonesia.

Vermillion, Douglas, ed. 1996. *The Privatization and Self-Management of Irrigation*, edited by Douglas Vermillion. Colombo, Sri Lanka: International Irrigation Management Institute.

Wade, Robert. 1987. *Village Republics: Economic Conditions for Collective Action in South India.* Cambridge: Cambridge University Press.

Wittfogel, Karl. 1957. *Oriental Despotism: A Comparative Study of Total Power.* New Haven, CT: Yale University Press.

Young, Oran R. 2002. *The Institutional Dimensions of Environmental Change: Fit, Interplay and Scale.* Cambridge, MA: MIT Press.

Chapter 14

ANTHROPOLOGICAL INSIGHTS INTO STAKEHOLDER PARTICIPATION IN WATER MANAGEMENT OF THE EDWARDS AQUIFER IN TEXAS

John M. Donahue

❧

Introduction

There was an almost audible rush of relief in the room when after three years the stakeholders unanimously agreed to the last of the items that might finally lead to a resolution of the seventeen-year debate over management of the Edwards Aquifer and protection of its endangered species. For three years, some eighty stakeholders representing a wide range of institutions and citizenry had met monthly to forge a process that would resolve a long-standing conflict on how best to preserve the habitat of the species and avoid a federal takeover of water management of the aquifer. Ad hoc committees, work groups, and an expert science subcommittee had met between the monthly meetings; over the three-year period, they had generated reports on various aspects of water management. The stakeholders contracted with biologists and hydrologists to address the technical aspects of groundwater management. They brought in professional facilitators to help them work through the recommendations of the committees and consultants. The agreements made that afternoon meeting in San Marcos, Texas, were not the end of the process, but were a significant watershed. Everyone had agreed on a plan to proceed, a result that was not ensured until that very day amid discussions, caucuses, and further clarification of the issues. The significance of that day and the lessons for water management generally can only be understood in the context of the debate that had begun some seventeen years earlier.

During the past forty years, conflicts over the use of the Edwards Aquifer in central Texas have pitted multiple interests against one other. Agricultural irrigators, municipal water systems, industrial users, river systems, and recreational users, to mention a few, all bring different cultural definitions of water to the debate. Whereas water is a hydrological given, it is also a cultural construction. Negotiations over water management often fail because of what Rothman (1995: 19) calls "the hidden grammar" of a group's values, priorities, motivations and patterns of behavior. A related factor when institutional actors are involved in water conflicts is what Nickum and Greenstadt (1998: 160) refer to as "project culture," where priority is given to institutional maintenance through the budgetary increases that certain water development policies provide. In such cases institution building may interfere with the original mission of the agency or organization. Both of these processes were potential threats to water management of the Edwards Aquifer. This chapter discusses how those obstacles were overcome and what lessons this case may have for similar conflicts over water management.

The Research Question and Methodology

The aquifer is a multivocal symbol that unites and divides people across political, economic, social, and geographical spectrums. Precisely in this context, the aquifer surfaces as a boundary object.[1] It has a solid presence but also morphs easily into disparate meanings for different interested parties. Research over the past four years has focused on an initiative of the U.S. Fish and Wildlife Service (USFWS) to facilitate a consensus-based forum of stakeholders who would solicit the best science on which to recommend a management policy that would comply with the Endangered Species Act of 1973. Several questions needed to be addressed, such as, how do the different stakeholders in the debate over management define the value of water? And, how is it possible to mediate among competing interests with different cultural meanings given to water? Furthermore, which power relations are served most by which cultural definitions? More specifically, what minimum spring flows at Comal and San Marcos would ensure the survival of endangered species? In another drought of record, such as occurred in the mid-1950s, to what extent would pumping from the aquifer need to be reduced to ensure adequate spring flow?

My research on water issues over the Edwards Aquifer spans twenty years (Donahue 1998; Donahue and Saunders 2001; Klaver and Donahue 2005). The research reported here began in 2007 with the inauguration of the Edwards Aquifer Recovery Implementation Program (EARIP, or RIP). Attendance at monthly meetings and numerous subcommittee meetings allowed me the opportunity to document the process of consensus building. Interviews at critical points during the following four years provided further perspective on how the actors were dealing with long-standing disagreements on contentious issues of aquifer management policy.

The process of consensus building can be understood as having three essential components (see figure 14.1). First, as suggested by Fisher and Ury (1991), the concerns among parties must include substantive issues. This was certainly the case with the EARIP process, as all participants agreed that the key question was how to protect the habitat of the endangered species in the springs while maintaining sufficient water for pumpers across the aquifer.[2] In addition, success in consensus building demands that all the stakeholders be present at the table. The EARIP was the first effort to establish this procedure among stakeholders who, for the previous seventeen years, had been at loggerheads. Finally, the third dimension in the process triangle is participation. All parties must be made to feel that their voice is being heard and that the decision-making process is transparent and based on consensus. Consensus may not mean that all agree, but that all parties can live with a decision of the group.

Before the EARIP process began in 2007 there had been widespread mistrust among the many stakeholders who depend on the aquifer. The level of trust increased significantly, however, over the first four years of its operation. Discussions in the meetings were civil, relaxed and obliging. In her interviews with stakeholders early in the process, Anna Muñoz (Muñoz and Donahue 2009) noted that several expressed skepticism that the effort would yield much, given the history of antagonism among the various interests over the previous twenty to thirty years. In her follow-up interviews two years later, she found that many expressed surprise at the working relationships that had developed. Before taking up the process itself, some background, hydrological and historical, is in order.

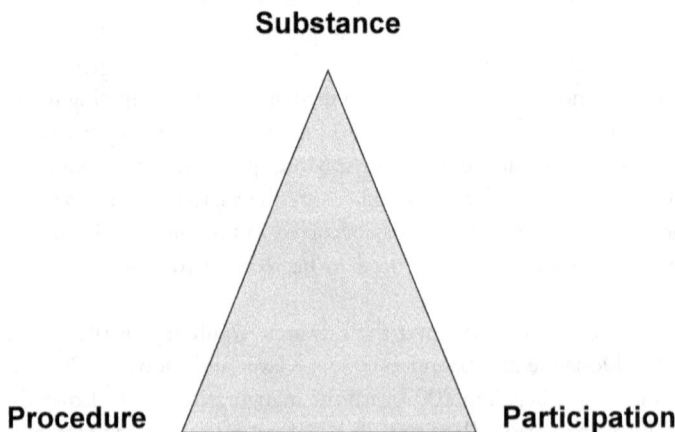

Figure 14.1. Process triangle

Background

In early Westerns, Texas is often portrayed as dry, devoid of vegetation except tumbleweed, and populated by more coyotes than people. This stereotype might fit the western Texas deserts, but the state enjoys a wide range of environmental zones, from the lush rice country and piney woods of the east to the Chihuahua Desert around El Paso in the west. Between these two major zones is the city of San Antonio, the third-largest metropolitan area in Texas after Houston and Dallas-Fort Worth, and now the eighth largest city in the United States, with 1.21 million people. Given its proximity to the border, San Antonio is culturally more like Monterrey, Mexico, 280 miles to the south, than it is like Austin, the state's capital, some eighty miles to the north. The influence of northern Mexico can be found in San Antonio's food, music, and architecture. More than half of the city's residents can trace their origins back to Mexico—partly for the simple reason that for a considerable period of its history San Antonio, in fact, was in Mexico.

In one important regard, however, San Antonio is unlike her neighbor to the south and more like other cities in the United States. Everywhere spacious lawns are carpeted with water-thirsty St. Augustine grass (*Stenotaphrum secundatum*). Imported from tropical Africa, this grass accounts—together with landscape watering—for some 30 to 40 percent of San Antonio's annual water usage, and symbolizes the ambivalences faced by South Texans in adapting to a semiarid environment. Yet, on the other side of the border, Mexicans long ago came to grips with living in a semiarid climate. Their homes normally front directly onto the street, obviating any need for a lawn. An inner courtyard yields a common green space for the enjoyment of all in the home, with minimal watering demands.

A major hydrological difference adds to the cultural contrasts between northern Mexico and southern Texas in watering landscapes. Whereas northern Mexico depends on the surface water of rainfall and riverine sources, San Antonio and the whole semiarid region that characterizes the area between the limestone hills of central Texas and the Gulf Coast is blessed with groundwater from a rechargeable, underground aquifer—the Edwards Aquifer (see figure 14.2).[3] The city relies for more than 99 percent of its municipal supply on the aquifer, giving San Antonio the dubious distinction of being the largest city in the country to rely so heavily on groundwater (Glennon 2002: 89).

Human adaptation to the geology of the Edwards Aquifer region has created a socioeconomic landscape that concentrates the agricultural and ranching economic sector in the west, the more industrial and metropolitan sector in the center, and a major recreational and service sector in the east, around the springs in Comal and Hayes Counties.

To the north of the aquifer lies an area of recharge where runoff from rainfall flows into the aquifer through porous limestone features and percolates downward to the reservoir. Because the aquifer varies in elevation from west to east, it flows

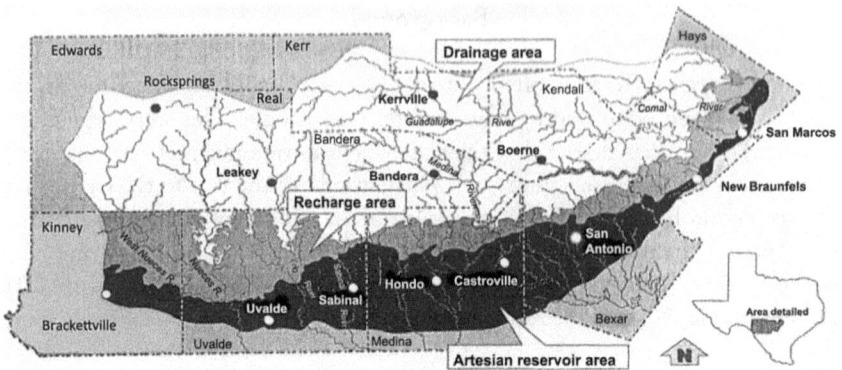

Figure 14.2. Edward's Aquifer region

eastward, emerging at various springs along the way (see figure 14.3). Estimates place the aquifer's capacity at between twenty-five and fifty-five million acre-feet of water, which is more than all the surface water in the state (Maclay and Land 1988).[4] The region's annual pumping is about 450,000 acre-feet, or 1.0 percent to 2.4 percent of the aquifer's capacity.[5]

Aquifer waters percolate to the surface at springs in the central (Bexar County) and eastern expanses of the aquifer (Hays and Comal Counties), which, in turn, have been sites of urban population growth. With advances in pumping technology during the early twentieth century, large tracts of irrigated farmland were opened up on the western expanses of the aquifer in Medina, Uvalde, and Kinney Counties. Currently farmers to the west invest heavily in irrigation to grow cash crops such as corn, cotton, sorghum, and Bermuda grass, as well as fall cabbage and spinach. All are water-intensive cash crops that depend increasingly on irrigation as one moves further west over the aquifer. On average, Medina and Uvalde Counties account for 38 percent of overall pumping—primarily for crop irrigation and ranching. Comal and Hays Counties account for 8 percent. The water then

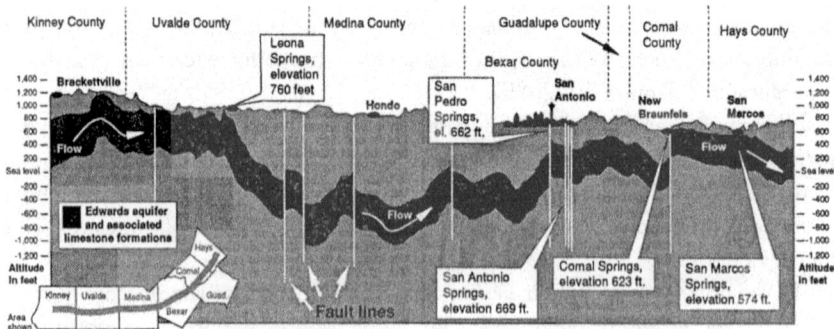

Figure 14.3. Cross-section of the Edward's Aquifer

flows south in the Comal and Guadalupe Rivers to petrochemical plants down-stream—where industrial effluent is discharged into the rivers. The major pumpers of aquifer water (44 percent) are the City of San Antonio and the farming communities in Bexar County.

Tourism provides important income for the eastern counties that boast of parks and natural springs. There, federally listed endangered species could suffer if spring flows diminish because of overdrafting the aquifer. Tourism in San Antonio is now a more significant source of income than are the three military bases. The River Walk alone delivers $3.5 billion a year in tourist dollars—removing the Alamo from its century-long status of tourist attraction number one in San Antonio (Glennon 2002: 87). Continued building of high-density, suburban residential areas over the aquifer's recharge zone will impact water quality and quantity. The few industrial plants in existence demand substantial amounts of water in the manufacturing process. Toyota now makes pickup trucks in a plant on the south side of the city, which has increased local demand for water. On average, it takes 400,000 liters (105,000 gallons) to make one car (Barlow and Clarke 2002: 8). Kelly Air Force Base, until it was closed in 1996 as part of a congressional cost-cutting effort, was the major depot for refurbishing cargo planes and jet engines. Large plumes of pollutants have been discovered under neighborhoods to the south of the base. Three large military bases, one Army and two Air Force, continue to be major players in the local economy—and in water issues.

Recent Conflict

The Edwards Aquifer has been a battleground for myriad competing interests—geographical, governmental, and corporate. Regional surface-water authorities, environmental groups, municipal water purveyors, irrigators, industrial water consumers, state and county legislative bodies, and federal environmental agencies have all been involved in a tug-of-war over the management of the aquifer (Donahue 1998).

The conflict came to a head in 1991 when the Lone Star Chapter of the Sierra Club and the Guadalupe/Blanco River Authority (GBRA) filed a lawsuit before Senior U.S. Judge Lucius Bunton in Midland (*Sierra Club v. Lujan* 1993). The plaintiffs contended that the USFWS was violating the Endangered Species Act by failing to implement a plan to preserve flows at the aquifer-fed springs at Comal and San Marcos.

In November, Judge Bunton announced that the hearing on the Sierra Club/GBRA lawsuit would be held before the end of the year and would be limited to four days to save money. There would be no closing arguments and the two sides would have two weeks in which to respond with written rebuttals or points. John Hall, the Texas Water Commissioner, asked the judge to hold his decision until the Texas legislature had time to act on proposed legislation in its spring 1993 session. The proposed legislation called for an initial cap of 450,000 acre-feet with successive reductions to 400,000 acre feet by 2008 and to whatever was necessary to protect spring flow by 2012. Some leeway was left in the bill to allow the Texas legislature to lift the pumping caps in the future as necessary.

Regional pumping was at 542,000 acre feet and environmentalists had demanded 165,000 to 250,000 acre feet pumping caps. Computer runs made in July 1992 at the Texas Water Development Board showed that reducing pumping to 350,000 acre feet could not prevent Comal Springs from drying up in a drought as severe as the drought of 1947–1957—"the drought of record" (Thorkildsen and McElhaney 1992).

On January 30, 1993, Judge Bunton ruled in favor of the Sierra Club/GBRA and urged the Texas Legislature, before the end of its current session on May 31, to reduce Edwards pumping in times of drought by at least 60 percent (to two hundred thousand acre-feet) to ensure the survival of endangered species. Failure to do so would result in a federal takeover of pumping from the aquifer. In April, the USFWS issued its best estimates on what pumping limits and spring flow requirements would ensure the survival of the species. By its own admission, these requirements were based on conservative estimates inasmuch as current data for modeling were unavailable. The Texas legislature took up the issue, and under threat of a federal takeover made a major change in the right of capture, which historically has been the rule for groundwater pumping. The right of capture holds that the owner of the property over a groundwater source holds rights to pump the water beneath the owner's property.

The Edwards Aquifer Authority

In the aftermath of the judge's ruling, the state legislature considered three bills that were eventually brought together in a compromise bill. The compromise called for replacing the elected Edwards Underground Water District with an appointed board, called the Edwards Aquifer Authority (EAA), with three representatives from Bexar County and three each from the eastern and western counties. A larger advisory committee would bring together the various river authorities and downstream industrial users. The authority would issue water-rights permits based on historical usage and alter pumping rights up or down depending on spring flow rates at certain test wells. The authority would issue a comprehensive water management plan for the region, including the development of alternative water supplies. The operations of the authority would be funded mostly from aquifer pumpers, but also a small portion from downstream users. Only the authority could buy water rights and landowners could sell up to 50 percent of their allocations.[6] The authority would install meters and be responsible for monitoring and protecting water quality (*San Antonio Express News*, May 27, 1993).

The compromise legislation, Senate Bill 1477, was passed without the support of two representatives from San Antonio, Karyne Conley and Ciro Rodriguez. Both objected to the severe pumping limits placed on San Antonio and to abolishing an elected board in favor of an appointed one (*San Antonio Express News,* May 29, 1993). Conley and Rodriguez were prophetic in their reservations regarding the abolition of an elected board in favor of an appointed one. The Mexican American Legal Defense and Educational Fund (MALDEF) asked the Federal Justice Department to rule on MALDEF's contention that the appointed board violated the federal Voting Rights Act, which prohibits any retrogression of minority voting rights (*San Antonio Express News,* June 2, 1993).

As the Justice Department weighed the merits of the suit, appointments to the new agency were made during the summer months. As Judge Bunton's September 1 deadline for EAA pumping regulations approached, he decided to delay any further action on his part until the Justice Department made its determination on the voting rights issue. On November 19, 2003, the Justice Department ruled that the state plan to create an appointed board in lieu of the existing elected one violated Section 5 of the Voting Rights Act. This ruling put the state's aquifer agency in limbo, and did not resolve the issue of pumping limits. That decision still stood in Judge Bunton's federal court. Judge Bunton declared Senate Bill 1477 effectively dead and indicated that new legislation would be needed to meet the Justice Department's demands. Responding to a motion from the Sierra Club, Judge Bunton appointed a water master who was to draw up an emergency water management plan. Subsequently, the Fifth Circuit Court of Appeals in New Orleans overturned Judge Bunton's declaration of a water emergency and a date for activation of a plan to restrict pumping. The Fifth Circuit ruled that the EAA had the means to ensure spring flow and protect the endangered species. The Texas Legislature removed the last obstacle by amending its earlier legislation, making the new EAA an elected, rather than an appointed body. For the next several years the EAA continued to face court actions from pumpers unhappy with their permitted allocations.

The Edwards Aquifer Recovery Implementation Program

While the creation of the EAA represented a major step forward in groundwater management, the issue of spring flows and pumping caps remained unresolved. In 1993, the Texas legislature had set pumping caps at 450,000 acre-feet. When the EAA finished issuing permits based on historical usage, the total amount permitted was 549,000 acre-feet, well above the limit set by Judge Bunton and the legislature. By 2005, neither the EAA nor the legislature had been able to determine the minimal continuous spring flows necessary to ensure survival of the endangered species. Animosity continued between aquifer pumpers and spring flow or downstream interests.

In late 2006, the newly appointed USFWS Texas state administrator for ecological services began to meet informally with stakeholders who represented the many parties with a stake in resolution of the conflict. She had been part of the Recovery Implementation Program on the Middle Rio Grande River in New Mexico while administrator there. She understood that a resolution to the management conflicts over the Edwards Aquifer would entail a close collaboration between the stakeholders and the USFWS. To that end, she invited any interested parties to a public meeting in San Marcos on February 16, 2007. Some eighty individuals attended, many from among the institutional stakeholders such as the San Antonio Water System and the GBRA. Also present were representatives of environmental groups, irrigators, and several legislators and their staff. Also present were four consultants in collaborative learning, two from out-of-state and two from Texas A&M University. Texas A&M had agreed to offer logistic and administrative support in the creation of the EARIP. The meeting had been well planned; when the offer was made to form a stakeholders' group to address with USFWS the issues of spring flow and pumping limits, the response was enthusiastic. For the next three years, monthly meetings would have between sixty and eighty participants.

The first major task of the group was to scope out the boundaries or geographical area of collaboration, identify the stakeholders, and agree on the first organizational steps. To address these tasks, two meetings were scheduled with the consultants in collaborative learning, the first at the offices of the EAA on February 27, 2007 and the second a week later in San Marcos. The workshops emphasized situational improvement rather than problem solving by looking at desirable and feasible change.[7] The process was to be one of conflict management rather than conflict resolution. Systems thinking or seeing the relationships among the several elements that make up the aquifer system would be critical for a successful process. For this type of thinking to occur, the participants would need to engage in considerable learning about the core values of the stakeholders, the science of the aquifer, and the language in which cultural values and scientific knowledge are expressed. Key to a successful collaborative learning experience is ongoing face-to-face communication and negotiation. Participants were invited to practice systems thinking through visualization, an exercise in which groups discussed and pictured the boundaries of the system (aquifer), the actors, the divergent views on its management, and the range of options open to managing the diverse uses of the groundwater system.

Reactions to the workshops were positive. One person noted,

> Stakeholder involvement is especially important. There are simply too many details involved with water resource issues to expect that any one person, or even a few people, can fully understand the range of implications. Legal issues, environmental concerns, engineering and scientific aspects, property rights, competing demands for water, economic considerations, and quality of life are just some of the elements that must be identified and evaluated (Muñoz and Donahue 2009).

The meeting of April 5, 2007, was held at the offices of the San Antonio Water System. Representatives of the USFWS spoke on minimum spring flows, the Habitat Conservation Plan (HCP) process, and the necessary memorandum of agreement (MOA) that the participants would need to sign and present to the USFWS. The MOA would be one part of a narrative that the stakeholders would need to construct. Other parts would include a governance structure, a statement of mission and goals, and a vision statement. The MOA went through several drafts between April and December 2007, when it was adopted (EARIP 2007). The MOA describes the EARIP as "a collaborative initiative among stakeholders to participate in efforts to contribute to the recovery of the Edwards Species, develop aquifer management measures, and develop conservation measures for the Edwards Aquifer," (EARIP 2007: 1). Of significance is that the MOA commits the stakeholders to the broader issue of groundwater management, which includes the recovery of the species.[8]

Almost from the first meeting, the stakeholders agreed to work with a consensus model of decision making. If consensus could not be reached on a given point, representatives of the opposing points of view would meet with an issues team, who would facilitate reaching a position that everyone could live with, even though the language might not reflect the original concerns.

By the time of the June 2007 meeting, a significant development threatened to undermine the sense of solidarity and consensus process that was growing among the stakeholders. In May 2007, the Texas legislature passed Senate Bill 3 which coopted several elements of the EARIP process. Senate Bill 3 directed the EAA to "cooperatively develop a recovery implementation program" through a consensus decision-making process that would involve a steering committee of twenty-one members, expandable to twenty-six (Legislature of the State of Texas 2007: 146). They were to hire a program manager and enter into an MOA by December 2007. They were to appoint a science subcommittee and agree to enter into an implementing agreement by December 2009, to be signed by September 2012. The larger stakeholder group was now faced with the question of whether consensus could be legislated. To its credit, the newly designated steering committee members did not feel the need to distinguish themselves from the larger group except for roll call and voting. All stakeholders had voice, which accounts for the sustained participation of sixty to eighty persons at every monthly meeting and at numerous subcommittee meetings. As it turned out, there was only one issue on which there was not consensus and for which a vote had to be taken (see "Conflict over EARIP Boundaries" below). To give the EARIP time to develop its program document, the HCP, Senate Bill 3 also raised the pumping caps to 572,000 acre-feet and adjusted the EAA pumping limits during a drought as part of a "critical period management plan" (Legislature of the State of Texas 2007: 142).

In the meantime, stakeholders finalized the "Program Operational Rules for RIP Steering Committee Members and RIP Participants"; those rules spelled out the consensus process and voting procedure (EARIP 2009)

Conflict over EARIP Boundaries

One outcome of the consensus-based approach has been general agreement on goals, objectives, and operating procedures for the EARIP. The single exception occurred in November 2009 when the issue of the scope of the EARIP was again raised.[9] The new studies subcommittee had recommended three additional studies be funded, in part, by the EARIP. The subcommittee reached consensus on recommending funding for two of the studies, but not the third. One stakeholder on the subcommittee opposed EARIP funds for a study of the impact of instream flows on the habitat of the whooping crane in the estuaries of San Antonio Bay.

According to the by-laws an Issues Committee brought together the parties for and against the whooping crane study. In favor of the study were the GBRA and the San Antonio River Authority. Opposed was the San Antonio Water System. When the issues committee was unable to forge a compromise, the matter was brought back to the full stakeholder committee for further discussion and a vote.

Reasons for and against the whooping crane study evolved around an earlier study that had cost $1.2 million. The originators of the proposed study, the GBRA and the San Antonio River Authority raised questions as to the validity of the earlier study and the need for a follow-up. Those opposed noted that the only dissent to the earlier study had been the GBRA. Underlying the discussion was a deeper issue as to the scope of the EARIP itself. The traditional division among aquifer pumpers, spring flow, and downstream interests emerged. Should the stakeholders limit EARIP funds to aquifer pumping and spring flow issues, or should funds be allocated for instream flows and estuary concerns throughout the year? Generally, river authorities, representatives from Hayes and Comal Counties, and environmental groups favored a broader boundary or scope of the EARIP. Irrigators over the aquifer and municipal water purveyors in Bexar and Medina Counties generally opposed the new study and favored a more limited scope for the EARIP. According to the by-laws, a super majority of 75 percent was needed to pass the motion to fund the whooping crane study. A friendly amendment was added that funding the study would not be a precedent for extending the scope of the EARIP. It failed seventeen in favor and nine against. With two exceptions, the voting followed the traditional divide among aquifer, springflow, and downstream interests.

Analysis of the case

Before the vote the project manager of the EARIP had reminded the participants that they were not speaking as individuals, but as representatives of their respective institutions. His remark and the vote pattern highlighted a continuing issue in the history of aquifer management conflicts: project culture. Nickim and Greenstadt (1998: 150–151) note that in the case of the Lake Biwa water management project in Japan the various governmental agencies proposed plans that "represented a clash of bureaucratic interests and cultures that to some extent reflected their different clienteles." In further elaborating on project culture, Donahue and Johnston (1998: 342–343) note that "bureaucracies are structured and defined by past events and agendas and carry with them a legacy of outdated goals and interests."

The challenge to the institutional stakeholders in the EARIP was to redefine their particular institutional and geographical interests in terms that would meet the legislative deadlines within a consensus model of decision making. The EARIP had less than a year to provide an HCP that the USFWS would accept. Without a plan, aquifer management to protect the endangered species could come under federal control. The whooping crane vote cast a doubt on whether a consensus could be reached in time.

Mediation Efforts

A second and more significant event took place in the same November 2009 meeting. The science subcommittee was made up of hydrologists and biologists whom the stakeholders had selected by consensus to study and offer the best science on spring flows. After months of deliberations and study, they issued their preliminary report, which came as a shock to stakeholders. The report indicated that in a drought, such as that in the 1950s, aquifer pumping would have to be cut by as much as 97 percent to ensure survival of the endangered species in the Comal and San Marcos Springs. The report dramatically changed the group dynamic. All stakeholders were aware that such a drastic reduction was politically untenable. The report stimulated intense discussion of alternative means to ensure survival of the species and still meet the needs of irrigators and municipal water systems.

The stakeholders had agreed in August 2009 to hire a consulting firm to assist them in developing a program document or HCP that would include spring flow requirements and recommended pumping caps. During September, facilitators met individually with the stakeholders on the steering committee and with any others who wished to share their views. The facilitators began to attend the monthly meetings and facilitated discussion of the dramatic science subcommittee report of November 2009.[10] In that same meeting, they witnessed the conflict over boundaries in the defeat of the proposed whooping crane study. The dynamics of the November meeting set the stage for the December retreat held at the Y.O. Ranch in the Texas hill country. The hope was that a neutral setting would allow the stakeholders to more easily address the issues of spring flow and pumping caps that had to be resolved. The rules for the retreat (table 14.1) are instructive of the challenges that the facilitators faced in reconciling the institutional interests of the stakeholders with the legal, hydrologic, and human demographic realities of the Edwards Aquifer.

Table 14.1. Proposed ground rules for the retreat

The Rules

1. Come prepared by reviewing past EARIP documents and work (particularly on background regarding possible covered actions).
2. Commit to work through dialogue (i.e., joint learning), not debate. Speak to explain, listen to understand.
3. Work to meet not only your own interests, but also the bona fide interests of the other stakeholders.
4. Come prepared to discuss substantive issues, choices, and how to make the best decisions.
5. Be able to brainstorm, develop ideas, and invent without committing. This is a time for exploratory ideas and considerations, not proposals and counteroffers.
6. Agree that stakeholders' willingness to discuss an option will not be considered an agreement to support that option—merely a willingness to discuss. Similarly, a stakeholder who suggests consideration of an option is free to later decide that the option is not suitable.
7. Recognize that part of this exploration is figuring out where we have commonality, where we do not, and why, and what possible additional work might be needed to bridge differences.
8. If applicable, loosen your grip on your favorite solution, and be open to considering other options or combinations of options.

A second retreat was held in late January 2010 with a facilitated discussion on how to get water to the species. Four strategies, pros and cons, were discussed. Options included recharge enhancement with dams that would hold floodwaters and increase water availability in the aquifer. Brush management, another enhancement strategy, involved the removal of thirsty Ashe juniper trees. A second strategy discussed was a storage option using quarries to hold floodwaters. A third strategy would involve aquifer storage and recovery. The idea was to pump or inject water into non–Edwards Aquifer formations that store water long term until it was needed for spring flow supplementation. A fourth strategy would be off-channel storage where water could be retained near the principal channel and could be delivered to the springs using recharge.

The remainder of the year was dedicated to review of the four options noted above. The facilitators assisted the stakeholders to address the work of four outside consultants: Two were hired as outside reviewers of the science subcommittee report.[11] Another was a hydrologist who looked into the amount of water that each of the above strategies or options might provide. The fourth was a firm whose task was to cost out the various options that the EARIP wished to be considered in a draft HCP.

A third two-day retreat in late July 2010 revealed further clarifications of steps that the stakeholders were considering. These included nonengineering as well as engineered options to ensure spring flow. Engineered solutions included new aquifer storage and recovery projects, quarries, and recharge or recirculation structures. Nonengineered approaches to maintaining spring flow envisioned a bottom-up approach, which included conservation measures, recharge enhancement structures (earthen dams), the currently operating aquifer storage and recovery in the Carrizo-Wilcox Aquifer in South Bexar County, brush management, and the dry year option.[12] Concerns centered around three issues: the high cost of the engineering solutions, the scientific uncertainty as to how much spring flow the species need to survive (avoid jeopardy), and thirdly, the risk that a "bottom-up approach" would not provide the needed spring flow during a record drought.[13]

In the September 2010 meeting, the two outside consultants who reviewed the science subcommittee report offered the flow numbers they concluded could be the basis for the HCP. In the discussion that followed, stakeholders used a number of language items with which they attempted to understand the meaning of the numbers. Were the flow numbers a wall or a cliff beyond which the species would be in jeopardy? Did the numbers suggest a buffer that would mitigate harm to the habitat? Were the flow numbers a trigger for instituting recovery measures? The flow numbers raised issues of biological survival, habitat recovery, risk, and the range and duration of flows over time.

The facilitators suggested that the stakeholders consider an adaptive management approach that would allow the incremental implementation of the bottom-up approaches before undertaking the more costly engineering options. The stakeholders continued the discussion in another retreat later that month. Given the expense and uncertainty of the proposed solutions, the stakeholders discussed a decision-making process that would involve adaptive management. The draft HCP could involve two phases. During the first five to seven years of the HCP permitting period, the nonengineering solutions would be put into place and closely monitored. In the second phase of eight to fifteen years, engineering solutions would be put into place if the spring flows were not at their necessary levels. All agreed to the approach with one exception: one stakeholder believed that his organization would object to such a drawn-out process. Further discussion revealed that even if a one-time approach was preferred to a phased approach, an adaptive management methodology of constant monitoring and reevaluation would be necessary.

With agreement on a phased approach, the stakeholders then turned to the ranges for spring flow that the consultants would use for modeling resulting impact on the species. This was a point of contention since some stakeholders such as municipal water purveyors would prefer lower numbers for required spring flow, thus allowing more pumping. Environmentalists and spring flow or downstream interests would prefer higher numbers to be modeled to protect the species even if that meant putting caps on pumping at earlier stages of a drought. Again, the concern of the former was that flows be such as not to put the species in jeopardy while the latter sought higher flows that would minimize take or alternation of the habitat.[14]

The fear of the former was that the higher flow numbers would be considered jeopardy numbers and call for drastic reductions in pumping. Representatives of the two sides informally caucused during and after lunch and finally agreed on a range of spring flow numbers to be used in modeling their impact on the species. All agreed that the ranges were for the purpose of computer modeling and not a commitment to a jeopardy number. Such a determination could better be made with the data provided by the three flow models than just one or two within a limited range.

With the spring flow modeling numbers agreed on, the final task was to address critical period management in which pumping would be reduced to a level that would ensure adequate spring flow. Currently there are four stages of reduced pumping (from 20 percent to 40 percent), depending on the level of the aquifer at two monitoring wells called J-13 and J-27 and two springs, Comal Spring and San Marcos Spring (EAA 2010: 4). Stakeholders agreed to a proposal that if the fourth stage does not ensure spring flow, there could be a fifth stage at which pumping would be limited to 320,000 acre-feet. The problem is such a low cap on pumping would affect groups across the aquifer differently. Those municipalities whose access to aquifer water would be eliminated at 320,000 acre-feet would need access to water from an aquifer storage and recovery facility. Stakeholders are aware that after the USFWS accepts the HCP, there will need to be an agency tasked to implement and monitor the entire process.

An important watershed in negotiations was reached when after three years, the stakeholders agreed on a phased, adaptive management approach, spring flow modeling numbers, and a critical period management plan. The process of implementation would continue, but with some relief among all the participants. Their agreement marked a way out of what for years had seemed to be an irreconcilable conflict.

Conclusion

Earlier I noted that that concerns among parties in conflict include not only the substantive issue(s) in question, but also procedural and psychological factors. While aquifer management has been a substantive issue for some forty years, strategies to address the procedural and psychological issues have been lacking. The EARIP represents an effort to bring the three together and reach agreement by consensus on water management policy. While much still has to be done to create an HCP that will be acceptable to the USFWS, significant obstacles have been overcome.

I argue that the time stakeholders spent in addressing procedural interests ensured that everyone felt that the rules were fair and that the process was transparent. The hope was that front-loading the group's decision-making process would make implementation of groundwater management rules at a later time better understood if not totally welcomed. Such seems to be the case.

Geographical spacing of monthly meetings was another factor that contributed to the growing feeling of legitimacy for and trust in the process. Representatives of previously antagonistic water management institutions now found themselves meeting in the headquarters of one another's agencies. Participants met in several venues across the aquifer, including San Antonio, and farther east in the Texas cities of New Braunfels, San Marcos, and Seguin. There were occasional meetings in Austin when the Texas legislature was in session. The success of social and cultural change relies as heavily on social relationships as it does on shared ideological or cultural norms. Success of this effort lies in great part in the opportunities provided for participants to listen to one another. Over time, participants learned to distinguish between a person and his or her position. This allowed individuals to change positions during the discussions and for new positions to emerge. During discussions, the program manager of the EARIP and the facilitators did not identify with one or another position, but restated positions for further clarification. They also provided summary statements from time to time as to the state of the discussion, and provided a timeline for making critical decisions so that participants could remain focused.

The efforts to create a consensus-based procedure allowed a new cultural discourse to emerge. Earlier cultural understandings of water and management remain, but monthly stakeholder meetings and frequent subcommittee meetings, open to all, allowed the emergence of a new cultural narrative of water management. This discourse is encoded in minutes, subcommittee reports, and invited presentations. Beyond a sharing of values such as the protection of endangered species, and optimal management of the aquifer, there is at work a microlevel process of consensus building as a means to achieve those ends. Within that context, the construction and sharing of documents has become a way of demonstrating process (consensus) and product (meeting legislative deadlines for policy options).

In his ethnographic study of nature and culture in the Amazon, Hugh Raffles (2002) discusses how actors manage multiple discourses about the environment. A similar process is under way among stakeholders in the Edwards Aquifer. As noted earlier, in addition to the traditional voices about aquifer management, new language items such as consensus, adaptive management, and structured decision making have emerged during the three years of discussion. These and other language items are codified in presentations, documents, and minutes that now constitute an emerging narrative of aquifer management.

Wolf (1990: 593) argues that "power is ... never external to signification—it inhabits meaning." Associated with those narratives and their meanings are varying degrees of power to enforce the priority of one over the others. There is no doubt that the narrative of endangered species is powerful in that the USFWS is federally mandated to enforce the law. To its credit, the EARIP has to date successfully negotiated the Scylla and Charybdis of litigation to find a win-win water management policy, or at least one that everyone can live with.

Coda

Since this chapter was completed in December 2010, several developments have taken place to which I would like to draw the reader's attention. Meetings during 2011 dealt with finalizing the HCP for presentation to the USFWS by the deadline of December 2011. Major responsibility for drafting the HCP fell to the program manager when the consultant who had been hired to do that task was unavailable. The program manager and the stakeholders worked through the many details and requirements, including the financing and implementation of the HCP.

Financing of the multimillion-dollar projects in the HCP was a major hurtle. There were two possible funding sources: a sales tax in the seventeen counties that would benefit from the HCP, or an increase in pumping fees. While the former would have been a more equitable arrangement including counties over the aquifer as well as downstream counties, the Texas legislature was in an antitax mood. The result was that the pumpers would have to bear the cost of the HCP at least initially.

The second issue was that of implementation. The signatories of the HCP were designated to be the EAA, which has legislative authority to regulate pumping from the aquifer. The San Antonio Water System, and the cities of New Braunfels and San Marcos would each seek incidental take protection for pumping that the EAA authorizes (see note 6). The cities of New Braunfels and San Marcos and the Texas State University would also seek incidental take permission as they manage the spring and river systems in their local areas for recreational use.

When the EARIP took a final vote on the HCP in December 2011, there was only one negative vote and one abstention among the twenty-six committee members. The EAA unanimously approved the HCP on December 28, 2011. USFWS approved the HCP on February 15, 2013 (USFWS 2003).

Notes

1. The art of negotiation is to reach practical, communally crafted decisions. Only when everyone's concerns are taken seriously can concessions be made. For everyone to understand the various concerns, these must be "translated" into one another's conceptual and practical vocabulary or framework. Boundary objects facilitate this translation: they have a specific meaning

for each group, but their structure is common enough to form a leverage or connection among the various groups. They could be seen as vehicles of translation that enable coherence across social worlds; in their capacity to cross boundaries, boundary objects create a meeting ground. In this evolution of cooperation, vested interests might shift, attracting potentially new players in the field. Cooperative management is, by definition, an ongoing process of education and negotiation (see Klaver and Donahue 2005: 109–110).

2. Federally recognized endangered species include four surface-dwelling species: the Fountain darter, Texas wild-rice, the San Marcos salamander, and the Comal Springs riffle beetle; and three subterranean species: the Comal Springs dryopid beetle, the Peck's cave amphipod, and the Texas blind salamander.

3. Both figures are used with permission of the *San Antonio Express News*.

4. An acre-foot of water is the amount of water that can stand in one acre of land at a depth of one foot. One acre-foot is equal to 325,851 gallons and can supply a family of five for a year. In 1993, the University of Texas Bureau of Economic Geology calculated that the aquifer holds up to 215 million acre-feet of water, four times that estimated in 1978 (*San Antonio Express News*, October 13, 1993; July 22, 1994). Glennon (2002: 91) estimates the capacity to be more than 250 million acre-feet. It should be noted, however, that not all of that water is accessible for human use.

5. Total aquifer discharge includes pumping but also springflows. On the average, springflows, primarily in Comal and Hays counties, account for 54 percent of total aquifer withdrawal.

6. The EAA issues permits to pump based on mutually agreed on historical pumping averages. Permit holders, be they individuals or corporations, may sell one-half of their annual allocations to other users. This has allowed municipal pumpers to buy or lease pumping rights from farmers. These are paper transfers and involve no infrastructure such as pipelines. Irrigators must use one-half of their allocation for agriculture.

7. The following material is from the collaborative workshop training. See also Daniels and Walker 2001.

8. Recovery in this case is the amount of spring flow necessary for the endangered species to survive and not be put in jeopardy of extinction. Take is the point at which a change in habitat affects the survival of some, but not all, of the endangered species. Take flows are significantly higher than jeopardy flows. For example, take flows at Comal Springs would begin at 150 cubic feet per second and at San Marcos at 110 cubic feet per second. Jeopardy flows must be higher than 30 cubic feet per second at Comal and 45 cubic feet per second at San Marcos in a HCP, but not as high as the take numbers.

9. The boundaries of the EARIP (scope) had been an issue since the early collaborative learning workshops in 2007. Spring flow and downstream interests argued that the Edwards ecosystem included the bays and estuaries at the mouth of the Guadalupe River. The blue crab, principal food of the endangered whooping crane, depended on freshwater inflows to reproduce.

10. The final science subcommittee report was sent to outside reviewers to ensure that the analysis was objective and reflected best practices.

11. Facilitators included the outside consultants and the program manager for the EARIP. The stakeholders held them in high regard for their expertise and their impartiality. They were adept at summarizing discussions and clarifying issues while holding the stakeholders to the legislative timeline.

12. The dry year option program seeks to reduce aquifer pumping demands during specified critical periods (or critical period management) in order to improve aquifer levels and protect spring flows, the protected species, and their habitats (see Edwards Aquifer Habitat Conservation Plan 2010).

13. According to the Endangered Species Act, jeopardy occurs "when an action is reasonably expected, directly or indirectly, to diminish a species' numbers, reproduction, or distribution so

that the likelihood of survival and recovery in the wild is appreciably reduced" (USFWS 2012, "What is Jeopardy?").

14. This difference of perspective was reflected in the science subcommittee report and that of the outside reviewers. The science subcommittee focused on flows that would result in take, while the reviewers were concerned with jeopardy. See note 8.

References

Barlow, Maude, and Tony Clarke. 2002. *Blue Gold: The Battle Against Corporate Theft of the World's Water.* London: Earthscan.

Daniels, Steven E., and Gregg B. Walker. 2001. *Working through Environmental Conflict: The Collaborative Learning Approach.* New York: Praeger.

Donahue, John M. 1998. Water Wars in South Texas: Managing the Edwards Aquifer. In *Water, Culture and Power: Local Struggles in a Global Context,* edited by John M. Donahue and Barbara Rose Johnston. Washington, DC: Island Press.

Donahue, John, and Barbara Rose Johnston. 1998. Conclusion. In *Water Culture and Power: Local Struggles in a Global Context,* edited by John M. Donahue and Barbara Rose Johnston (340–341). Covelo, CA: Island Press.

Donahue, John M., and Jon Q. Sanders. 2001. Sitting Down at the Table: Mediation Efforts in Resolving Water Conflicts. In *San Antonio: An Environmental History,* edited by Char Miller. Pittsburgh, PA: University of Pittsburgh Press.

Edwards Aquifer Authority (EAA). 2009. Critical Period: the Evolution of Drought Management (1989–present). Presented to Edwards Aquifer RIP, June 29. http://www.eahcp.org/files/uploads/06-29-10EAACPMPresentation.pdf

Edwards Aquifer Habitat Conservation Plan. 2010. *Dry Year Option Program for the Edwards Aquifer Implementation Program (EARIP).* October 8 addendum. http://www.eahcp.org/files/uploads/10-15-~4.PDF

Edwards Aquifer Recovery Implementation Program (EARIP or RIP). 2007. Memorandum of Agreement for the Edwards Aquifer Recovery Implementation Program. December 13. http://www.edwardsaquifer.org/files/HCP_Appendix_A.pdf

———. 2009. Program Operational Rules for RIP Steering Committee Members and RIP Participants as Amended May 14, 2009. http://www.eahcp.org/files/uploads/05-14-09Attachment6_ProposedRevisionstoRules.pdf

Fisher, Roger, and William Ury. 1991. *Getting to Yes: Negotiating Agreement Without Giving In.* New York: Viking Penguin.

Glennon, Robert. 2002. *Water Follies: Groundwater Pumping and the Fate of America's Fresh Waters.* Washington, DC: Island Press.

Klaver, Irene, and John M. Donahue. 2005. Whose Water Is It Anyway? Boundary Negotiations on the Edwards Aquifer in Texas. In *Globalization, Water and Health: Resource Management in Times of Scarcity,* edited by Linda and Scott Whiteford. Santa Fe, NM: School of American Research.

Legislature of the State of Texas. 2007. Senate Bill SB3. http://www.legis.state.tx.us/tlodocs/80R/billtext/pdf/SB00003F.pdf#navpanes=0

Maclay, R.W., and R.F. Land. 1988. Simulation of flow in the Edwards aquifer, San Antonio region, Texas, and refinement of storage and flow concepts: U.S. Geological Survey Water-Supply Paper 2336-A.

Muñoz, Anna, and John M. Donahue. 2009. Constructing the Common Ground: Consensus Building among Stakeholders in Protecting Endangered Species. Paper presented in the panel Managing Water Conflicts: Consensus Building among Stakeholders in Protecting Endangered Species. Annual Meetings of the Society for Applied Anthropology. Santa Fe, NM, March 19.

Nickim, James, and Daniel Greenstadt. 1998. Transacting a Commons: The Lake Biwa Comprehensive Development Plan, Singha Prefecture, Japan. In *Water Culture and Power: Local Struggles in a Global Context,* edited by John M. Donahue and Barbara Rose Johnston, 150–151. Covelo, CA: Island Press.

Raffles, Hugh. 2002. In *Amazonia: A Natural History.* Princeton, NJ: Princeton University Press.

Sierra Club v. Lujan 1993 WL 151353 (W.D.Tex.1993).

Thorkildsen, D., and McElhaney, P.D. 1992. Model refinement and applications for the Edwards (Balcones Fault Zone) aquifer in the San Antonio region. Texas: Texas Water Development Board Report 340.

U.S. Fish and Wildlife Service (USFWS). 2012. Section 7 Consultation: A Brief Explanation. October 24. http://www.fws.gov/midwest/Endangered/section7/section7.html

———. 2013. Service Approves Edwards Aquifer Recovery Implementation Program Incidental Take Permit. http://www.fws.gov/southwest/es/AustinTexas/ESA_C_HCP_news.html#EA RIP_final

Wolf, Eric. 1990. Distinguished Lecture: Facing Power—Old Insights, New Questions. *American Anthropologist* 92: 586–596.

CONTRIBUTORS

Jeannette Armstrong is an indigenous author and activist. Her published works include literary titles and academic writing on a wide variety of indigenous issues. She was awarded the 2004 Eco Trust Buffet Award for Indigenous Leadership, and is distinguished with honorary doctorates from the University of St. Thomas, the University of British Columbia Okanagan, and the University of Queens. She holds the Okanagan Lifetime Fellow Award. She is the executive director of the En'owkin Centre, the cultural research and education facility of the Okanagan Nation, and in 2013 was appointed Canada Research Chair in Indigenous Studies at University of British Columbia Okanagan. She has a PhD from the University of Greifswald in Germany, interdisciplinary in environmental ethics and Syilx indigenous literatures. She is an Okanagan indigenous Syilx language and culture specialist and traditional knowledge keeper. She currently serves on Environment Canada's Aboriginal Traditional Knowledge (ATK) Sub-committee of the Committee on the Status of Endangered Wildlife in Canada (COSEWIC).

Rita Brara teaches at the Department of Sociology, University of Delhi, India. She is the author of *Shifting Landscapes: The Making and Remaking of Village Commons in India* (Oxford University Press, 2006). Besides ecology, she writes on kinship and popular culture.

Alexandra Brewis is a biocultural anthropologist interested in the connections between human, cultural, and biological variation. She is currently executive director of the School of Human Evolution and Social Change at Arizona State University (ASU), and directs ASU's Center for Global Health. Over the past three decades, she has conducted fieldwork in Mexico, Micronesia, Polynesia, and the United States, and has addressed a variety of complicated health and environmental problems, including reproductive health, obesity, ADHD, and water insecurity.

Bryan Bruns received a bachelor of arts in anthropology from Beloit College and a doctorate in development sociology from Cornell University. He coedited *Negotiating Water Rights* and *Water Rights Reform: Lessons for Institutional Design* (2005, International Food Policy Research Institute). His papers on community participation in irrigation and water resources management are available at www.bryanbruns.com. He worked as sociologist and water user association specialist on the Earthquake and Tsunami Emergency Support Project in Nanggroe Aceh Darussalam and Nias Island, Sumatra, which was funded by a grant from the Asian Development Bank. He worked on a variety of projects related to participatory irrigation development in Indonesia beginning in 1988.

Hugo De Burgos is assistant professor at the University of British Columbia Okanagan. He was born and raised in El Salvador but moved to Canada to attend university, earning a bachelor of arts at McGill University, a master of arts from the University of Toronto, and a doctorate from the University of Alberta. He has published three books of anthropology and history focusing on three colonial cities in El Salvador in a collection titled *City and Memory* (1999, Dirección Nacional de Publicaciones e Impresos). He has also published several medical anthropology articles and produced and directed an ethnographic film on lead poisoning in El Salvador.

John M. Donahue is professor emeritus of anthropology in the Department of Sociology and Anthropology at Trinity University in San Antonio, Texas, where he began teaching in 1974. Dr. Donahue's research on water issues dates back to the 1980s. He coedited *Water, Culture and Power: Local Struggles in a Global Context* with Barbara Rose Johnston (1998, Island Press). He subsequently conducted research on conflict resolution and more recently on cultural definitions of water rights in Texas. He is an active stakeholder in the Edwards Aquifer Recovery Implementation Program (http://earip.tamu.edu/index.cfm). He is currently working on a coedited volume with Henry Casey Walsh (University of California–Santa Barbara), and Irene Klaver (UNT, *Infrastructures of Water and the Mexico–U.S. Borderlands.*

Issaka Kanton Osumanu is a lecturer in the Department of Environment and Resource Studies, University for Development Studies, Wa Campus, Ghana. He holds a doctor of philosophy degree in geography and resource development from the University of Ghana. Research interests include urban environmental health, environmental change and resource management, management of poverty alleviation, food security, and environmental change.

Liam Leonard is series editor of the Advances in Ecopolitics Book Series with Emerald Group Publishing. He currently works as a lecturer in sociology, criminology, and human rights at the Institute of Technology, Sligo, Ireland. He is a member of the executive committee of the Sociology Association of Ireland, and is the chair of the Green Link West think tank, as well as founder and editor of the *Journal of Social Criminology.*

Fabiana Li is assistant professor of anthropology at the University of Manitoba, Canada. Her research on conflicts over mining brings together interests in environmental politics, social movements, Latin America, and the anthropology of science and technology.

Lyla Mehta is a research fellow at the Institute of Development Studies at the University of Sussex, and an adjunct professor at Noragric, Norwegian University of Life Sciences. Her work focuses on water and sanitation, forced displacement and resistance, scarcity, rights and access, and the politics of environment or development and sustainability. She has extensive field research in rural India studying the politics of water scarcity and the linkages between gender, displacement, and resistance. Additionally, she has worked on the right to water in South Africa and studied the cultural and institutional aspects of sanitation in Bangladesh, Ethiopia, India, and Indonesia.

Nefissa Naguib (doctor of philosophy) is an anthropologist and senior researcher at Chr. Michelsen Institute (CMI) in Bergen, and Prof II at the Department of Social Anthropology, University of Bergen. Her specialization lies in globalization and transformations, gender, social aspects of water and food, religious minorities, and the politics of memory and aesthetics. Naguib is the research director of a project titled "Muslim Devotional Practices, Aesthetics, and Cultural Formation" and serves as the coordinator of "Cultures and Politics of Faith" thematic research group. Her regions of interest are the Middle East (Egypt, Syria, Palestine) and Turkey.

Marlowe Sam is an indigenous political and social activist currently pursuing a doctoral degree at the University of British Columbia Okanagan, an institution where he also received his bachelor of arts degree in indigenous studies, and his master of arts degree in interdisciplinary studies. The primary focus of his doctoral studies is on aboriginal water rights in Canada, which is a continuation of his undergraduate and master's research, which was primarily focused on the social, political, and cultural history of water for the Okanagan Nation.

Sveinn Sigurdsson is a graduate student in sociocultural anthropology at Arizona State University's School of Human Evolution and Social Change. His research interests include the health effects of resource insecurity, access to health care, and cross-cultural patterns of treatment-seeking behavior. He has conducted ethnographic fieldwork in Guatemala, Iceland, and the United States.

Sarah C. Smith obtained a master of public health degree from Tulane University and a doctorate in anthropology from the School of Social Science, University of Queensland. She has conducted research in Cambodia and in the United States.

Her research interests include infectious diseases, human–environment interaction, and global heath policy. She now works for EnCompass LLC, a small consulting firm in Maryland. She is a quality improvement adviser for research and evaluation on the U.S. Agency for International Development (USAID) Health Care Improvement Project (HCI), which aims to improve the quality of health care around the world. In this role, Smith provides training and technical assistance on qualitative research to country program staff in Africa, Asia, Latin America, and the Middle East.

Rhian Stotts is a graduate student in anthropology at Arizona State University's School of Human Evolution and Social Change. Her research focuses on archaeology of the eastern Mediterranean Bronze Age, primarily on the island of Cyprus. She studies issues surrounding the emergence of social complexity using ceramic analysis, landscape modeling, and social network analysis.

Veronica Strang is professor of anthropology and executive director of the Institute of Advanced Study at Durham University, Durham, England. Prior to studying anthropology, she worked as a writer and researcher on environmental issues, and contributed to The Brundtland Report. She received her doctorate at the University of Oxford in 1994, and has written extensively on water, land, and resource issues in Australia and the United Kingdom. She is the author of *Uncommon Ground: Cultural Landscapes and Environmental Values* (1997, Berg); *The Meaning of Water* (2004, Berg); *Gardening the World: Agency, Identity, and the Ownership of Water* (2009, Berghahn Books); and an edited volume (with Mark Busse), *Ownership and Appropriation* (2010, Berg). In 2007, she was named as one of UNESCO's Lumières de l'Eau (water's leading lights).

Swathi Veeravalli completed a bachelor of arts degree from The George Washington University in international affairs and African politics, and then worked for several years in the Washington, DC, metropolitan area on issues related to international development and grassroots advocacy. She earned a master of science degree in water science, policy, and management from the University of Oxford. With fieldwork experience in Australia, Botswana, Kenya, South Africa, and the United Kingdom, she has continued to focus on the relationship between identity, water access, and the impacts of conflict and climate variability in structuring those relationships. Currently, she is working as a social science researcher with the U.S. Army Corps of Engineers examining the relationship between water and the surrounding landscape.

John Richard Wagner is associate professor of anthropology at the University of British Columbia Okanagan. After receiving his bachelor of arts and master of arts degrees from the University of Victoria, he worked for several years as a research consultant to indigenous communities in British Columbia. He received his doctoral degree from McGill University in 2002, after conducting fieldwork in Papua New Guinea on issues of resource management, conservation, and development. He joined the Social Research for Sustainable Fisheries project at St. Francis Xavier University in Antigonish, Nova Scotia, for two years as a postdoctoral fellow, and has coauthored several journal articles with Anthony Davis, the director of that project. Since taking up his current position at the University of British Columbia Okanagan in 2003, he has focused his attention on water management issues in the Okanagan Valley. He continues to conduct research on language and local ecological knowledge in Papua New Guinea.

Amber Wutich is a cultural anthropologist on the faculty of Arizona State University's School of Human Evolution and Social Change. Her research examines human adaptability to resource insecurity, with a focus on water and food insecurity in urban environments. She directs the Global Ethnohydrology Study, a multiyear, multicountry comparative study of local water knowledge, and has conducted ethnographic fieldwork in Bolivia, Mexico, Paraguay, and the United States.

Abigail York is a political scientist on the faculty of Arizona State University's School of Human Evolution and Social Change. She investigates the emergence, evolution, and impact of rules, norms, and policies on communities and landscapes. Her work examines governance of social-ecological systems in the modern, historical, and ancient contexts through collaboration with archaeologists, geographers, historians, and anthropologists.

INDEX

www.ingramcontent.com/pod-product-compliance
Lightning Source LLC
Chambersburg PA
CBHW060027030426
42334CB00019B/2204